Python Data Cleaning and Preparation Best Practices

A practical guide to organizing and handling data from various sources and formats using Python

Maria Zervou

Python Data Cleaning and Preparation Best Practices

Group Product Manager: Apeksha Shetty
Publishing Product Managers: Deepesh Patel and Chayan Majumdar
Book Project Manager: Hemangi Lotlikar
Senior Content Development Editor: Manikandan Kurup
Technical Editor: Kavyashree K S
Copy Editor: Safis Editing
Proofreader: Manikandan Kurup
Indexer: Hemangini Bari
Production Designer: Joshua Misquitta
Senior DevRel Marketing Executive: Nivedita Singh

First published: September 2024
Production reference: 1190924

Published by Packt Publishing Ltd.
Grosvenor House
11 St Paul's Square
Birmingham
B3 1RB, UK.

ISBN 978-1-83763-474-3
www.packtpub.com

I want to extend my deepest thanks to those who have been by my side throughout the journey of writing this book while managing work in parallel. I am immensely grateful to everyone who has cheered me on, offered feedback, and inspired me to keep going. A special thanks to my family, for their unwavering support and for teaching me the power of determination. To my mentors, friends, and partner, who have guided me over the years and helped me see the bigger picture, and from whom I have learned so much! This accomplishment is as much yours as it is mine. Thank you for being part of this journey!

– Maria Zervou

Contributors

About the author

Maria Zervou is a Generative AI and machine learning expert, dedicated to making advanced technologies accessible. With over a decade of experience, she has led impactful AI projects across industries and mentored teams on cutting-edge advancements. As a machine learning specialist at Databricks, Maria drives innovative AI solutions and industry adoption. Beyond her role, she democratizes knowledge through her YouTube channel, featuring experts on AI topics. A recognized thought leader and finalist in the Women in Tech Excellence Awards, Maria advocates for responsible AI use and contributes to open source projects, fostering collaboration and empowering future AI leaders.

About the reviewers

Mohammed Kamil Khan is currently a scientific programmer at UTHealth Houston's McWilliams School of Biomedical Informatics, wherein he works on data preprocessing, GWAS, and post-GWAS analysis of imaging data. He has a master's degree from the **University of Houston – Downtown** (**UHD**), having majored in data analytics. With an unwavering passion for democratizing knowledge, Kamil strives to make complex concepts accessible to all. Moreover, Kamil's commitment to sharing his expertise led him to publish articles on platforms such as DigitalOcean, Open Source For You magazine, and Red Hat's opensource.com. These articles explore a diverse range of topics, including pandas DataFrames, API data extraction, SQL queries, and much more.

Ashish Shukla is a seasoned professional with 12 years of experience, specializing in Azure technologies, particularly Azure Databricks, for the past 9 years. Formerly associated with Microsoft, Ashish has been instrumental in leading numerous successful projects leveraging Azure Databricks. Currently serving as an associate manager of data operations at PepsiCo India, he brings extensive expertise in cloud-based data solutions, ensuring robust and innovative data operations strategies.

Beyond his professional roles, Ashish is an active contributor to the Azure community through his technical blogs and engagements as a speaker on Azure technologies, where he shares valuable insights and best practices in data management and cloud computing.

Krishnan Raghavan is an IT professional with over 20 years of experience in software development and delivery excellence across multiple domains and technologies, including C++, Java, Python, Angular, Golang, and data warehouses.

When not working, Krishnan likes to spend time with his wife and daughter, as well as reading fiction, nonfiction, and technical books and participating in Hackathons. Krishnan tries to give back to the community by being part of the GDG – Pune volunteer group.

You can connect with Krishnan at `mailtokrishnan@gmail.com` or via LinkedIn.

I'd like to thank my wife, Anita, and daughter, Ananya, for giving me the time and space to review this book.

Table of Contents

3

Data Profiling – Understanding Data Structure, Quality, and Distribution 61

4

Cleaning Messy Data and Data Manipulation 95

5

Data Transformation – Merging and Concatenating 109

6

Data Grouping, Aggregation, Filtering, and Applying Functions 137

7

Data Sinks 155

Part 2: Downstream Data Cleaning – Consuming Structured Data

8

Detecting and Handling Missing Values and Outliers 209

9

Normalization and Standardization 255

10

Handling Categorical Features 267

11

Consuming Time Series Data 289

Part 3: Downstream Data Cleaning – Consuming Unstructured Data

12

13

Preface

In today's fast-paced data-driven world, it's easy to be dazzled by the headlines about **artificial intelligence** (**AI**) breakthroughs and advanced **machine learning** (**ML**) models. But ask any seasoned data scientist or engineer, and they'll tell you the same thing: *the true foundation of any successful data project is not the flashy algorithms or sophisticated models—it's the data itself, and more importantly, how that data is prepared.*

Throughout my career, I have learned that data preprocessing is the unsung hero of data science. It's the meticulous, often complex process that turns raw data into a reliable asset, ready for analysis, modeling, and ultimately, decision-making. I've seen firsthand how the right preprocessing techniques can transform an organization's approach to data, turning potential challenges into powerful opportunities.

Yet, despite its importance, data preprocessing is often overlooked or undervalued. Many see it as a tedious step, a bottleneck that slows down the exciting work of building models and delivering insights. But I've always believed that this phase is where the most critical work happens. After all, even the most sophisticated algorithms can't make up for poor-quality data. That's why I've dedicated much of my professional journey to mastering this art—exploring the best tools, techniques, and strategies to make preprocessing more efficient, scalable, and aligned with the ever-evolving landscape of AI.

This book aims to demystify the data preprocessing process, offering both a solid grounding in traditional methods and a forward-looking perspective on emerging techniques. We'll explore how Python can be leveraged to clean, transform, and organize data more effectively. We'll also look at how the advent of **large language models** (**LLMs**) is redefining what's possible in this space. These models are already proving to be game changers, automating tasks that were once manual and time-consuming, and providing new ways to enhance data quality and usability.

Throughout the pages, I'll share insights from my experiences, the challenges faced, and the lessons learned along the way. My hope is to provide you with not just a technical roadmap but also a deeper understanding of the strategic importance of data preprocessing in today's data ecosystem. I strongly believe in the philosophy of "learning by doing," so this book includes a wealth of code examples for you to follow along with. I encourage you to try these examples, experiment with the code, and challenge yourself to apply the techniques to your own datasets.

By the end of this book, you'll be equipped with the knowledge and skills to approach data preprocessing not just as a necessary step but also as a critical component of your overall data strategy.

So, whether you're a data scientist, engineer, analyst, or simply someone looking to enhance their understanding of data processes, I invite you to join me on this journey. Together, we will explore how to harness the power of data preprocessing to unlock the full potential of your data.

Who this book is for

This book is for readers with a working knowledge of Python, a good grasp of statistical concepts, and some experience in manipulating data. This book will not start from scratch but will rather build on existing skills, introducing you to sophisticated preprocessing strategies, hands-on code examples, and practical exercises that require a degree of familiarity with the core principles of data science and analytics.

What this book covers

Chapter 1, Data Ingestion Techniques, provides a comprehensive overview of the data ingestion process, emphasizing its role in collecting and importing data from various sources into storage systems for analysis. You will explore different ingestion methods such as batch and streaming modes, compare real-time and semi-real-time ingestion, and understand the technologies behind data sources. The chapter highlights the advantages, disadvantages, and practical applications of these methods.

Chapter 2, Importance of Data Quality, emphasizes the critical role data quality plays in business decision-making. It highlights the risks of using inaccurate, inconsistent, or outdated data, which can lead to poor decisions, damaged reputations, and missed opportunities. You will explore why data quality is essential, how to measure it across different dimensions, and the impact of data silos on maintaining data quality.

Chapter 3, Data Profiling – Understanding Data Structure, Quality, and Distribution, explores data profiling and focuses on scrutinizing and validating datasets to understand their structure, patterns, and quality. You will learn how to perform data profiling using tools such as the pandas Profiler and Great Expectations and understand when to use each tool. Additionally, the chapter covers tactics for handling large data volumes and compares profiling methods to improve data validation.

Chapter 4, Cleaning Messy Data and Data Manipulation, focuses on the key strategies for cleaning and manipulating data, enabling efficient and accurate analysis. It covers techniques for renaming columns, removing irrelevant or redundant data, fixing inconsistent data types, and handling date and time formats. By mastering these methods, you will learn how to enhance the quality and reliability of your datasets.

Chapter 5, Data Transformation – Merging and Concatenating, explores techniques for transforming and manipulating data through merging, joining, and concatenating datasets. It covers methods to combine multiple datasets from various sources, handle duplicates effectively, and improve merging performance. The chapter also provides practical tricks to streamline the merging process, ensuring efficient data integration for insightful analysis.

Chapter 6, Data Grouping, Aggregation, Filtering, and Applying Functions, covers the essential techniques of data grouping and aggregation, which are vital for summarizing large datasets and generating meaningful insights. It discusses methods to handle missing or noisy data by aggregating values, reducing data volume, and enhancing processing efficiency. The chapter also focuses on grouping data by various keys, applying aggregate and custom functions, and filtering data to create valuable features for deeper analysis or ML.

Chapter 7, Data Sinks, focuses on the critical decisions involved in data processing, particularly the selection of appropriate data sinks for storage and processing needs. It delves into four essential pillars: choosing the right data sink, selecting the correct file type, optimizing partitioning strategies, and understanding how to design a scalable online retail data platform. The chapter equips you with the tools to enhance efficiency, scalability, and performance in data processing pipelines.

Chapter 8, Detecting and Handling Missing Values and Outliers, delves into techniques for identifying and managing missing values and outliers. It covers a range of methods, from statistical approaches to advanced ML models, to address these issues effectively. The key areas of focus include detecting and handling missing data, identifying univariate and multivariate outliers, and managing outliers in various datasets.

Chapter 9, Normalization and Standardization, covers essential preprocessing techniques such as feature scaling, normalization, and standardization, which ensure that ML models can effectively learn from data. You will explore different techniques, including scaling features to a range, Z-score scaling, and using a robust scaler, to address various data challenges in ML tasks.

Chapter 10, Handling Categorical Features, addresses the importance of managing categorical features, which represent non-numerical information in datasets. You will learn various encoding techniques, including label encoding, one-hot encoding, target encoding, frequency encoding, and binary encoding, to transform categorical data for ML models.

Chapter 11, Consuming Time Series Data, delves into the fundamentals of time series analysis, covering key concepts, methodologies, and applications across various industries. It includes understanding the components and types of time series data, identifying and handling missing values, and techniques for analyzing trends and patterns over time. The chapter also addresses dealing with outliers and feature engineering to enhance predictive modeling with time series data.

Chapter 12, Text Preprocessing in the Era of LLMs, focuses on mastering text preprocessing techniques that are essential for optimizing the performance of LLMs. It covers methods for cleaning text, handling rare words and spelling variations, chunking, and tokenization strategies. Additionally, it addresses the transformation of tokens into embeddings, highlighting the importance of adapting preprocessing approaches to maximize the potential of LLMs.

Chapter 13, Image and Audio Preprocessing with LLMs, examines preprocessing techniques for unstructured data, particularly images and audio, to extract meaningful information. It includes methods for image preprocessing, such as **optical character recognition** (**OCR**) and image caption generation with the BLIP model. The chapter also explores audio data handling, including converting audio to text using the Whisper model, providing a comprehensive overview of working with multimedia data in the context of LLMs.

To get the most out of this book

To benefit fully from this book, you should have a good knowledge of Python and a grasp of data engineering and data science basics.

Software/hardware covered in the book	Operating system requirements
Python 3	Windows, macOS, or Linux
Visual Studio Code (or your preferred IDE)	

If you are using the digital version of this book, we advise you to type the code yourself or access the code from the book's GitHub repository (a link is available in the next section). Doing so will help you avoid any potential errors related to the copying and pasting of code.

The GitHub repository follows the chapters of the book, and all the scripts are numbered according to the sections within each chapter. Each script is independent of the others, so you can move ahead without having to run all the scripts beforehand. However, it is critically recommended to follow the flow of the book so that you don't miss any necessary information.

Download the example code files

You can download the example code files for this book from GitHub at `https://github.com/PacktPublishing/Python-Data-Cleaning-and-Preparation-Best-Practices`. If there's an update to the code, it will be updated in the GitHub repository.

We also have other code bundles from our rich catalog of books and videos available at `https://github.com/PacktPublishing/`. Check them out!

Conventions used

There are a number of text conventions used throughout this book.

`Code in text`: Indicates code words in text, database table names, folder names, filenames, file extensions, pathnames, dummy URLs, user input, and Twitter handles. Here is an example: "The delete_entry() function is used to remove an entry, showing how data can be deleted from the store"

A block of code is set as follows:

```
def process_in_batches(data, batch_size):
    for i in range(0, len(data), batch_size):
        yield data[i:i + batch_size]
```

When we wish to draw your attention to a particular part of a code block, the relevant lines or items are set in bold:

```
user_satisfaction_scores = [random.randint(1, 5) for _ in range(num_
users)]
```

Any command-line input or output is written as follows:

```
$ mkdir data
pip install pandas
```

Bold: Indicates a new term, an important word, or words that you see onscreen. For instance, words in menus or dialog boxes appear in **bold**. Here is an example: "It involves storing data on remote servers accessed from anywhere via the internet, **rather than on local devices**"

> **Tips or important notes**
> Appear like this.

Get in touch

Feedback from our readers is always welcome.

General feedback: If you have questions about any aspect of this book, email us at `customercare@packtpub.com` and mention the book title in the subject of your message.

Errata: Although we have taken every care to ensure the accuracy of our content, mistakes do happen. If you have found a mistake in this book, we would be grateful if you would report this to us. Please visit `www.packtpub.com/support/errata` and fill in the form.

Piracy: If you come across any illegal copies of our works in any form on the internet, we would be grateful if you would provide us with the location address or website name. Please contact us at `copyright@packt.com` with a link to the material.

If you are interested in becoming an author: If there is a topic that you have expertise in and you are interested in either writing or contributing to a book, please visit `authors.packtpub.com`.

Share your thoughts

Once you've read *Python Data Cleaning and Preparation Best Practices*, we'd love to hear your thoughts! Scan the QR code below to go straight to the Amazon review page for this book and share your feedback.

`https://packt.link/r/1-837-63474-2`

Your review is important to us and the tech community and will help us make sure we're delivering excellent quality content.

Download a free PDF copy of this book

Thanks for purchasing this book!

Do you like to read on the go but are unable to carry your print books everywhere?

Is your eBook purchase not compatible with the device of your choice?

Don't worry, now with every Packt book you get a DRM-free PDF version of that book at no cost.

Read anywhere, any place, on any device. Search, copy, and paste code from your favorite technical books directly into your application.

The perks don't stop there, you can get exclusive access to discounts, newsletters, and great free content in your inbox daily

Follow these simple steps to get the benefits:

1. Scan the QR code or visit the link below

`https://packt.link/free-ebook/9781837634743`

2. Submit your proof of purchase
3. That's it! We'll send your free PDF and other benefits to your email directly

Part 1: Upstream Data Ingestion and Cleaning

This part focuses on the foundational stages of data processing, starting from data ingestion to ensuring its quality and structure for downstream tasks. It guides readers through the essential steps of importing, cleaning, and transforming data, which lay the groundwork for effective data analysis. The chapters explore various methods for ingesting data, maintaining high-quality datasets, profiling data for better insights, and cleaning messy data to make it ready for analysis. Further, it covers advanced techniques like merging, concatenating, grouping, and filtering data, along with choosing appropriate data destinations or sinks to optimize processing pipelines. Each chapter in this part equips readers with the knowledge to handle raw data and turn it into a clean, structured, and usable form.

This part has the following chapters:

- *Chapter 1, Data Ingestion Techniques*
- *Chapter 2, Importance of Data Quality*
- *Chapter 3, Data Profiling – Understanding Data Structure, Quality, and Distribution*
- *Chapter 4, Cleaning Messy Data and Data Manipulation*
- *Chapter 5, Data Transformation – Merging and Concatenating*
- *Chapter 6, Data Grouping, Aggregation, Filtering, and Applying Functions*
- *Chapter 7, Data Sinks*

1

Data Ingestion Techniques

Data ingestion is a critical component of the data life cycle and sets the foundation for subsequent data transformation and cleaning. It involves the process of collecting and importing data from various sources into a storage system where it can be accessed and analyzed. Effective data ingestion is crucial for ensuring data quality, integrity, and availability, which directly impacts the efficiency and accuracy of data transformation and cleaning processes. In this chapter, we will dive deep into the different types of data sources, explore various data ingestion methods, and discuss their respective advantages, disadvantages, and real-world applications.

In this chapter, we'll cover the following topics:

- Ingesting data in batch mode
- Ingesting data in streaming mode
- Real-time versus semi-real-time ingestion
- Data sources technologies

Technical requirements

You can find all the code for the chapter in the following GitHub repository:

`https://github.com/PacktPublishing/Python-Data-Cleaning-and-Preparation-Best-Practices/tree/main/chapter01`

You can use your favorite IDE (VS Code, PyCharm, Google Colab, etc.) to write and execute your code.

Ingesting data in batch mode

Batch ingestion is a data processing technique whereby large volumes of data are collected, processed, and loaded into a system at scheduled intervals, rather than in real-time. This approach allows organizations to handle substantial amounts of data efficiently by grouping data into batches, which are then processed collectively. For example, a company might collect customer transaction data throughout the day and then process it in a single batch during off-peak hours. This method is particularly useful for organizations that need to process high volumes of data but do not require immediate analysis.

Batch ingestion is beneficial because it optimizes system resources by spreading the processing load across scheduled times, often when the system is underutilized. This reduces the strain on computational resources and can lower costs, especially in cloud-based environments where computing power is metered. Additionally, batch processing simplifies data management, as it allows for the easy application of consistent transformations and validations across large datasets. For organizations with regular, predictable data flows, batch ingestion provides a reliable, scalable, and cost-effective solution for data processing and analytics.

Let's explore batch ingestion in more detail, starting with its advantages and disadvantages.

Advantages and disadvantages

Batch ingestion offers several notable advantages that make it an attractive choice for many data processing needs:

- **Efficiency** is a key benefit, as batch processing allows for the handling of large volumes of data in a single operation, optimizing resource usage and minimizing overhead
- **Cost-effectiveness** is another benefit, reducing the need for continuous processing resources and lowering operational expenses.
- **Simplicity** makes it easier to manage and implement periodic data processing tasks compared to real-time ingestion, which often requires more complex infrastructure and management
- **Robustness**, as batch processing is well-suited for performing complex data transformations and comprehensive data validation, ensuring high-quality, reliable data

However, batch ingestion also comes with certain drawbacks:

- There is an inherent delay between the generation of data and its availability for analysis, which can be a critical issue for applications requiring real-time insights.
- Resource spikes can occur during batch processing windows, leading to high resource usage and potential performance bottlenecks
- Scalability can also be a concern, as handling very large datasets may require significant infrastructure investment and management

- Lastly, maintenance is a crucial aspect of batch ingestion; it demands careful scheduling and ongoing maintenance to ensure the timely and reliable execution of batch jobs

Let's look at some common use cases for ingesting data in batch mode.

Common use cases for batch ingestion

Any data analytics platform such as data warehouses or data lakes requires regularly updated data for **Business Intelligence** (**BI**) and reporting. Batch ingestion is integral as it ensures that data is continually updated with the latest information, enabling businesses to perform comprehensive and up-to-date analyses. By processing data in batches, organizations can efficiently handle vast amounts of transactional and operational data, transforming it into a structured format suitable for querying and reporting. This supports BI initiatives, allowing analysts and decision-makers to generate insightful reports, track **Key Performance Indicators** (**KPIs**), and make data-driven decisions.

Extract, Transform, and Load (**ETL**) processes are a cornerstone of data integration projects, and batch ingestion plays a crucial role in these workflows. In ETL processes, data is extracted from various sources, transformed to fit the operational needs of the target system, and loaded into a database or data warehouse. Batch processing allows for efficient handling of these steps, particularly when dealing with large datasets that require significant transformation and cleansing. This method is ideal for periodic data consolidation, where data from disparate systems is integrated to provide a unified view, supporting activities such as data migration, system integration, and master data management.

Batch ingestion is also widely used for backups and archiving, which are critical processes for data preservation and disaster recovery. Periodic batch processing allows for the scheduled backup of databases, ensuring that all data is captured and securely stored at regular intervals. This approach minimizes the risk of data loss and provides a reliable restore point in case of system failures or data corruption. Additionally, batch processing is used for data archiving, where historical data is periodically moved from active systems to long-term storage solutions. This not only helps in managing storage costs but also ensures that important data is retained and can be retrieved for compliance, auditing, or historical analysis purposes.

Batch ingestion use cases

Batch ingestion is a methodical process involving several key steps: data extraction, data transformation, data loading, scheduling, and automation. To illustrate these steps, let's explore a use case involving an investment bank that needs to process and analyze trading data for regulatory compliance and performance reporting.

Batch ingestion in an investment bank

An investment bank needs to collect, transform, and load trading data from various financial markets into a central data warehouse. This data will be used for generating daily compliance reports, evaluating trading strategies, and making informed investment decisions.

Data extraction

The first step is identifying the sources from which data will be extracted. For the investment bank, this includes trading systems, market data providers, and internal risk management systems. These sources contain critical data such as trade execution details, market prices, and risk assessments. Once the sources are identified, data is collected using connectors or scripts. This involves setting up data pipelines that extract data from trading systems, import real-time market data feeds, and pull risk metrics from internal systems. The extracted data is then temporarily stored in staging areas before processing.

Data transformation

The extracted data often contains inconsistencies, duplicates, and missing values. Data cleaning is performed to remove duplicates, fill in missing information, and correct errors. For the investment bank, this ensures that trade records are accurate and complete, providing a reliable foundation for compliance reporting and performance analysis. After cleaning, the data undergoes transformations such as aggregations, joins, and calculations. For example, the investment bank might aggregate trade data to calculate daily trading volumes, join trade records with market data to analyze price movements, and calculate key metrics such as **Profit and Loss** (**P&L**) and risk exposure. The transformed data must be mapped to the schema of the target system. This involves aligning the data fields with the structure of the data warehouse. For instance, trade data might be mapped to tables representing transactions, market data, and risk metrics, ensuring seamless integration with the existing data model.

Data loading

The transformed data is processed in batches, which allows the investment bank to handle large volumes of data efficiently, performing complex transformations and aggregations in a single run. Once processed, the data is loaded into the target storage system, such as a data warehouse or data lake. For the investment bank, this means loading the cleaned and transformed trading data into their data warehouse, where it can be accessed for compliance reporting and performance analysis.

Scheduling and automation

To ensure that the batch ingestion process runs smoothly and consistently, scheduling tools such as Apache Airflow or Cron jobs are used. These tools automate the data ingestion workflows, scheduling them to run at regular intervals, such as every night or every day. This allows the investment bank to have up-to-date data available for analysis without manual intervention. Implementing monitoring is crucial to track the success and performance of batch jobs. Monitoring tools provide insights into job execution, identifying any failures or performance bottlenecks. For the investment bank, this ensures that any issues in the data ingestion process are promptly detected and resolved, maintaining the integrity and reliability of the data pipeline.

Batch ingestion with an example

Let's have a look at a simple example of a batch processing ingestion system written in Python. This example will simulate the ETL process. We'll generate some mock data, process it in batches, and load it into a simulated database.

You can find the code for this part in the GitHub repository at `https://github.com/PacktPublishing/Python-Data-Cleaning-and-Preparation-Best-Practices/blob/main/chapter01/1.batch.py`. To run this example, we don't need any bespoke library installation. We just need to ensure that we are running it in a standard Python environment (Python 3.x):

1. We create a `generate_mock_data` function that generates a list of mock data records:

```
def generate_mock_data(num_records):
    data = []
    for _ in range(num_records):
        record = {
            'id': random.randint(1, 1000),
            'value': random.random() * 100
        }
        data.append(record)
    return data
```

   Each record is a dictionary with two fields:

   - `id`: A random integer between 1 and 1000
   - `value`: A random float between 0 and 100

   Let's have a look at what the data looks like:

```
print("Original data:", data)
{'id': 449, 'value': 99.79699336555473}
{'id': 991, 'value': 79.65999078145887}
```

   A list of dictionaries is returned, each representing a data record.

2. Next, we create a batch processing function:

```
def process_in_batches(data, batch_size):
    for i in range(0, len(data), batch_size):
        yield data[i:i + batch_size]
```

   This function takes the data, which is a list of data records to process, and `batch_size`, which represents the number of records per batch, as parameters. The function uses a `for` loop to iterate over the data in steps of `batch_size`. The `yield` keyword is used to generate batches of data, each of the `batch_size` size. A generator that yields batches of data is returned.

3. We create a `transform_data` function that transforms each record in the batch:

```
def transform_data(batch):
    transformed_batch = []
    for record in batch:
        transformed_record = {
            'id': record['id'],
            'value': record['value'],
            'transformed_value': record['value'] * 1.1
        }
        transformed_batch.append(transformed_record)
return transformed_batch
```

This function takes as an argument the batch, which is a list of data records to be transformed. The transformation logic is simple: a new `transformed_value` field is added to each record, which is the original value multiplied by 1.1. At the end, we have a list of transformed records. Let's have a look at some of our transformed records:

```
{'id': 558, 'value': 12.15160339587219, 'transformed_value':
13.36676373545941}
{'id': 449, 'value': 99.79699336555473, 'transformed_value':
109.77669270211021}
{'id': 991, 'value': 79.65999078145887, 'transformed_value':
87.62598985960477}
```

4. Next, we create a `load_data` function to load the data. This function simulates loading each transformed record into a database:

```
def load_data(batch):
    for record in batch:
        # Simulate loading data into a database
        print(f"Loading record into database: {record}")
```

This function takes the batch as a parameter, which is a list of transformed data records that is ready to be loaded. Each record is printed to the console to simulate loading it into a database.

5. Finally, we create a `main` function. This function calls all the aforementioned functions:

```
def main():
    # Parameters
    num_records = 100 # Total number of records to generate
    batch_size = 10 # Number of records per batch

    # Generate data
```

```
    data = generate_mock_data(num_records)
    # Process and load data in batches
    for batch in process_in_batches(data, batch_size):
        transformed_batch = transform_data(batch)
        print("Batch before loading:")
        for record in transformed_batch:
            print(record)
        load_data(transformed_batch)
        time.sleep(1) # Simulate time delay between batches
```

This function calls `generate_mock_data` to create the mock data and uses `process_in_batches` to divide the data into batches. For each batch, the function does the following:

- Transforms the batch using `transform_data`
- Prints the batch to show its contents before loading
- Simulates loading the batch using `load_data`

Now, let's transition from batch processing to a streaming paradigm. In streaming, data is processed as it arrives, rather than in predefined batches.

Ingesting data in streaming mode

Streaming ingestion is a data processing technique whereby data is collected, processed, and loaded into a system in real-time, as it is generated. Unlike batch ingestion, which accumulates data for processing at scheduled intervals, streaming ingestion handles data continuously, allowing organizations to analyze and act on information immediately. For instance, a company might process customer transaction data the moment it occurs, enabling real-time insights and decision-making. This method is particularly useful for organizations that require up-to-the-minute data analysis, such as in financial trading, fraud detection, or sensor data monitoring.

Streaming ingestion is advantageous because it enables immediate processing of data, reducing latency and allowing organizations to react quickly to changing conditions. This is particularly beneficial in scenarios where timely responses are critical, such as detecting anomalies, personalizing user experiences, or responding to real-time events. Additionally, streaming can lead to more efficient resource utilization by distributing the processing load evenly over time, rather than concentrating it into specific batch windows. In cloud-based environments, this can also translate into cost savings, as resources can be scaled dynamically to match the real-time data flow. For organizations with irregular or unpredictable data flows, streaming ingestion offers a flexible, responsive, and scalable approach to data processing and analytics. Let's look at some of its advantages and disadvantages.

Advantages and disadvantages

Streaming ingestion offers several distinct advantages, making it an essential choice for specific data processing needs:

- One of the primary benefits is the ability to obtain real-time insights from data. This immediacy is crucial for applications such as fraud detection, real-time analytics, and dynamic pricing, where timely data is vital.
- Streaming ingestion supports continuous data processing, allowing systems to handle data as it arrives, thereby reducing latency and improving responsiveness.
- This method is highly scalable, as well as capable of managing high-velocity data streams from multiple sources without significant delays.

However, streaming ingestion also presents some challenges:

- Implementing a streaming ingestion system can be *complex*, requiring sophisticated infrastructure and specialized tools to manage data streams effectively.
- Continuous processing demands constant computational resources, which can be costly and resource-intensive.
- Ensuring data consistency and accuracy in a streaming environment can be difficult due to the constant influx of data and the potential for out-of-order or duplicate records

Let's look at common use cases for ingesting data in batch mode.

Common use cases for streaming ingestion

While batch processing is well-suited for periodic, large-scale data updates and transformations, streaming data ingestion is crucial for real-time data analytics and applications that require immediate insights. Here are some common use cases for streaming data ingestion.

Real-time fraud detection and security monitoring

Financial institutions use streaming data to detect fraudulent activities by analyzing transaction data in real-time. Immediate anomaly detection helps prevent fraud before it can cause significant damage. Streaming data is used in cybersecurity to detect and respond to threats immediately. Continuous monitoring of network traffic, user behavior, and system logs helps identify and mitigate security breaches as they occur.

IoT and sensor data

In manufacturing, streaming data from sensors on machinery allows for predictive maintenance. By continuously monitoring equipment health, companies can prevent breakdowns and optimize maintenance schedules.

Another interesting application in the IoT and sensors space is **smart cities**. Streaming data from various sensors across a city (traffic, weather, pollution, etc.) helps in managing city operations in real-time, improving services such as traffic management and emergency response.

Online recommendations and personalization

Streaming data enables e-commerce platforms to provide real-time recommendations to users based on their current browsing and purchasing behavior. This enhances user experience and increases sales. Platforms such as Netflix and Spotify use streaming data to update recommendations as users interact with the service, providing personalized content suggestions in real-time.

Financial market data

Stock traders rely on streaming data for up-to-the-second information on stock prices and market conditions to make informed trading decisions. Automated trading systems use streaming data to execute trades based on predefined criteria, requiring real-time data processing for optimal performance.

Telecommunications

Telecommunication companies use streaming data to monitor network performance and usage in real-time, ensuring optimal service quality and quick resolution of issues. Streaming data also helps in tracking customer interactions and service usage in real-time, enabling personalized customer support and improving the overall experience.

Real-time logistics and supply chain management

Streaming data from GPS devices allows logistics companies to track vehicle locations and optimize routes in real-time, improving delivery efficiency. Real-time inventory tracking helps businesses maintain optimal stock levels, reducing overstock and stockouts while ensuring timely replenishment.

Streaming ingestion in an e-commerce platform

Streaming ingestion is a methodical process involving several key steps: data extraction, data transformation, data loading, and monitoring and alerting. To illustrate these steps, let's explore a use case involving an e-commerce platform that needs to process and analyze user activity data in real-time for personalized recommendations and dynamic inventory management.

An e-commerce platform needs to collect, transform, and load user activity data from various sources such as website clicks, search queries, and purchase transactions into a central system. This data will be used for generating real-time personalized recommendations, monitoring user behavior, and managing inventory dynamically.

Data extraction

This is the first step is identifying the sources from which data will be extracted. For the e-commerce platform, this includes web servers, mobile apps, and third-party analytics services. These sources contain critical data such as user clicks, search queries, and transaction details. Once the sources are identified, data is collected using streaming connectors or APIs. This involves setting up data pipelines that extract data from web servers, mobile apps, and analytics services in real-time. The extracted data is then streamed to processing systems such as Apache Kafka or AWS Kinesis.

Data transformation

The extracted data often contains inconsistencies and noise. Real-time data cleaning is performed to filter out irrelevant information, handle missing values, and correct errors. For the e-commerce platform, this ensures that user activity records are accurate and relevant for analysis. After cleaning, the data undergoes transformations such as parsing, enrichment, and aggregation. For example, the e-commerce platform might parse user clickstream data to identify browsing patterns, enrich transaction data with product details, and aggregate search queries to identify trending products. The transformed data must be mapped to the schema of the target system. This involves aligning the data fields with the structure of the real-time analytics system. For instance, user activity data might be mapped to tables representing sessions, products, and user profiles, ensuring seamless integration with the existing data model.

Data loading

The transformed data is processed continuously using tools such as Apache Flink or Apache Spark Streaming. Continuous processing allows the e-commerce platform to handle high-velocity data streams efficiently, performing transformations and aggregations in real-time. Once processed, the data is loaded into the target storage system, such as a real-time database or analytics engine, where it can be accessed for personalized recommendations and dynamic inventory management.

Monitoring and alerting

To ensure that the streaming ingestion process runs smoothly and consistently, monitoring tools such as Prometheus or Grafana are used. These tools provide real-time insights into the performance and health of the data ingestion pipelines, identifying any failures or performance bottlenecks. Implementing alerting mechanisms is crucial to promptly detect and resolve any issues in the streaming ingestion process. For the e-commerce platform, this ensures that any disruptions in data flow are quickly addressed, maintaining the integrity and reliability of the data pipeline.

Streaming ingestion with an example

As we said, in streaming, data is processed as it arrives rather than in predefined batches. Let's modify the batch example to transition to a streaming paradigm. For simplicity, we will generate data continuously, process it immediately upon arrival, transform it, and then load it:

1. The `generate_mock_data` function generates records *continuously* using a generator and simulates a delay between each record:

```
def generate_mock_data():
    while True:
        record = {
            'id': random.randint(1, 1000),
            'value': random.random() * 100
        }
        yield record
        time.sleep(0.5)  # Simulate data arriving every 0.5
seconds
```

2. The `process_stream` function processes each record as it arrives from the data generator, without waiting for a batch to be filled:

```
def process_stream(run_time_seconds=10):
    start_time = time.time()
    for record in generate_mock_data():
        transformed_record = transform_data(record)
        load_data(transformed_record)
        # Check if the run time has exceeded the limit
        if time.time() - start_time > run_time_seconds:
            print("Time limit reached. Terminating the stream
processing.")
            break
```

3. The `transform_data` function transforms each record individually as it arrives:

```
def transform_data(record):
    transformed_record = {
        'id': record['id'],
        'value': record['value'],
        'transformed_value': record['value'] * 1.1  # Example
transformation
    }
    return transformed_record
```

4. The `load_data` function simulates loading data by processing each record as it arrives, instead of processing each record within a batch as before:

```
def load_data(record):
    print(f"Loading record into database: {record}")
```

Let's move from real-time to semi-real-time processing, which you can think it as batch processing over short intervals. It is usually called micro-batch processing.

Real-time versus semi-real-time ingestion

Real-time ingestion refers to the process of collecting, processing, and loading data almost instantaneously as it is generated, as we have discussed. This approach is critical for applications that require immediate insights and actions, such as fraud detection, stock trading, and live monitoring systems. Real-time ingestion provides *the lowest latency*, enabling businesses to react to events as they occur. *However, it demands robust infrastructure and continuous resource allocation, making it complex and potentially expensive to maintain.*

Semi-real-time ingestion, on the other hand, also known as near real-time ingestion, involves processing data with *minimal delay*, typically in seconds or minutes, rather than instantly. This approach *strikes a balance between real-time and batch processing*, providing timely insights while reducing the resource intensity and complexity associated with true real-time systems. Semi-real-time ingestion is suitable for applications such as social media monitoring, customer feedback analysis, and operational dashboards, where near-immediate data processing is beneficial but not critically time-sensitive.

Common use cases for near-real-time ingestion

Let's look at some of the common use cases wherein we can use near-real-time ingestion.

Real-time analytics

Streaming enables organizations to continuously monitor data as it flows in, allowing for real-time dashboards and visualizations. This is critical in industries such as finance, where stock prices, market trends, and trading activities need to be tracked live. It also allows for instant report generation, facilitating timely decision-making and reducing the latency between data generation and analysis.

Social media and sentiment analysis

Companies track mentions and sentiments on social media in real-time to manage brand reputation and respond to customer feedback promptly. Streaming data allows for the continuous analysis of public sentiment towards brands, products, or events, providing immediate insights that can influence marketing and PR strategies.

Customer experience enhancement

Near-real-time processing allows support teams to access up-to-date information on customer issues and behavior, enabling quicker and more accurate responses to customer inquiries. Businesses can also use near-real-time data to update customer profiles and trigger personalized marketing messages, such as emails or notifications, shortly after a customer interacts with their website or app.

Semi-real-time mode with an example

Transitioning from real-time to semi-real-time data processing involves adjusting the example to introduce a more structured approach to handling data updates, rather than processing each record immediately upon arrival. This can be achieved by batching data updates over short intervals, which allows for more efficient processing while still maintaining a responsive data processing pipeline. Let's have a look at the example and as always, you can find the code in the GitHub repository `https://github.com/PacktPublishing/Python-Data-Cleaning-and-Preparation-Best-Practices/blob/main/chapter01/3.semi_real_time.py`:

1. For generating mock data continuously, there are no changes from the previous example. This continuously generates mock data records with a slight delay (`time.sleep(0.1)`).
2. For processing in semi-real-time, we can use a deque to buffer incoming records. This function processes records when either the specified time interval has elapsed, or the buffer reaches a specified size (`batch_size`). Then, it converts the deque to a list (`list(buffer)`) before passing it to `transform_data`, ensuring the data is processed in a batch:

```
def process_semi_real_time(batch_size, interval):
    buffer = deque()
    start_time = time.time()

    for record in generate_mock_data():
        buffer.append(record)
```

3. Check whether the interval has elapsed, or the buffer size has been reached:

```
        if (time.time() - start_time) >= interval or len(buffer)
>= batch_size:
```

4. Process and clear the buffer:

```
            transformed_batch = transform_data(list(buffer))  #
Convert deque to list
            print(f"Batch of {len(transformed_batch)} records
before loading:")
            for rec in transformed_batch:
                print(rec)
```

```
            load_data(transformed_batch)
            buffer.clear()
            start_time = time.time()  # Reset start time
```

5. Then, we transform each record in the batch. There are no changes from the previous example and we load the data.

When you run this code, it continuously generates mock data records. Records are buffered until either the specified time interval (`interval`) has elapsed, or the buffer reaches the specified size (`batch_size`). Once the conditions are met, the buffered records are processed as a batch, transformed, and then "loaded" (printed) into the simulated database.

When discussing the different types of data sources that are suitable for batch, streaming, or semi-real-time streaming processing, it's essential to consider the diversity and characteristics of these sources. Data can originate from various sources, such as databases, logs, IoT devices, social media, or sensors, as we will see in the next section.

Data source solutions

In the world of modern data analytics and processing, the diversity of data sources available for ingestion spans a wide spectrum. From traditional file formats such as CSV, JSON, and XML to robust database systems encompassing both SQL and NoSQL variants, the landscape expands further to include dynamic APIs such as REST, facilitating real-time data retrieval. Message queues such as Kafka offer scalable solutions for handling event-driven data while streaming services such as Kinesis and pub/sub enable continuous data flows crucial for applications demanding immediate insights. Understanding and effectively harnessing these diverse data ingestion sources is fundamental to building robust data pipelines that support a broad array of analytical and operational needs.

Let's start with event processing.

Event data processing solution

In a real-time processing system, data is ingested, processed, and responded to almost instantaneously, as we've discussed. Real-time processing systems often use message queues to handle incoming data streams and ensure that data is processed in the order it is received, without delays.

The following Python code demonstrates a basic example of using a message queue for processing messages, which is a foundational concept in both real-time and semi-real-time data processing systems. The `Queue` class from Python's `queue` module is used to create a queue—a data structure that follows the **First-in-First-out** (**FIFO**) principle. In this context, a queue is used to manage messages or tasks that need to be processed. The code simulates an event-based system where messages (in this case, strings such as `message 0`, `message 1`, etc.) are added to a queue. This mimics a scenario wherein events or tasks are generated and need to be processed in the order they arrive. Let's have a look at each part of the code. You can find the code file at `https://github.com/`

PacktPublishing/Python-Data-Cleaning-and-Preparation-Best-Practices/blob/main/chapter01/4.work_with_queue.py:

1. The `read_message_queue()` function initializes a queue object `q` using the `Queue` class from the `queue` module:

   ```
   def read_message_queue():
       q = Queue()
   ```

2. This loop adds 10 messages to the queue. Each message is a string in the format message `i`, where `i` ranges from 0 to 9:

   ```
   for i in range(10): # Mocking messages
       q.put(f"message {i}")
   ```

3. This loop continuously retrieves and processes messages from the queue until it is empty. `q.get()` retrieves a message from the queue, and `q.task_done()` signals that the retrieved message has been processed:

   ```
   while not q.empty():
       message = q.get()
       process_message(message)
       q.task_done() # Signal that the task is done
   ```

4. The following function takes a message as input and prints it to the console, simulating the processing of the message:

   ```
   def process_message(message):
       print(f"Processing message: {message}")
   ```

5. Call the `read_message_queue` function:

   ```
   read_message_queue()
   ```

Here, the `read_message_queue` function reads messages from the queue and processes them one by one using the `process_message` function. This demonstrates how event-based systems handle tasks—by placing them in a queue and processing each task as it becomes available.

The `while not q.empty()` loop ensures that each message is processed in the exact order it was added to the queue. This is crucial in many real-world applications where the order of processing matters, such as in handling user requests or processing logs.

The `q.task_done()` method signals that a message has been processed. This is important in real-world systems where tracking the completion of tasks is necessary for ensuring reliability and correctness, especially in systems with multiple workers or threads.

In real-world applications, message queues are often integrated into more sophisticated data streaming platforms to ensure scalability, fault tolerance, and high availability. For instance, in real-time data processing, platforms such as Kafka and AWS Kinesis come into play.

Ingesting event data with Apache Kafka

There are different technologies to ingest and handle event data. One of the technologies we will discuss is Apache Kafka. Kafka is an open source distributed event streaming platform first developed by LinkedIn and later donated to the Apache Software Foundation. It is designed to handle large amounts of data in real-time and provides a scalable and fault-tolerant system for processing and storing streams.

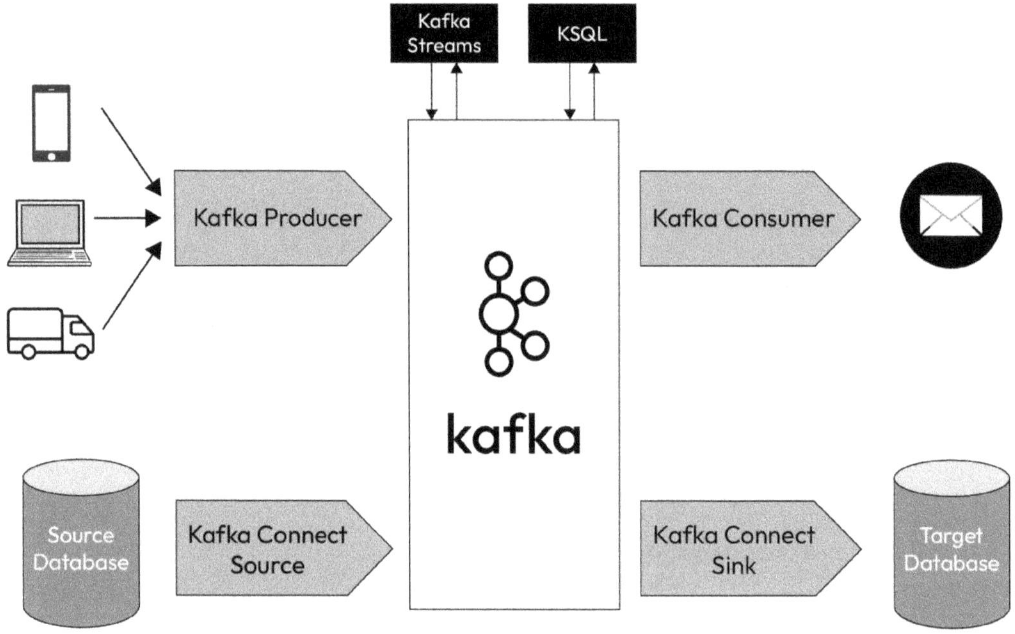

Figure 1.1 – Components of Apache Kafka

Let's see the different components of Apache Kafka:

- **Ingestion**: Data streams can be ingested into Kafka using Kafka producers. Producers are applications that write data to Kafka topics, which are logical channels that can hold and organize data streams.
- **Processing**: Kafka can process streams of data using Kafka Streams, a client library for building real-time stream processing applications. Kafka Streams allows developers to build custom stream-processing applications that can perform transformations, aggregations, and other operations on data streams.

- **Storage***:* Kafka stores data streams in distributed, fault-tolerant clusters called Kafka brokers. Brokers store the data streams in partitions, which are replicated across numerous brokers for fault tolerance.
- **Consumption***:* Data streams can be consumed from Kafka using Kafka consumers. Consumers are applications that read data from Kafka topics and process it as needed.

Several libraries can be used to interact with Apache Kafka in Python; we will explore the most popular ones in the next section.

Which library should you use for your use case?

`Kafka-Python` is a pure Python implementation of Kafka's protocol, offering a more Pythonic interface for interacting with Kafka. It is designed to be simple and easy to use, making it particularly appealing for beginners. One of its primary advantages is its simplicity, making it easier to install and use compared to other Kafka libraries. Kafka-Python is flexible and well-suited for small to medium-sized applications, providing the essential features needed for basic Kafka operations without the complexity of additional dependencies. Its pure Python nature means that it does not rely on any external libraries beyond Python itself, streamlining the installation and setup process.

`Confluent-kafka-python` is a library developed and maintained by Confluent, the original creator of Kafka. It stands out for its high-performance and low-latency capabilities, leveraging the `librdkafka` C library for efficient operations. The library offers extensive configuration options akin to the Java Kafka client and closely aligns with Kafka's feature set, often pioneering support for new Kafka features. It is particularly well-suited for production environments where both performance and stability are crucial, making it an ideal choice for handling high-throughput data streams and ensuring reliable message processing in critical applications.

Transitioning from event data processing to databases involves shifting focus from real-time data streams to persistent data storage and retrieval. While event data processing emphasizes handling continuous streams of data in near real-time for immediate insights or actions, databases are structured repositories designed for storing and managing data over the long term.

Ingesting data from databases

Databases, whether relational or non-relational, serve as foundational components in data management systems. Classic databases and NoSQL databases are two different types of database management systems that differ in architecture and characteristics. A classic database, also known as a relational database, stores data in tables with a fixed schema. Classic databases are ideal for applications that require complex querying and transactional consistency, such as financial systems or enterprise applications.

On the other hand, NoSQL databases do not store data in tables with a fixed schema. They use a document-based approach to store data in a flexible schema format. They are designed to be scalable and handle large amounts of data, with a focus on high-performance data retrieval. NoSQL databases are well-suited for applications that require high performance and scalability, such as real-time analytics, content management systems, and e-commerce platforms.

Let's start with relational databases.

Performing data ingestion from a relational database

Relational databases are useful for batch ETL processes where structured data from various sources needs consolidation, transformation, and loading into a data warehouse or analytical system. SQL-based operations are efficient for joining and aggregating data before processing. Let's try to understand how SQL databases represent data in tables with rows and columns using a code example. We'll simulate a basic SQL database interaction using Python dictionaries to represent tables and rows. You can see the full code example at `https://github.com/PacktPublishing/Python-Data-Cleaning-and-Preparation-Best-Practices/blob/main/chapter01/5.sql_databases.py`:

1. We create a `read_sql` function that simulates reading rows from a SQL table, represented here as a list of dictionaries where each dictionary corresponds to a row in the table:

```
def read_sql():
# Simulating a SQL table with a dictionary
    sql_table = [
        {"id": 1, "name": "Alice", "age": 30},
        {"id": 2, "name": "Bob", "age": 24},
    ]
    for row in sql_table:
        process_row(row)
```

2. The `process_row` function takes a row (dictionary) as input and prints its contents, simulating the processing of a row from a SQL table:

```
def process_row(row):
    print(f"Processing row: id={row['id']}, name={row['name']},
age={row['age']}")

read_sql()
```

3. Let's print our SQL table in the proper format:

```
print(f"{'id':<5} {'name':<10} {'age':<3}")
print("-" * 20)
# Print each row
for row in sql_table:
    print(f"{row['id']:<5} {row['name']:<10} {row['age']:<3}")
```

This will print the following output:

```
id   name      age
------------------
```

```
1    Alice    30
2    Bob      24
```

The key to learning from the previous example is understanding how SQL databases structure and manage data through tables composed of rows and columns, and how to efficiently retrieve and process these rows programmatically. This knowledge is crucial because it lays the foundation for effective database management and data manipulation in any application.

In real-world applications, this interaction is often facilitated by libraries and drivers such as **Java Database Connectivity** (JDBC) or **Open Database Connectivity** (**ODBC**), which provide standardized methods for connecting to and querying databases. These libraries are typically wrapped by higher-level frameworks or libraries in Python, making it easier for developers to ingest data from various SQL databases without worrying about the underlying connectivity details. Several libraries can be used to interact with SQL databases using Python; we will explore the most popular ones in the following section.

Which library should you use for your use case?

Let's explore the different libraries available for interacting with SQL databases in Python, and understand when to use each one:

- **SQLite** (sqlite3) is ideal for small to medium-sized applications, local storage, and prototyping. Its zero-configuration, serverless architecture makes it perfect for lightweight, embedded database needs and quick development cycles. It is especially useful in scenarios where the overhead of a full-fledged database server is unnecessary. Avoid using sqlite3 for applications requiring high concurrency or extensive write operations, or where multiple users need to access the database simultaneously. It is not suitable for large-scale applications or those needing robust security features and advanced database functionalities.
- **SQLAlchemy** is suitable for applications requiring a high level of abstraction over raw SQL, support for multiple database engines, and complex queries and data models. It is ideal for large-scale production environments that need flexibility, scalability, and the ability to switch between different databases with minimal code changes. Avoid using SQLAlchemy for small, lightweight applications where the overhead of its comprehensive ORM capabilities is unnecessary. If you need direct, low-level access to a specific database's features and are comfortable writing raw SQL queries, a simpler database adapter such as sqlite3, Psycopg2, or MySQL Connector/Python might be more appropriate.
- **Psycopg2** is the go-to choice for interacting with PostgreSQL databases, making it suitable for applications that leverage PostgreSQL's advanced features, such as ACID compliance, complex queries, and extensive data types. It is ideal for production environments requiring reliability and efficiency in handling PostgreSQL databases. Avoid using Psycopg2 if your application does not interact with PostgreSQL. If you need compatibility with multiple database systems or a higher-level abstraction, consider using SQLAlchemy instead. Also, it might not be the best choice for lightweight applications where the overhead of a full PostgreSQL setup is unnecessary.

- **MySQL Connector/Python** (`mysql-connector-python`) is great for applications that need to interact directly with MySQL databases. It is suitable for environments where compatibility and official support from Oracle are critical, as well as for applications leveraging MySQL's features such as transaction management and connection pooling. Do not use MySQL Connector/Python if your application requires compatibility with multiple database systems or a higher-level abstraction. For simpler applications where the overhead of a full MySQL setup is unnecessary, or where MySQL's features are not specifically needed, consider other lightweight alternatives.

After understanding the various libraries and their use cases for interacting with SQL databases, it's equally important to explore alternatives for scenarios where the traditional relational model of SQL databases may not be the best fit. This brings us to NoSQL databases, which offer flexibility, scalability, and performance for handling unstructured or semi-structured data. Let's delve into the key Python libraries for working with popular NoSQL databases and examine when and how to use them effectively.

Performing data ingestion from the NoSQL database

Non-relational databases can be used for storing and processing large volumes of semi-structured or unstructured data in batch operations. They are particularly effective when the schema can evolve or when handling diverse data types in a consolidated manner. NoSQL databases excel in streaming and semi-real-time workloads due to their ability to handle high throughput and low-latency data ingestion. They are commonly used for capturing and processing real-time data from IoT devices, logs, social media feeds, and other sources that generate continuous streams of data.

The provided Python code mocks a NoSQL database with a dictionary and processes each key-value pair. Let's have a look at each part of the code:

1. The `process_entry` function takes a key and its associated value from the data store and prints a formatted message showing the processing of that key-value pair. It provides a simple way to view or handle individual entries, highlighting how data is accessed and manipulated based on its key:

   ```
   def process_entry(key, value):
       print(f"Processing key: {key} with value: {value}")
   ```

2. The following function prints the entire `data_store` dictionary in a tabular format:

   ```
   def print_data_store(data_store):
       print(f"{'Key':<5} {'Name':<10} {'Age':<3}")
       print("-" * 20)
       for key, value in data_store.items():
           print(f"{key:<5} {value['name']:<10} {value['age']:<3}")
   ```

   It starts by printing column headers for `Key`, `Name`, and `Age`, followed by a separator line for clarity. It then iterates over all key-value pairs in the `data_store` dictionary, printing each entry's key, name, and age. This function helps visualize the current state of the data store. The initial state of the data is as follows:

```
Initial Data Store:
Key   Name      Age
-----------------------
1     Alice     30
2     Bob       24
```

3. This function adds a new entry to the `data_store` dictionary:

```
def create_entry(data_store, key, value):
    data_store[key] = value
    return data_store
```

It takes a key and a value, then inserts the value into `data_store` under the specified key. The updated `data_store` dictionary is then returned. This demonstrates the ability to add new data to the store, showcasing the creation aspect of **Create, Read, Update, and Delete (CRUD)** operations.

4. The `update_entry` function updates an existing entry in the `data_store` dictionary:

```
def update_entry(data_store, key, new_value):
    if key in data_store:
        data_store[key] = new_value
    return data_store
```

It takes a key and `new_value`, and if the key exists in the `data_store` dictionary, it updates the corresponding value with `new_value`. The updated `data_store` dictionary is then returned. This illustrates how existing data can be modified, demonstrating the update aspect of CRUD operations.

5. The following function removes an entry from the `data_store` dictionary:

```
def delete_entry(data_store, key):
    if key in data_store:
        del data_store[key]
    return data_store
```

It takes a key, and if the key is found in the `data_store` dictionary, it deletes the corresponding entry. The updated `data_store` dictionary is then returned.

6. The following function wraps all the process together:

```
def read_nosql():
    data_store = {
        "1": {"name": "Alice", "age": 30},
        "2": {"name": "Bob", "age": 24},
    }
```

```
    print("Initial Data Store:")
    print_data_store(data_store)

    # Create: Adding a new entry
    new_key = "3"
    new_value = {"name": "Charlie", "age": 28}
    data_store = create_entry(data_store, new_key, new_value)

    # Read: Retrieving and processing an entry
    print("\nAfter Adding a New Entry:")
    process_entry(new_key, data_store[new_key])

    # Update: Modifying an existing entry
    update_key = "1"
    updated_value = {"name": "Alice", "age": 31}
    data_store = update_entry(data_store, update_key, updated_
value)

    # Delete: Removing an entry
    delete_key = "2"
    data_store = delete_entry(data_store, delete_key)

    # Print the final state of the data store
    print("\nFinal Data Store:")
    print_data_store(data_store)
```

This code illustrates the core principles of NoSQL databases, including schema flexibility, key-value pair storage, and basic CRUD operations. It begins with the `read_nosql()` function, which simulates a NoSQL database using a dictionary, `data_store`, where each key-value pair represents a unique identifier and associated user information. Initially, the `print_data_store()` function displays the data in a tabular format, highlighting the schema flexibility inherent in NoSQL systems. The code then demonstrates CRUD operations. It starts by adding a new entry with the `create_entry()` function, showcasing how new data is inserted into the store. Following this, the `process_entry()` function retrieves and prints the details of the newly added entry, illustrating the read operation. Next, the `update_entry()` function modifies an existing entry, demonstrating the update capability of NoSQL databases. The `delete_entry()` function is used to remove an entry, showing how data can be deleted from the store. Finally, the updated state of the `data_store` dictionary is printed again, providing a clear view of how the data evolves through these operations.

7. Let's execute the whole process:

```
read_nosql()
```

This returns the final datastore:

```
Final Data Store:
Key   Name      Age
----------------------
1     Alice     31
2     Charlie   28
```

In the preceding example, we demonstrated an interaction with a *mocked* NoSQL system using Python so that we can showcase the core principles of NoSQL databases such as schema flexibility, key-value pair storage, and basic CRUD operations. We can now better grasp how NoSQL databases differ from traditional SQL databases in terms of data modeling and handling unstructured or semi-structured data efficiently.

There are several libraries that can be used to interact with NoSQL databases. In the next section, we will explore the most popular ones.

Which library should you use for your use case?

Let's explore the different libraries available for interacting with NoSQL databases in Python, and understand when to use each one:

- `pymongo` is the official Python driver for MongoDB, a popular NoSQL database known for its flexibility and scalability. `pymongo` allows Python applications to interact seamlessly with MongoDB, offering a straightforward API to perform CRUD operations, manage indexes, and execute complex queries. `pymongo` is particularly favored for its ease of use and compatibility with Python's data structures, making it suitable for a wide range of applications from simple prototypes to large-scale production systems.
- `cassandra-driver` (Cassandra): The `cassandra-driver` library provides Python applications with direct access to Apache Cassandra, a highly scalable NoSQL database designed for handling large amounts of data across distributed commodity servers. Cassandra's architecture is optimized for write-heavy workloads and offers tunable consistency levels, making it suitable for real-time analytics, IoT data, and other applications requiring high availability and fault tolerance.

Transitioning from databases to file systems involves shifting the focus from structured data storage and retrieval mechanisms to more flexible and versatile storage solutions.

Performing data ingestion from cloud-based file systems

Cloud storage is a service model that allows data to be remotely maintained, managed, and backed up over the internet. It involves storing data on remote servers accessed from anywhere via the internet, *rather than on local devices*. Cloud storage has revolutionized the way we store and access data. It provides a flexible and scalable solution for individuals and organizations, enabling them to store large amounts of data *without investing in physical hardware*. This is particularly useful for ensuring that data is always accessible and can be shared easily.

Amazon S3, Microsoft Azure Blob Storage, and Google Cloud Storage are all cloud-based object storage services that allow you to store and retrieve files in the cloud. Cloud-based file systems are becoming increasingly popular for several reasons.

Firstly, they provide a flexible and scalable storage solution that can easily adapt to the changing needs of an organization. This means that as the amount of data grows, additional storage capacity can be added without the need for significant capital investment or physical infrastructure changes. Thus, it can help reduce capital expenditures and operational costs associated with maintaining and upgrading on-premises storage infrastructure.

Secondly, cloud-based file systems offer high levels of accessibility and availability. With data stored in the cloud, users can access it from anywhere with an internet connection, making it easier to collaborate and share information across different teams, departments, or locations. Additionally, cloud-based file systems are designed with redundancy and failover mechanisms, ensuring that data is always available even in the event of a hardware failure or outage. Finally, they provide enhanced security features to protect data from unauthorized access, breaches, or data loss. Cloud service providers typically have advanced security protocols, encryption, and monitoring tools to safeguard data and ensure compliance with data privacy regulations.

Files in cloud-based storage systems are essentially the same as those on local devices, but they are stored on remote servers and accessed over the internet. However, how are these files organized in these cloud storage systems? Let's discuss that next.

Organizing files in cloud storage systems

One of the primary methods of organizing files in cloud storage is by using folder structures, similar to local file systems. Users can create folders and subfolders to categorize and store files systematically. Let's have a look at some best practices:

- Creating a logical and intuitive hierarchy that reflects how you work or how your projects are structured is essential. This involves designing a folder structure that mimics your workflow, making it easier to locate and manage files. For instance, you might create main folders for different departments, projects, or clients, with subfolders for specific tasks or document types. This hierarchical organization not only saves time by reducing the effort needed to find files but also enhances collaboration by providing a clear and consistent framework that team members can easily navigate.
- Using *consistent naming conventions* for folders and files is crucial for ensuring easy retrieval and maintaining order within your cloud storage. A standardized naming scheme helps avoid confusion, reduces errors, and speeds up the process of locating specific documents. For example, adopting a format such as `YYYY-MM-DD_ProjectName_DocumentType` can provide immediate context and make sorting and searching more straightforward. Consistent

naming also facilitates automation and integration with other tools, as predictable file names can be more easily processed by scripts and applications.

- Grouping files by project or client is an effective way to keep related documents together and streamline project management. This method involves creating dedicated folders for each project or client, where all relevant files, such as contracts, communications, and deliverables, are stored.
- Many cloud storage systems allow tagging files with keywords or metadata, which significantly enhances file categorization and searchability. Tags are essentially labels that you can attach to files, making it easier to group and find documents based on specific criteria. Metadata includes detailed information, such as the author, date, project name, and file type, which provides additional context and aids in more precise searches. By using relevant tags and comprehensive metadata, you can quickly filter and locate files, regardless of their location within the folder hierarchy. This practice is particularly useful in large storage systems where traditional folder structures might become cumbersome.

From discussing cloud storage systems, the focus now shifts to exploring the capabilities and integration opportunities offered by APIs.

APIs

APIs have become increasingly popular in recent years due to their ability to enable seamless communication and integration between different systems and services. APIs provide developers with a standardized and flexible way to access data and functionality from other systems, allowing them to easily build new applications and services that leverage existing resources. APIs have become a fundamental building block for modern software development and are widely used across a wide range of industries and applications.

Now that we understand what APIs represent, let's move on to the `requests` Python library with which developers can programmatically access and manipulate data from remote servers.

The requests library

When it comes to working with APIs in Python, the `requests` library is the go-to Python library for making HTTP requests to APIs and other web services. It makes it easy to send HTTP/1.1 requests using Python, and it provides many convenient features for working with HTTP responses.

Run the following command to install the `requests` library:

```
pip install requests==2.32.3
```

Let's have a quick look at how we can use this library:

1. Import the `requests` library:

   ```
   import requests
   ```

2. Specify the API endpoint URL:

```
url = "https://jsonplaceholder.typicode.com/posts"
```

3. Make a GET request to the API endpoint:

```
response = requests.get(url)
```

4. Get the response content:

```
print(response.content)
```

Here, we're making a GET request to the API endpoint at `https://jsonplaceholder.typicode.com/posts` and storing the `response` object in the `response` variable. We can then print the response content using the `content` attribute of the `response` object. The `requests` library provides many other methods and features for making HTTP requests, including support for POST, PUT, DELETE, and other HTTP methods, handling headers and cookies, and handling redirects and authentication.

Now that we've explained the `requests` library, let's move on to a specific example of retrieving margarita cocktail data from the `Cocktail DB` API, which can illustrate how practical web requests can be in accessing and integrating real-time data sources into applications.

Learn how to make a margarita!

The use case demonstrates retrieving cocktail data from the `Cocktail DB` API using Python. If you want to improve your bartending skills and impress your friends, you can use an open API to get real-time information on the ingredients required for any cocktail. For this, we will use the `Cocktail DB` API and the `request` library to see which ingredients we need for a margarita:

1. Define the API endpoint URL. We are making a request to the Cocktail DB API endpoint to search for cocktails with the `margarita` name:

```
url = "https://www.thecocktaildb.com/api/json/v1/1/search.
php?s=margarita"
```

2. Make the API request. We define the API endpoint URL as a string and pass it to the `requests.get()` function to make the GET request:

```
response = requests.get(url)
```

3. Check whether the request was successful (status code `200`) and get the data. The API response is returned as a JSON string, which we can extract by calling the `response.json()` method. We then assign this JSON data to a variable called `data`:

```
if response.status_code == 200:
    # Extract the response JSON data
```

```
    data = response.json()
    # Check if the API response contains cocktails data
    if 'drinks' in data:
        # Create DataFrame from drinks data
        df = pd.DataFrame(data['drinks'])
        # Print the resulting DataFrame
        print(df.head())
    else:
        print("No drinks found.")
```

4. If the request was not successful, print this error message:

```
else:
    print(f"Failed to retrieve data from API. Status code:
{response.status_code}")
```

You can replace the `margarita` search parameter with any other cocktail name or ingredient to get data for different drinks.

With this, we come to the end of our first chapter. Let's summarize what we have learned so far.

Summary

Throughout this chapter, we covered essential technologies in modern computing and data management. We began by discussing batch ingestion, a method whereby large volumes of data are collected and processed at scheduled intervals, offering efficiency and cost-effectiveness for organizations with predictable data flows. In contrast, we explored streaming ingestion, which allows data to be processed in real-time, enabling immediate analysis and rapid response to changing conditions. We followed with streaming services such as Kafka for real-time data processing. We moved to SQL and NoSQL databases—such as PostgreSQL, MySQL, MongoDB, and Cassandra—highlighting their strengths in structured and flexible data storage, respectively. We explored APIs such as REST for seamless system integration. Also, we delved into file systems, file types, and attributes, alongside cloud storage solutions such as Amazon S3 and Google Cloud Storage, emphasizing scalability and data management strategies. These technologies collectively enable robust, scalable, and efficient applications in today's digital ecosystem.

In the upcoming chapter, we will dive deep into the critical aspects of data quality and its significance in building reliable data products. We'll explore why ensuring high data quality is paramount for making informed business decisions, enhancing customer experiences, and maintaining operational efficiency.

2

Importance of Data Quality

Did you know that data serves as the backbone of many important business decisions? Without accurate, complete, and consistent information, companies risk making faulty judgments that could potentially damage their reputation, client relationships, and business overall. Consistency issues across different datasets can cause confusion and prevent meaningful analysis from happening. Irrelevant or outdated data can misguide the judgment of decision-makers, resulting in suboptimal choices. On the other hand, building high-quality data products serves as a powerful asset, empowering organizations to make informed decisions, uncover valuable insights, identify trends, mitigate risks, and gain a competitive edge.

In this chapter, we will dive deep into the following topics:

- Why data quality is important
- Different dimensions to measure data quality in your data products
- The impact of data silos on data quality

Technical requirements

You can find all the code for the chapter in the following GitHub repository:

`https://github.com/PacktPublishing/Python-Data-Cleaning-and-Preparation-Best-Practices/tree/main/chapter02`

Why data quality is important

Allow me to unveil the reasons why data quality matters:

- **Accurate data can give you a competitive advantage**: Organizations depend on data to determine patterns, trends, preferences, and other vital aspects governing their ecosystem. If your data quality is subpar, the resulting analysis and conclusions may be skewed, resulting in wrong moves that could jeopardize your entire business.

- **Complete data is the backbone of cost optimization**: Data forms the foundation of automation and optimization, which can drive up productivity and lower expenses when executed properly. Incomplete or low-quality data can cause bottlenecks and increase costs. Imagine countless man-hours wasted on fixing errors that would have been avoided if only there had been higher standards set for data entry.
- **Top-notch data can lead to satisfied customers who stick around long term**: The heartbeat of every business depends on satisfied clients, whose loyalty can ensure sustained growth over time. Wrong data about customers might translate into personalized experiences that don't match customer characteristics or even into inaccurate billings and unfulfilled requests. These disappointed customers may take their business elsewhere, leaving your company struggling to survive.
- **Compliant data is a requirement to avoid unwanted legal consequences**: Several industries must follow specific rules concerning data precision, safety, and privacy. Adherence requires a high level of data quality to satisfy strict guidelines and prevent legal penalties along with the potential loss of consumer confidence.
- **To avoid data silos, you need to trust your data**: When various entities within an enterprise must work together to leverage data, ensuring its integrity is essential. Incompatibility or discrepancies may obstruct cooperation and can hinder integration efforts and create data silos.
- **Data quality actually means trust**: Data quality directly impacts the trust and credibility stakeholders place in an organization. By maintaining high-quality data, organizations can foster trust among customers, partners, and investors.

Now that we have a clearer picture of why data quality is important, let's move to the next section, where we will dive deep into the different dimensions of data quality.

Dimensions of data quality

As emphasized earlier, superior data quality forms the foundation upon which informed decisions and strategic insights are built. With this in mind, let us now examine which **Key Performance Indicators (KPIs)** we could use to measure the data quality of our assets.

Completeness

Completeness measures the extent to which data is complete and lacks missing values or fields. KPIs can include metrics such as the percentage of missing data or missing data points per record.

The following code will output the completeness percentages for each column in your dataset. A higher percentage indicates a higher level of completeness, while a lower percentage suggests more missing values:

1. We'll start by importing the `pandas` library to work with the dataset:

   ```
   import pandas as pd
   ```

2. Next, we create a sample dataset with the following columns: `Name`, `Age`, `Gender`, and `City`. Some values are intentionally missing (represented as `None`):

   ```
   data = {
      'Name': ['Alice', 'Bob', 'Charlie', 'David', 'Eve'],
      'Age': [25, 30, None, 28, 22],
      'Gender': ['Female', 'Male', 'Male', 'Male', 'Female'],
      'City': ['New York', 'Los Angeles', 'Chicago', None, 'San
   Francisco']
   }
   ```

3. Then, we create a pandas DataFrame:

   ```
   df = pd.DataFrame(data)
   ```

4. We'll then use the `isnull()` function to identify missing values in each column and use the `sum()` function to count the total number of missing values for each column:

   ```
   completeness = df.isnull().sum()
   ```

5. Next, we'll calculate the completeness percentage:

   ```
   total_records = len(df)
   completeness_percentage = (1- completeness / total_records) *
   100
   ```

   This will print the following output:

   ```
   Completeness Check:
   Name      0
   Age       1
   Gender    0
   City      1

   Completeness Percentage:
   Name      100.0
   Age        80.0
   Gender    100.0
   City       80.0
   ```

The completeness check shows the number of missing values for each column, and the completeness percentage indicates the proportion of missing values in each column relative to the total number of records. This output indicates that the `Name` and `Gender` columns have 100% completeness, whereas the `Age` and `City` columns have 80%.

> **Note – completeness percentage**
>
> The completeness percentage is calculated by dividing the count of missing values by the total number of records in the dataset and then multiplying by 100. It represents the proportion of missing values relative to the size of the dataset.
>
> **The higher the percentage of completeness is, the better!**

Accuracy

Accuracy assesses the correctness of data by comparing it to a trusted source or benchmark.

The following code will output the accuracy percentage based on the comparison between the actual and expected values:

1. We'll start by loading the required libraries:

```
import pandas as pd
```

2. Next, we create a sample dataset named `data` and a reference dataset named `reference_data`. Both datasets have the same structure (columns: `Name`, `Age`, `Gender`, and `City`):

```
data = {
  'Name': ['Alice', 'Bob', 'Charlie', 'David', 'Eve'],
  'Age': [25, 30, 28, 28, 22],
  'Gender': ['Female', 'Male', 'Male', 'Male', 'Female'],
  'City': ['New York', 'Los Angeles', 'Chicago', 'New York',
'San Francisco']
}

# Reference dataset for accuracy comparison
reference_data = {
  'Name': ['Alice', 'Bob', 'Charlie', 'David', 'Eve'],
  'Age': [25, 30, 29, 28, 22],
  'Gender': ['Female', 'Male', 'Male', 'Male', 'Female'],
  'City': ['New York', 'Los Angeles', 'Chicago', 'New York',
'San Francisco']
}
```

3. We create two pandas DataFrames, named `df` and `reference_df`, using the sample data and the reference data, respectively:

```
df = pd.DataFrame(data)
reference_df = pd.DataFrame(reference_data)
```

4. We create an `accuracy_check` variable and assign the result of the comparison between `df` and `reference_df` to it. This comparison is done using the == operator, which returns `True` for matching values and `False` for non-matching values:

```
accuracy_check = df == reference_df
```

We can compare the column with actual values to the column with expected values using the == operator.

5. We then calculate the accuracy percentage by taking the mean of the `accuracy_check` DataFrame for each column. The `mean` operation treats `True` as `1` and `False` as `0`, so it effectively calculates the percentage of matching values in each column:

```
accuracy_percentage = accuracy_check.mean() * 100
```

6. Finally, we print the results:

```
# Display the accuracy results
print("Accuracy Check:")
print(accuracy_check)
print("\nAccuracy Percentage:")
print(accuracy_percentage)
```

The output is as follows:

```
Accuracy Check:
   Name    Age  Gender  City
0  True   True    True  True
1  True   True    True  True
2  True  False    True  True
3  True   True    True  True
4  True   True    True  True

Accuracy Percentage:
Name      100.0
Age        80.0
Gender    100.0
City      100.0
```

The accuracy check shows `True` where the data matches the reference dataset and `False` where it doesn't. The accuracy percentage indicates the proportion of matching values in each column relative to the total number of records. This output indicates that the `Age` column is the only one that needs some more attention in this case. All the others are 100% accurate.

> **Note – accuracy percentage**
>
> An accuracy percentage can be calculated by taking the mean (average) of the comparison results for all columns and multiplying by 100. This represents the proportion of matching data values relative to the total number of data points.
>
> The higher the percentage of accuracy is, the better!

Wondering how to build the ground truth dataset?

The ground truth must be representative of the task you are trying to solve. This means that depending on where you are in the data life cycle, the ground truth is built differently and plays a different role as well.

Building a ground truth dataset is essential across various data personas, including data engineers, data analysts, and machine learning practitioners. For data engineers, ground truth is critical for data validation and testing. The good thing is that in most cases, data engineers can build the ground truth labels from historical data with the following:

- **Data validation rules**: Establish validation rules and constraints to verify the accuracy of the data as it flows through the pipeline.
- **Manual inspection**: Manually inspect data samples to identify inconsistencies or errors and create a dataset of validated and corrected data. This can serve as the ground truth dataset.

Data analysts rely on ground truth data to validate the accuracy of their findings, which can be acquired through expert annotations, historical data, and user feedback, ensuring that analytical insights reflect real-world phenomena:

- **Expert annotations**: If working with unstructured or text data, domain experts can manually annotate data samples with the correct labels or categories, serving as ground truth for analysis.
- **Historical data**: Use historical data with well-documented accuracy to serve as ground truth. This can be valuable when analyzing trends, patterns, or historical events.
- **Surveys and user feedback**: Collect data from surveys or user feedback to validate insights and conclusions drawn from the data. These can serve as qualitative ground truth.

Lastly, in the context of machine learning, ground truth data forms the backbone for model training and evaluation:

- **Manual labeling**: Annotate data samples manually to create a labeled dataset. This is common for tasks such as image classification, sentiment analysis, or object detection.

- **Crowdsourcing**: Use crowdsourcing platforms to collect annotations from multiple human workers, who collectively establish ground truth data.
- **Existing datasets**: Many machine learning tasks benefit from established benchmark datasets that have been widely used in the research community. You can use and update these datasets to your needs.
- **Domain expert labels**: Consult domain experts to provide labels or annotations for data, especially when domain-specific knowledge is required.
- **Synthetic data generation**: Create synthetic data with known ground truth labels to develop and test machine learning models. This is especially useful in the absence of real-world labeled data.

Regardless of the persona, it is vital to create, maintain, and continually assess the quality of ground truth data, being mindful of potential biases and limitations. This is because it significantly influences the effectiveness of data engineering, analysis, and machine learning efforts.

Timeliness

Timeliness evaluates how quickly data is captured, processed, and made available for use. Timeliness KPIs may include metrics such as data latency (time elapsed between data capture and availability) or adherence to data refresh schedules.

Measuring timeliness in data involves assessing the freshness or currency of the data with respect to a specific timeframe or event. Let's look at an example:

1. We start by importing the required libraries:

```
import pandas as pd
import numpy as np
from datetime import datetime, timedelta
```

2. We then generate a random dataset with timestamps and values. The timestamps are randomly distributed within a given time range to simulate real-world data:

```
np.random.seed(0) # For reproducibility
n_samples = 100
start_time = datetime(2023, 10, 25, 9, 0, 0)
end_time = datetime(2023, 10, 25, 16, 0, 0)

timestamps = [start_time + timedelta(minutes=np.random.
randint(0, (end_time - start_time).total_seconds() / 60)) for _
in range(n_samples)]
values = np.random.randint(50, 101, n_samples)

df = pd.DataFrame({'Timestamp': timestamps, 'Value': values})
```

3. We define a reference timestamp to compare the dataset's timestamps to:

```
reference_timestamp = datetime(2023, 10, 25, 12, 0, 0)
```

4. We set a timeliness threshold of 30 minutes. Data with timestamps within 30 minutes of the reference timestamp will be considered timely:

```
timeliness_threshold = 30
```

5. We calculate the timeliness for each record in the dataset by computing the time difference in minutes between the reference timestamp and each record's timestamp. We also create a Boolean column to indicate whether the record is timely based on the threshold:

```
df['Timeliness'] = (reference_timestamp - df['Timestamp']).
dt.total_seconds() / 60
df['Timely'] = df['Timeliness'] <= timeliness_threshold
```

6. Finally, we calculate the average timeliness of the dataset and display the results:

```
average_timeliness = df['Timeliness'].mean()
```

7. This will display the following output:

```
Dataset with Timestamps:
            Timestamp  Value  Timeliness  Timely
0 2023-10-25 11:52:00     71         8.0    True
1 2023-10-25 09:47:00     98       133.0   False
2 2023-10-25 10:57:00     99        63.0   False
3 2023-10-25 12:12:00     55       -12.0    True
4 2023-10-25 14:23:00     91      -143.0    True

Average Timeliness (in minutes): -23.8
Percentage of Timely Records: 61.0 %
```

This example provides a more realistic simulation of timeliness in a dataset with randomly generated timestamps and a timeliness threshold. *The average timeliness represents the average time deviation from the reference timestamp for the dataset.*

Good timeliness

A low average timeliness and a high percentage of timely records suggest that the dataset is current and aligns well with the reference timestamp. This is desirable in real-time applications or scenarios where up-to-date data is critical.

An important consideration here is how to define the reference timestamp. The reference timestamp is the point in time to which the dataset's timestamps are compared. It represents the expected or desired time for the data. For example: a record was created when a loyalty card was scanned at a retail store, and the reference time is when the record was logged in the database. So, we are calculating the time it took from the real creation of the event to a new entry to be stored in the database.

The smaller the reference threshold, the more *real-time* the application needs to be. On the other hand, the bigger the reference threshold is, the more time it takes to bring the data into your system (**batch application**). The choice between real-time and batch processing depends on the specific requirements of your application:

- **Real-time processing**: Choose real-time processing when immediate responses, low latency, and the ability to act on data as it arrives are crucial. It's suitable for applications where time-sensitive decisions are made.
- **Batch processing**: Select batch processing when low latency is not a strict requirement, and you can tolerate some delay in data processing. Batch processing is often more cost-effective and suitable for tasks that can be scheduled and automated.

How timeliness definitions change across different data personas

Timeliness is an essential aspect of data quality with applications across various roles, from data engineers to data analysts and machine learning practitioners. Here's how each role can leverage timeliness in the real world:

- **Data engineers**:
  - **Data pipeline monitoring**: Data engineers can use timeliness as a key metric for monitoring data pipelines. They can set up automated alerts or checks to ensure that data is arriving on time, identifying, and addressing delays in data ingestion.
  - **Data validation**: Data engineers can incorporate timeliness checks as part of their data validation processes, ensuring that data meets specified timing criteria before it is used for downstream processes.
- **Data analysts**:
  - **Real-time analytics**: Analysts in domains such as finance or e-commerce rely on real-time data to make informed decisions. They need to ensure that the data they analyze is up to date and reflects the current state of affairs.
  - **KPI monitoring**: Timeliness is essential in monitoring KPIs. Analysts use timely data to track and assess the performance of various business metrics.
- **Machine learning practitioners**:
  - **Feature engineering**: Timeliness plays a role in feature engineering for machine learning models. It's important to keep features as up-to-date as possible because this has a direct impact on model training and scoring.

- **Model training and evaluation**: In real-time predictive models, model training and evaluation rely on timely data. Practitioners must ensure that the training data is current to build effective models or perform real-time inference.
- **Concept drift detection**: Timeliness is critical in detecting concept drift, which occurs when the relationships within the data change over time. Machine learning models need to adapt to these changes, and timely data is necessary to monitor and detect such drift.

Here are some real applications of timeliness:

- **Finance**: In the financial sector, timeliness is crucial for stock trading, fraud detection, and risk management, where timely data can lead to better decisions and reduced risks
- **Healthcare**: Timeliness is vital for healthcare data, particularly in patient monitoring and real-time health data analysis
- **E-commerce**: Timely data is essential for e-commerce companies to monitor sales, customer behavior, and inventory in real-time
- **Transportation and L=logistics**: In supply chain management and logistics, real-time tracking and timely data are essential for route optimization and inventory management

Let's move on to consistency.

Consistency

Consistency measures the degree of consistency within the data and involves ensuring that data follows established rules, standards, and formats throughout a dataset. In more detail, we should check for the following:

- **Same format**: Data should follow the same format, structure, and standards across all records or columns. This uniformity ensures that data can be easily processed and compared.
- **Adherence to standards**: Data should adhere to predefined rules, guidelines, naming conventions, and reference data. For example, if a dataset contains product names, consistency would require that all product names follow a standardized naming convention.
- **Data type and format**: Consistency checks include verifying that data types (e.g., text, numbers, and dates) and data formats (e.g., date formats and numerical representations) are consistent.

Let's get a better understanding with an example:

1. We start by importing the `pandas` library:

```
import pandas as pd
```

2. We then create a sample dataset with product information, including product names. In this example, we'll check whether all product names start with PROD as per the naming convention:

```
data = {
  'ProductID': [1, 2, 3, 4, 5],
  'ProductName': ['PROD001', 'PROD002', 'Product003', 'PROD004',
'PROD005'],
}

df = pd.DataFrame(data)
```

3. Let's define the expected prefix:

```
expected_prefix = "PROD"
```

4. We check the consistency of the `ProductName` column by ensuring that all product names start with PROD. Inconsistent names will be flagged:

```
inconsistent_mask = ~df['ProductName'].str.startswith(expected_
prefix)

df['Consistency'] = ~inconsistent_mask
```

5. Then we calculate the percentage of consistent rows:

```
consistent_percentage = (df['Consistency'].sum() / len(df)) *
100
```

6. Finally, we display the results, including the dataset with the consistency check results:

```
print("Dataset with Consistency Check:")
print(df)
```

Here's the final output:

```
Dataset with Consistency Check:
   ProductID ProductName  Consistency
0          1     PROD001         True
1          2     PROD002         True
2          3  Product003        False
3          4     PROD004         True
4          5     PROD005         True
Percentage of Consistent Rows: 80.00%
```

In this specific dataset, three out of the five product names are consistent with the naming convention, resulting in an 80% consistency rate. The `Product003` entry is marked as inconsistent because it does not start with PROD.

This type of consistency check can be useful for ensuring that data adheres to specific standards or conventions, and the calculated percentage provides a quantitative measure of how many records meet the criteria.

> **Note**
>
> Higher consistency percentages imply more uniformity and conformity in the values within the respective columns. If there is a very low percentage, then we have a lot of distinct values in the dataset, which is not a mistake as long as we understand what the column represents and there is a meaning behind all the unique values.

Are you wondering about what other consistency checks you could apply?

Data consistency type	Example data	Code example

Format consistency	Phone numbers	`import re`
		`phone_numbers = ['(123) 456-7890', '123-456-7890', '1234567890']`
		`pattern = re.compile(r'^\(?(\d{3})\)?[-.\s]?(\d{3})[-.\s]?(\d{4})$')`
		`consistent_numbers = [number for number in phone_numbers if pattern.match(number)]`
		`print("Consistent Phone Numbers:", consistent_numbers)`

Naming conventions	Employee names	`employee_names = ['John Doe', 'Jane Smith', 'Bob Johnson', 'susan brown']`
		`consistent_names = [name.title() for name in employee_names]`
		`print("Consistent Employee Names:", consistent_names)`

Date and time consistency	Date formats	`import datetime`
		`dates = ['01/15/2022', '2022-01-15', 'Jan 15, 2022']`
		`consistent_dates = [date for date in dates if any(format in date for format in ('%m/%d/%Y', '%Y-%m-%d'))]`
		`print("Consistent Dates:", consistent_dates)`

Data consistency type	Example data	Code example

Code and identifier consistency	Product Stock Keeping Units (SKUs)	`product_skus = ['SKU-001', 'SKU002', 'SKU-003', 'SKU-004']`
		`consistent_skus = [sku if sku.startswith('SKU-') else 'SKU-' + sku for sku in product_skus]\`
		`print("Consistent Product SKUs:", consistent_skus)`

Reference data consistency	Country names	`reference_countries = ['USA', 'Canada', 'UK', 'Australia']`
		`data_countries = ['USA', 'Canada', 'US', 'Australia']`
		`consistent_countries = [country for country in data_countries if country in reference_countries]`
		`print("Consistent Countries:", consistent_countries)`

Unit of measurement consistency	Temperature units	`temperatures = ['25°C', '77°F', '298K', '25C']`
		`consistent_temperatures = [temp for temp in temperatures if temp.endswith('°C')]`
		`print("Consistent Temperatures:", consistent_temperatures)`

Currency consistency	Financial data	`financial_data = ['$1000', '€500', '500 USD', '£750']`
		`consistent_currencies = [data for data in financial_data if '$' in data]`
		`print("Consistent Currency Data:", consistent_currencies)`

Table 2.1 – Different consistency check options

Let's discuss uniqueness next.

Uniqueness

Uniqueness measures the presence of unique values in a dataset. It can help identify anomalies such as duplicated keys.

The code will output the validity results for each column, indicating whether the values in each column conform to the defined validity rules:

1. We import the `pandas` library:

   ```
   import pandas as pd
   ```

2. We then create a sample dataset containing email addresses. We want to check whether all email addresses in the dataset are unique:

   ```
   # Create a sample dataset
   data = {
       'Email': ['john.doe@example.com', 'jane.smith@example.com',
       'james.doe@example.com', 'susan.brown@example.com'],
   }

   df = pd.DataFrame(data)
   ```

3. We check the uniqueness of the `Email` column to ensure that no email address is duplicated in the dataset:

   ```
   # Check uniqueness and create a Boolean mask for duplicated
   email addresses
   duplicated_mask = df['Email'].duplicated(keep='first')

   # Create a new column to indicate uniqueness
   df['Uniqueness'] = ~duplicated_mask
   ```

4. Next, we calculate the uniqueness percentage:

   ```
   unique_percentage = (df['Uniqueness'].sum() / len(df)) * 100
   ```

5. Finally, we display the results, including the dataset with the uniqueness check results. Here's the output:

   ```
   Dataset with Uniqueness Check:
                        Email  Uniqueness
   0     john.doe@example.com        True
   1   jane.smith@example.com        True
   2    james.doe@example.com        True
   3  susan.brown@example.com        True
   Percentage of Unique Records: 100.00%
   ```

This output indicates that the values in the dataset are all unique.

Uniqueness checks are commonly performed in various industries and use cases. Here are some common examples of uniqueness checks in real-world scenarios:

- **Customer IDs**: In a customer database, each customer should have a unique identifier (customer ID) to prevent duplicate customer records
- **Product SKUs**: In inventory and e-commerce databases, each product should have a unique SKU to identify and manage products without duplication
- **Email Addresses**: Email addresses should be unique in a mailing list or user database to avoid sending duplicate messages or creating multiple accounts with the same email
- **Employee IDs**: In HR databases, each employee typically has a unique employee ID to differentiate between employees and manage their records effectively
- **Vehicle Identification Numbers (VINs)**: In the automotive industry, VINs are unique identifiers for each vehicle to track their history and ownership
- **Barcodes and QR codes**: In retail and logistics, barcodes and QR codes provide unique identifiers for products, packages, and items for tracking and inventory management
- **Usernames and user IDs**: In online platforms and applications, usernames and user IDs are unique to each user to distinguish them and manage user accounts
- **Serial Numbers**: In manufacturing, products often have unique serial numbers to identify and track individual items
- **Transaction IDs**: In financial systems, each transaction is assigned a unique transaction ID to prevent duplication and ensure proper record-keeping

When you have non-unique records in a dataset, it means that there are duplicate entries or records with identical key attributes (e.g., IDs, names, or email addresses) in the dataset. Non-unique records can lead to data quality issues and potentially cause errors in data analysis and reporting. To fix non-unique records in a dataset using Python, you can use various methods, including removing duplicates, aggregating data, or resolving conflicts based on your specific data and requirements. We'll discuss those strategies in a later chapter.

Duplication

Duplication assesses the presence of duplicate or redundant data within the dataset. Having duplicates in your data means that you have the same piece of information or record repeated multiple times within your dataset.

Example – customer database

Imagine you work for a company with a customer database that tracks information about each customer, including their contact details, purchases, and interactions.

Issue – duplicate customer records

You discover that there are duplicate customer records in the database. These duplicates may have occurred due to data entry errors, system issues, or other reasons. For instance, a customer named John Smith has two separate records with slight variations in contact details. One record has his email address as `john.smith@email.com`, and the other has `jsmith@email.com`.

This is generally considered undesirable for several reasons:

- **Data accuracy**: When you have multiple copies of the same information, it becomes challenging to determine which version is correct or up-to-date. This can lead to data inconsistency and confusion.
- **Storage efficiency**: Duplicated records consume unnecessary storage space. This is particularly important when dealing with large datasets, as it can lead to increased storage costs and longer data retrieval times.
- **Data integrity**: Duplicates can compromise data integrity. In situations where data relationships are critical, duplicated records can disrupt the integrity of your data model.
- **Efficient data processing**: Analyzing, querying, and processing datasets with fewer duplicates is more efficient. Data processing times are shorter, and results are more meaningful when you're not dealing with repetitive information.
- **Data analysis**: When performing data analysis or running statistical models, duplicated records can skew results and lead to incorrect conclusions. Reducing duplicates is crucial for accurate and meaningful analysis.
- **Cost savings**: Storing and managing duplicates incurs additional costs in terms of storage infrastructure and data management efforts. Eliminating duplicates can lead to cost savings.

Now imagine what happens if we extrapolate the same problem to a company that manages millions of customers. Could you see how expensive and confusing introducing duplicates could be?

While it's generally a best practice to minimize duplicate records in a dataset, there are some specific scenarios where accepting or allowing duplicates might be a reasonable choice:

Scenario	Data representation
Customer order history	Each row represents a separate order made by a customer. Multiple rows with the same customer ID are allowed to show order history.
Service requests	Records represent service requests, including multiple requests from the same customer or location over time. Duplicates are allowed to maintain a detailed history.
Sensor data	Each row contains sensor readings, which can include multiple entries with the same data values over time. Duplicates are allowed for tracking each reading.
Logging and audit trails	Log entries record events or actions, and some events may generate duplicate entries. Duplicates are preserved for detailed audit trails.
User interaction data	Records capture user interactions with a website or application. Duplicates can represent repeated interactions for analysis of user behavior.
Change history	Data versions or document changes result in multiple records, including duplicates that capture historical revisions. Duplicates are maintained for version history.

Table 2.2 – Duplicated records scenarios

In these scenarios, allowing duplicates serves specific data management or analysis goals, such as preserving historical data, maintaining a record of changes, or capturing detailed user interactions. The data representation aligns with these objectives, and duplicates are *intentionally* retained to support these use cases.

Let's see how to track duplicates with a code example. The code will output the number of duplicate records found in your dataset:

1. First, we import `pandas`:

   ```
   import pandas as pd
   ```

2. Next, we create a sample dataset with employee information. We'll intentionally introduce duplicate employee IDs to demonstrate the identification of duplicates:

   ```
   data = {
       'EmployeeID': [101, 102, 103, 101, 104, 105, 102],
       'FirstName': ['Alice', 'Bob', 'Charlie', 'David', 'Eve',
   'Frank', 'Bob'],
       'LastName': ['Smith', 'Johnson', 'Brown', 'Davis', 'Lee',
   'White', 'Johnson'],
   }

   df = pd.DataFrame(data)
   ```

3. We'll use pandas to identify and mark duplicate records based on the `EmployeeID` column. The `duplicated()` function is used to create a Boolean mask, where `True` indicates a duplicated record:

```
duplicated_mask = df.duplicated(subset='EmployeeID',
keep='first')
```

The `subset='EmployeeID'` argument specifies the column on which the duplication check is performed. `keep='first'` marks duplicates as `True` except for the first occurrence. You can change this parameter to `last` or `False` based on your requirements.

4. We then create a new column called `'IsDuplicate'` in the DataFrame to indicate whether each record is a duplicate or not:

```
df['IsDuplicate'] = duplicated_mask
```

5. We calculate the percentage of duplicate records by dividing the number of duplicate records (those marked as `True` in the `'IsDuplicate'` column) by the total number of records and then multiplying by 100 to express it as a percentage:

```
duplicate_percentage = (df['IsDuplicate'].sum() / len(df)) * 100
```

6. Finally, we display the dataset with the `IsDuplicate` column to see which records are duplicates. Here's the final output:

```
   EmployeeID FirstName LastName  IsDuplicate
0         101     Alice    Smith        False
1         102       Bob  Johnson        False
2         103   Charlie    Brown        False
3         101     David    Davis         True
4         104       Eve      Lee        False
5         105     Frank    White        False
6         102       Bob  Johnson         True
Percentage of Duplicate Records: 28.57%
```

This output indicates that 28.57% of records are duplicated in the dataset.

> Note
>
> **The fewer duplicated records, the better!**
>
> The threshold for what is considered an acceptable or "good" level of duplicated records in a dataset can vary depending on the specific context and the goals of your data management or analysis. There is no one-size-fits-all answer to this question, as it depends on factors such as the type of data, the purpose of the dataset, and industry standards.

Data usage

Data usage assesses the extent to which data is effectively utilized within the organization. Data usage KPIs can include metrics such as data utilization rates, the number of data requests or queries, or user satisfaction surveys regarding data availability and quality.

Scenario – corporate business intelligence dashboard

Imagine a large corporation that relies on data-driven decision-making to optimize its operations, marketing strategies, and financial performance. The corporation has a centralized **Business Intelligence** (**BI**) dashboard that provides various data analytics and insights to different departments and teams. This dashboard is crucial for monitoring the company's performance and making informed decisions.

In this scenario, assessing data usage metrics can be vital for optimizing the BI dashboard's effectiveness and ensuring it meets the organization's data needs. Here's what we will track in the code example:

- **Data utilization rates**: By tracking data utilization rates for different departments and teams, the organization can assess how often the dashboard is accessed and how extensively the data within it is used. For example, the marketing department might have a high data utilization rate, indicating a heavy reliance on the dashboard for campaign performance analysis. This metric can help identify areas of the organization where data-driven insights are most critical.
- **Number of data requests or queries**: Monitoring the number of data requests or queries made by users provides insights into the volume of data analysis conducted through the dashboard. High data request numbers may indicate a strong appetite for data-driven decision-making. This metric can also help identify peak usage times and popular data sources.
- **User satisfaction scores**: Collecting user satisfaction scores through surveys can gauge how well the BI dashboard meets user expectations. A lower average user satisfaction score may signal that the dashboard's features or user experience need improvement. Feedback from users can guide dashboard enhancements.
- **Organization data utilization rate**: Calculating the overall data utilization rate for the entire organization helps assess the dashboard's relevance and its effectiveness in serving the broader business goals. It also provides a benchmark to measure against in terms of data utilization improvements.

To calculate the number of data requests for the last month, you would need to log the data requests of your application with the associated timestamps. Let's see an example:

1. First, we import the `random` library:

   ```
   import random
   ```

2. Next, we create a function to simulate data usage metrics. In this function, we set the number of users in the organization to 500 users, but in a real-world scenario, you would replace this with the actual number of users in your organization. Let's have a look at the following function:

```
def simulate_data_usage():
    num_users = 500
    data_utilization_rates = [random.uniform(20, 90) for _ in
range(num_users)]
    data_requests = [random.randint(1, 100) for _ in
range(num_users)]
    organization_data_utilization_rate = sum(data_utilization_
rates) / num_users
    total_data_requests = sum(data_requests)
    user_satisfaction_scores = [random.randint(1, 5) for _ in
range(num_users)]
    avg_user_satisfaction_score = sum(user_satisfaction_scores)
/ num_users
    return {
        "data_utilization_rates": data_utilization_rates,
        "organization_data_utilization_rate": organization_data_
utilization_rate,
        "data_requests": data_requests,
        "total_data_requests": total_data_requests,
        "user_satisfaction_scores": user_satisfaction_scores,
        "avg_user_satisfaction_score": avg_user_satisfaction_
score,
}
```

The main goal of this function is to simulate data utilization rates for each user in the organization. The `random.uniform(20, 90)` function generates a random floating-point number between 20 and 90. We do this for each user, resulting in a list of utilization rates. Similarly, we simulate the number of data requests or queries made by each user. Here, we use `random.randint(1, 100)` to generate a random integer between 1 and 100 for each user, representing the number of data requests. Next, we calculate two organization-level metrics. The first is the average data utilization rate for the entire organization, and the second is the total number of data requests or queries across all users. We simulate user satisfaction scores using a scale from 1 to 5. Each user receives a random satisfaction score. We calculate the average user satisfaction score for the entire organization based on the simulated satisfaction scores.

3. We call the `simulate_data_usage()` function to run the simulation and store the results in the `data_usage_metrics` variable:

```
data_usage_metrics = simulate_data_usage()
```

4. Finally, we display the simulated data usage metrics. The output is as follows:

```
Organization Data Utilization Rate:
54.83%
```

```
Total Number of Data Requests or Queries:
25794

Average User Satisfaction Score:
2.93
```

Capturing the data usage of different data products is crucial for several reasons, especially in the context of organizations that rely on data for decision-making and operational effectiveness:

- **Optimizing resources**: By understanding how data products are used, organizations can allocate resources effectively. This includes identifying which data sources are heavily utilized and which may be underused. It helps in optimizing data storage, processing, and infrastructure resources.
- **Improving data quality**: Monitoring data usage can highlight data quality issues. For example, if certain data products are rarely accessed, it may indicate that the data quality is poor or that the data is no longer relevant. Capturing usage can trigger data quality improvements.
- **Identifying trends and patterns**: Data usage patterns can reveal insights into how data is consumed and what types of analyses or reports are most valuable to users. This input can inform data product development and enhancement strategies.
- **Cost management**: Knowing which data products are in high demand helps manage data-related costs. It enables organizations to invest resources wisely and avoid unnecessary expenses on maintaining or storing less-used data.
- **Security and compliance**: Tracking data usage is crucial for data security and compliance. Organizations can identify unauthorized access or unusual usage patterns that may indicate security breaches. It also helps in complying with data privacy regulations by demonstrating control over data access.
- **User satisfaction**: Understanding how data products are used and whether they meet user needs is vital for user satisfaction. It allows organizations to tailor data products to user requirements, resulting in better user experiences.
- **Capacity planning**: Capturing usage data helps in capacity planning for data infrastructure. It ensures that there is enough capacity to handle data traffic during peak usage periods, preventing performance bottlenecks.
- **Return on Investment (ROI) measurement**: For organizations investing in data products, tracking usage is essential for measuring the ROI. It helps determine whether the resources spent on data collection, processing, and presentation are justified by their impact on decision-making and business outcomes.

Next, let's discuss data compliance.

Data compliance

Data compliance evaluates the extent to which data adheres to regulatory requirements, industry standards, or internal data governance policies. Compliance KPIs may involve metrics such as the number of non-compliant data records, the percentage of data complying with specific regulations, or the results of data compliance audits. Data compliance is crucial for several important reasons, especially in today's data-driven and highly regulated business environment as presented in the following table.

Consequence/challenge	Description
Legal and regulatory consequences	Non-compliance can lead to legal actions, fines, and penalties
Reputational damage	Negative publicity and loss of trust among customers and stakeholders
Financial impact	Costs associated with fines, legal fees, data breach notifications, and so on
Data breaches	Increased risk of security breaches and unauthorized access
Data quality issues	Inaccurate or incomplete data impacts decision-making and efficiency
Loss of customers	Customers discontinuing relationships with non-compliant organizations
Legal liability	Potential legal liability for individuals and organizations
Additional monitoring and oversight	Imposition of stricter regulatory monitoring and oversight
Difficulty in expanding internationally	Hindrance in global expansion due to international non-compliance

Table 2.3 – Consequences of neglecting data compliance

Here's a Python example to illustrate a simplified scenario where we check the compliance of data records with specific regulations using randomly generated data:

1. We first import the `random` library:

   ```
   import random
   ```

2. Next, we create a function to simulate a dataset with compliance checks for a given number of data records:

   ```
   def simulate_data_compliance(num_records):
     data_records = []
     compliant_count = 0  # Counter for compliant records
   ```

3. Each data record consists of attributes such as `Age` and `Consent Given`, which are randomly generated:

```
    for _ in range(num_records):
        # Generate a random record (e.g., containing age and consent
fields)
        age = random.randint(18, 100)
        consent_given = random.choice([True, False])
```

4. We define compliance rules for these attributes based on a simplified scenario where, for instance, individuals must be 18 or older to provide consent:

```
        age_rule = age >= 18
        consent_rule = age >= 18 and consent_given
```

5. We check compliance with specific regulations, and for each data record, we report whether it complies with `Age` and `Consent` requirements:

```
        age_compliant = "Age Compliant" if age_rule else "Age
Non-Compliant"
        consent_compliant = "Consent Compliant" if consent_rule else
"Consent Non-Compliant"

        # Define overall compliance status
        compliance_status = "Compliant" if age_rule and consent_rule
else "Non-Compliant"
```

6. We introduce a `compliant_count` variable to keep track of the number of compliant records:

```
        # Count compliant records
        if compliance_status == "Compliant":
          compliant_count += 1

        data_records.append({
            "Age": age,
            "Consent Given": consent_given,
            "Age Compliance": age_compliant,
            "Consent Compliance": consent_compliant,
            "Overall Compliance Status": compliance_status
        })
```

Inside the loop that generates data records, we increment `compliant_count` whenever a record is compliant with the defined rules.

7. After generating all records, we calculate the percentage of compliant records as (compliant_count / num_records) * 100 and store it in the percentage_compliant variable:

```
    # Calculate the percentage of compliant records
    percentage_compliant = (compliant_count / num_records) * 100

    return data_records, percentage_compliant
```

8. We define the number of records we would like to simulate and start simulating the compliance checks by calling our simulate_data_compliance function:

```
# Define the number of data records to simulate
num_records = 100

# Simulate data compliance checks
data_records, percentage_compliant = simulate_data_compliance(num_records)
```

9. Finally, we display the results:

```
# Display the results for a sample of data records and the percentage of compliance
sample_size = 10
for record in data_records[:sample_size]:
    print(record)

print(f"\nPercentage of Compliant Records: {percentage_compliant:.2f}%")
```

This will display the following output:

```
Percentage of Compliant Records: 49.00%
```

Here's a table summarizing popular compliance checks along with examples:

Compliance check	Description and examples
Data privacy compliance	Ensuring the protection of **Personally Identifiable Information** (**PII**); an example is the secure storage of customer names and addresses.
GDPR compliance	Complying with the GDPR; an example is handling user data access and deletion requests.
HIPAA compliance	Ensuring healthcare data protection in accordance with HIPAA; an example is secure handling of **Electronic Protected Health Information** (**ePHI**)
PCI DSS compliance	Complying with the PCI DSS; an example is encrypting credit card information during payment processing

Data retention compliance	Managing data retention periods and secure archiving or deletion
Consent compliance	Verifying explicit user consent for data collection and processing; an example is opt-in consent for email marketing
Accuracy and completeness compliance	Regularly checking and correcting data for accuracy and completeness
Data classification and handling compliance	Labeling data by sensitivity and enforcing access controls; an example is classifying data as **Confidential** with restricted access.
Data encryption compliance	Encrypting sensitive data in transit and at rest
Access control compliance	Implementing role-based access control to restrict data access
Auditing and logging compliance	Maintaining audit logs of data access and changes
Data masking and anonymization Compliance	Protecting sensitive data through masking or anonymization
Data life cycle management compliance	Managing data from creation to disposal in accordance with policies
Data ethics and ethical compliance	Ensuring that data practices align with ethical standards
Non-discrimination compliance	Avoiding discriminatory uses of data; an example is fair lending practices in financial services

Table 2.4 – Key compliance checks

In practice, organizations may choose to calculate data quality KPIs, including completeness, on a daily, weekly, monthly, or quarterly basis. It is important to strike a balance between the frequency of measurement and the resources required to perform the assessments effectively. Regular monitoring and adjustments to the frequency of calculation can help ensure that data quality is continuously evaluated and maintained in accordance with business needs.

If the data has frequent updates, the monitoring metrics on the data should also be frequent. This ensures that any changes or updates to the data are captured in a timely manner and that the quality metrics remain up to date.

The more critical the data, the more frequent the update on the monitoring metrics should be. See here what critical data means:

Characteristic	Description
Vital for core operations	Essential for daily organizational functions
Key to decision-making	Instrumental in strategic, tactical, and operational decisions
High value and impact	Associated with significant financial value and operational impact
Sensitive and confidential	Often includes sensitive and confidential information
Business continuity and disaster recovery	Crucial for continuity planning and recovery measures
Customer trust and satisfaction	Directly impacts trust and satisfaction
Competitive advantage	May provide a competitive edge
Strategic asset	Recognized as a strategic resource

Table 2.5 – Critical data definition

If the data being measured for quality is vital for critical decision-making processes or sensitive operations, it may be necessary to calculate the quality KPIs on a more frequent basis.

Now that we understand how to evaluate our data products against different quality KPIs, let's see at which point in the data life cycle we need to apply those.

Implementing quality controls throughout the data life cycle

Data quality should be a fundamental consideration throughout the entire life cycle of data. From data ingestion to utilization by downstream analytic teams, data undergoes various changes, and ensuring its quality at each step is paramount. Here is a diagram of the quality check life cycle:

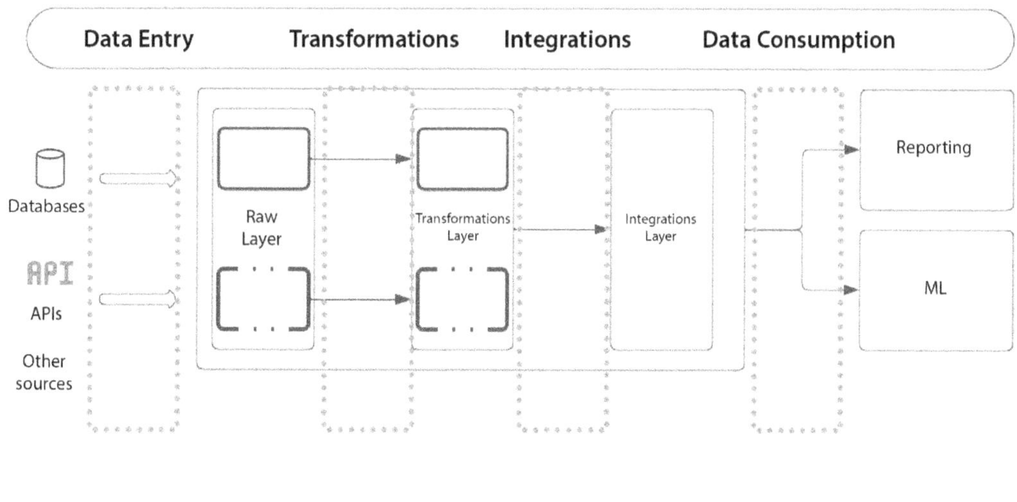

Figure 2.1 – Quality check life cycle

Let's understand more deeply what needs to happen at each step:

- **Data entry/ingestion**: Validating the data sources and ensuring that data is captured accurately and consistently while entering the system can limit errors in the downstream processes.

  Data persona --> data engineer

- **Data transformation**: By incorporating quality checks into the data transformation layer, organizations ensure that data remains reliable, accurate, and consistent throughout its journey from raw sources to its final destination.

  Data persona --> data engineer

- **Data integration**: When combining data from multiple sources or systems, data integration can introduce errors and inconsistencies. Applying quality checks at this level helps prevent data quality issues from propagating throughout the data ecosystem and supports a high level of confidence in the integrated data.

  Data persona --> data engineer and data scientist

- **Data consumption**: Analytics and machine learning models heavily depend on the quality of input data. This is particularly crucial in today's data-driven landscape, where the quality of data directly impacts an organization's success and competitive advantage.

  Data persona --> data scientist, analyst

As you can see in the preceding list, data flows in the system. Different teams collaborate on defining the quality metrics and applying the quality controls. Now, let's see what would happen if there was no collaboration between the different teams.

Data silos and the impact on data quality

Data silos, also known as isolated data repositories, are prevalent in many organizations today. Data silos refer to the practice of storing and managing data in isolated or disconnected systems or departments within an organization. These isolated data repositories have evolved over time, with various departments or business units maintaining the data separately and making it too complex to integrate. Organizations are increasingly aware of the limitations posed by data silos. They recognize that these silos hinder data-driven decision-making and operational efficiency. As a result, efforts to break down data silos and promote data integration and quality initiatives are on the rise, aimed at leveraging the full potential of data resources.

These silos pose challenges in maintaining data quality across the dimensions we've already discussed:

- **Not sharing data with the rest of the organization hinders its competitive advantage**: Data silos slow down decision-making by requiring employees to spend time searching for data from disparate sources, diverting their focus from gaining insights and taking action. Usually, data silos are associated with duplicate work as teams perform similar tasks independently, lacking efficient collaboration and information sharing. Conflicting interpretations of metrics frequently arise, causing confusion and disagreements among teams relying on different data sources. Misaligned assumptions and perspectives prevent progress and direction. Establishing clear communication guidelines and enforcing standardized methodologies is essential to align expectations and facilitate comprehension throughout the organization.
- **Not sharing data with the rest of the organization is very costly**: Data silos increase costs due to the maintenance of multiple scattered systems across the organization. Maintaining these disparate systems requires dedicated resources, both in terms of personnel and infrastructure (such as redundant storage in multiple places). Retrieving relevant information becomes time-consuming due to scattered data repositories, resulting in delays. Manually combining data from different sources introduces potential errors.

Now, let's summarize what we've learned in this chapter.

Summary

In this chapter, we discussed the critical role of high-quality data, providing a solid foundation for analytics, machine learning, and informed decision-making. To ensure data quality, organizations implement a series of checks and measures at various stages of the data pipeline:

- Data entry/ingestion: Data sources are validated to ensure accurate and consistent data capture, primarily overseen by data engineers
- Data transformation: Quality checks are incorporated into the transformation layer to maintain data reliability and accuracy, typically managed by data engineers
- Data integration: Checks prevent data quality issues from propagating and support confidence in integrated data, involving data engineers and data scientists
- Data consumption: Quality data input is vital for analytics and machine learning, impacting user trust and competitive advantage, and is driven by data scientists and analysts

These quality checks ensure that data adheres to defined standards, meets regulatory requirements, and is fit for its intended purpose. By implementing these checks, organizations maintain data accuracy, reliability, and transparency, facilitating better decision-making and ensuring data-driven success.

In the next chapter, we will explore how profiling tools can be used to continuously and automatically monitor data quality.

3

Data Profiling – Understanding Data Structure, Quality, and Distribution

Data profiling refers to scrutinizing, understanding, and validating datasets to learn more about their underlying structure, patterns, and quality. It is a critical step in data management and ingestion as it can enhance data quality and accuracy and ensure compliance with regulatory standards. In this chapter, you will learn how to perform profiling with different tools and how to change your tactics as the data volume increases.

In this chapter, we will deep dive into the following topics:

- Understanding data profiling
- Data profiling with the pandas profiler
- Data validation with Great Expectations
- Comparing Great Expectations and the pandas profiler – when to use what
- How to profile big data volumes

Technical requirements

For this chapter, you will need to install a Python interpreter that can be downloaded and installed using the instructions given here: `https://www.python.org/downloads/`.

You can find all the code for the chapter in the following GitHub repository: `https://github.com/PacktPublishing/Python-Data-Cleaning-and-Preparation-Best-Practices/tree/main/chapter03`.

Understanding data profiling

If you have never heard of **data profiling** before starting this chapter, it is a comprehensive process that involves analyzing and examining data from various sources to gain insights into the structure, quality, and overall characteristics of a dataset. Let's start by describing the main goals of data profiling.

Identifying goals of data profiling

Data profiling helps us understand the structure and quality of the data. As a result, we can get a better idea of the best way to organize the different datasets, identify potential data integration challenges, assess data quality, and identify and address issues that may affect the reliability and trustworthiness of the data.

Let's deep dive into the three main goals of data profiling.

Data structure

One of the main goals of data profiling is to understand the data's structure. This entails examining the data types, formats, and relationships between different data fields.

Here's an example of a simple table structure. Consider a table named `Employee` that stores information about employees in a company:

EmployeeID	FirstName	LastName	Position	Department	Salary
1	John	Doe	Software Eng	IT	75000
2	Jane	Smith	Data Analyst	Analytics	60000
3	Bob	Johnson	Project Manager	Project Management	85000

Let's break this table down:

- `EmployeeID`: A unique identifier for each employee
- `FirstName` and `LastName` are columns storing the first and last names of employees
- `Position`: The job title or position of the employee
- `Department`: The department in which the employee works
- `Salary`: The salary of the employee

This table structure is organized into rows and columns. Each row represents a specific employee, and each column represents a different attribute or piece of information about the employee. The table structure allows for easy querying, filtering, and joining of data. The values in each column adhere to a specific data type (e.g., integer, string, etc.), and relationships between tables can be established using keys.

This is a simplified example, but in real-world scenarios, tables can have more columns and complex relationships.

Data quality

Data quality involves evaluating the overall reliability and trustworthiness of the data. Through data profiling, we can identify various data quality problems, including duplicate records, incorrect or inconsistent values, missing values, and outliers. By quantifying these issues, organizations gain an understanding of the extent to which the data can be trusted and relied upon for analysis.

Data distribution

Understanding data distribution within each field or column is another key objective of data profiling. By analyzing data distribution, organizations gain insights into patterns, frequencies, and anomalies present in the data.

Let's imagine that we are working for an e-commerce company and we are collecting daily sales revenue data. By examining the distribution, we can identify trends in sales:

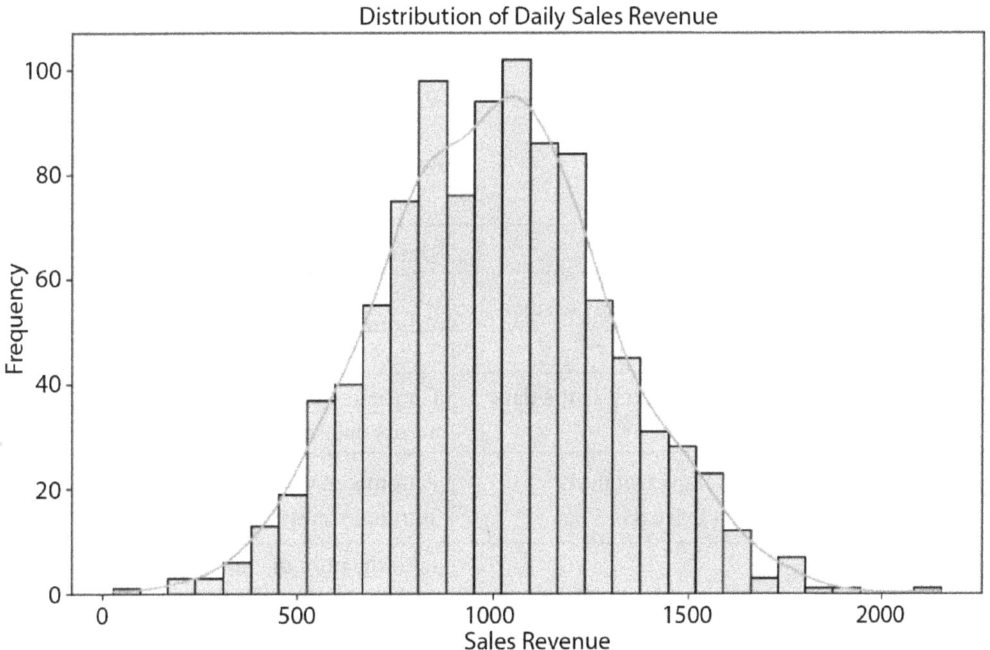

Figure 3.1 – Distribution of daily sales revenue

In this histogram, we can see that the sales data follows a normal distribution, indicating that data near the mean is more frequent in occurrence than data far from the mean. In this way, we can understand the mean daily sales we can expect on a regular day.

Now that we understand what challenges data profiling can deal with, let's see the different ways you can go about performing data profiling.

Exploratory data analysis options – profiler versus manual

When performing **exploratory data analysis** (**EDA**), there are different approaches you can take to understand your data, including conducting manual analysis or using a profiler.

Manual EDA involves writing custom code or using general-purpose data analysis libraries (e.g., pandas in Python) to explore the data. It gives you more flexibility and control over the analysis process. You can customize the analysis based on your specific requirements and questions. Manual EDA allows for more in-depth exploration, including custom calculations, feature engineering, and advanced visualizations. It can be beneficial when dealing with complex data or when you have specific domain knowledge that you want to apply to the analysis.

A **profiler** is a tool or library specifically designed for analyzing and summarizing data. It automates many EDA tasks and provides quick insights into the data's structure, summary statistics, missing values, data types, and distributions. It can save you time by automating repetitive tasks and providing a comprehensive overview of the dataset.

Let's see more in detail when to use what:

	Manual EDA	**Data Profiling**
Pros	Flexibility to explore data based on specific needs	Automated process for quick insights
	In-depth understanding of the data through custom code	Consistent and standardized analysis across datasets
	Greater control over analysis techniques and visuals	Automated visualizations and summary statistics
		Identification of data quality issues and anomalies
Cons	A time-consuming process requiring manual effort and repetitive	Limited customization options in generated reports
	Higher likelihood of human errors or biases	May not capture complex relationships or patterns
	Lack of standardization across analysts or teams	Less flexibility compared to manual analysis

	Manual EDA	Data Profiling
		Reliance on predefined algorithms and techniques

Table 3.1 – Comparison between Manual EDA versus using a profiler tool

As the data grows, manual EDA becomes increasingly time-consuming and prone to human error, leading to inconsistent results and potentially overlooked data issues. Manual efforts also lack scalability and reproducibility, making it difficult to handle large datasets and collaborate effectively. That is why for the rest of the chapter, we will focus on how you can use different profiling tools to perform EDA on the data; however, in practice, a combined approach is often implemented. We will also provide some insights on how to change your tools given the size of your data.

Profiling data with pandas' ydata_profiling

Let's see an example in Python that showcases data profiling using the `ProfileReport` class from the `ydata-profiling` library.

Let's start with installing a few libraries first:

```
pip install pandas
pip install ydata-profiling
pip install ipywidgets
```

In the following code example, we will use the `iris` dataset from the `seaborn` library, which is an open source dataset.

Next, we are going to read the dataset and perform some initial EDA with *minimal code*!

1. We'll start by importing the libraries and loading the dataset directly from its URL using the `read_csv()` function from pandas:

```
import pandas as pd
import ydata_profiling as pp
```

2. Load the `iris` dataset from the `seaborn` library:

```
iris_data = pd.read_csv('https://raw.githubusercontent.com/
mwaskom/seaborn-data/master/iris.csv')
```

3. Next, we'll perform data profiling by creating a profile report using the `ProfileReport()` function from `pandas_profiling`:

```
profile = pp.ProfileReport(iris_data)
```

4. We'll generate an HTML report using the `to_file()` method, which exports the profiling results to an HTML file for easy sharing and further analysis:

   ```
   profile.to_file("data_profile_report.html")
   ```

5. Optionally, we can embed the report in the notebook:

   ```
   profile.to_notebook_iframe()
   ```

6. Writing the report to a JSON file is optional, but a best practice:

   ```
   profile.to_file(output_path+"/pandas_profiler.json")
   ```

Let's explore the results of the profiler one by one.

Overview

The first section in the profiling report is the **Overview** section. In the **Overview** section, you have multiple tabs, as shown in the following figure:

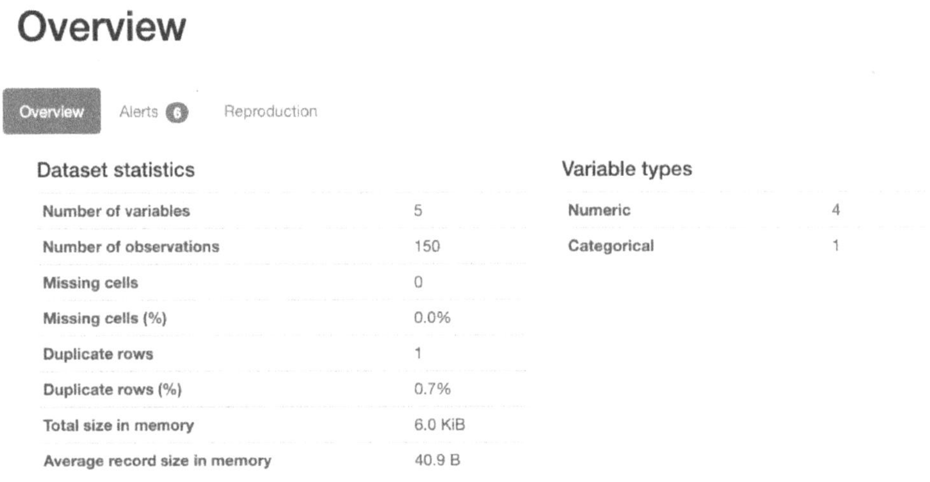

Overview

Overview | Alerts 6 | Reproduction

Dataset statistics

Dataset statistics	
Number of variables	5
Number of observations	150
Missing cells	0
Missing cells (%)	0.0%
Duplicate rows	1
Duplicate rows (%)	0.7%
Total size in memory	6.0 KiB
Average record size in memory	40.9 B

Variable types	
Numeric	4
Categorical	1

Figure 3.2 – Overview of pandas profiler results

In the **Overview** tab of the profiler results, we can see the following:

- **Number of variables**: The iris dataset has five variables – sepal length, sepal width, petal length, petal width, and species
- **Number of observations**: The dataset contains 150 rows
- **Missing cells**: No missing values are present in the iris dataset
- **Duplicate rows**: There is one duplicate row

Then, we have the **Alerts** tab, as shown in the following screenshot:

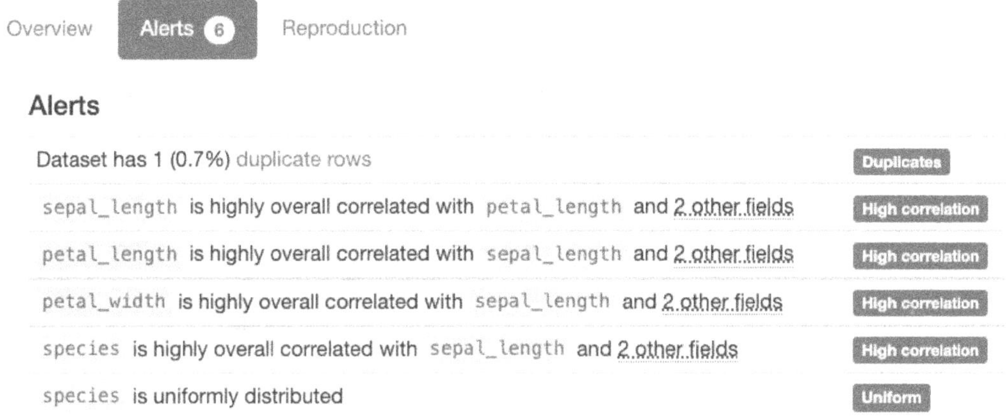
Overview
Alerts 6
Reproduction
Alerts
Dataset has 1 (0.7%) duplicate rows
Duplicates
sepal_length is highly overall correlated with petal_length and 2 other fields
High correlation
petal_length is highly overall correlated with sepal_length and 2 other fields
High correlation
petal_width is highly overall correlated with sepal_length and 2 other fields
High correlation
species is highly overall correlated with sepal_length and 2 other fields
High correlation
species is uniformly distributed
Uniform

Figure 3.3 – The Alerts tab of the ydata_profiling profiler

In the **Alerts** tab, we can see all the potential quality issues in our dataset, from high correlation to missing values. The most amazing thing is that those alerts are being created by the profiler automatically when you perform the profile creation. Never forget to check the **Alerts** tab as it shows the summary of all the potential issues in your dataset. The following screenshot shows the profiling performed for *each* numerical feature in the dataset. Let's see an example with `sepal_length`:

sepal_length
Real number (ℝ)

Distinct	35	**Minimum**	4.3
Distinct (%)	23.3%	**Maximum**	7.9
Missing	0	**Zeros**	0
Missing (%)	0.0%	**Zeros (%)**	0.0%
Infinite	0	**Negative**	0
Infinite (%)	0.0%	**Negative (%)**	0.0%
Mean	5.8433333	**Memory size**	1.3 KiB

5 6 7 8

Figure 3.4 – Numerical feature profiling

In the `sepal_length` part of the profile page, we can get more details about the specific numeric feature. A similar analysis is performed for all the other numeric features in the dataset. We can see that this feature has 35 different values and there are no missing values in the dataset for this feature. All the values are positive, which makes sense as this feature represents the length of the sepal and these values can range from 4.3 to 7.9. The histogram shows the distribution for the feature.

species
Categorical

HIGH CORRELATION UNIFORM

Distinct	3
Distinct (%)	2.0%
Missing	0
Missing (%)	0.0%
Memory size	1.3 KiB

setosa	50
versicolor	50
virginica	50

Figure 3.5 – Categorical feature profiling

In the `species` part of the profile page, we can get more details about the specific categorical feature. A similar analysis is performed for all the other categorical features in the dataset. We can see that this feature has three different values (`sectosa`, `versicolor`, and `virginica`) and there are no missing values in the data for this feature. From the graph, we can see that we have the same number of records for each value of the feature (50).

Interactions

Another section in the profiling report is an **Interactions** section, which visualizes the relationships and potential interactions between different columns in the dataset. These charts are particularly useful for identifying potential correlations or dependencies between variables presented as *scatter plots*.

The following figure shows the interactions between different variables. This chart can be created for every different combination of *numeric* variables. Let's see an example for petal length and petal width.

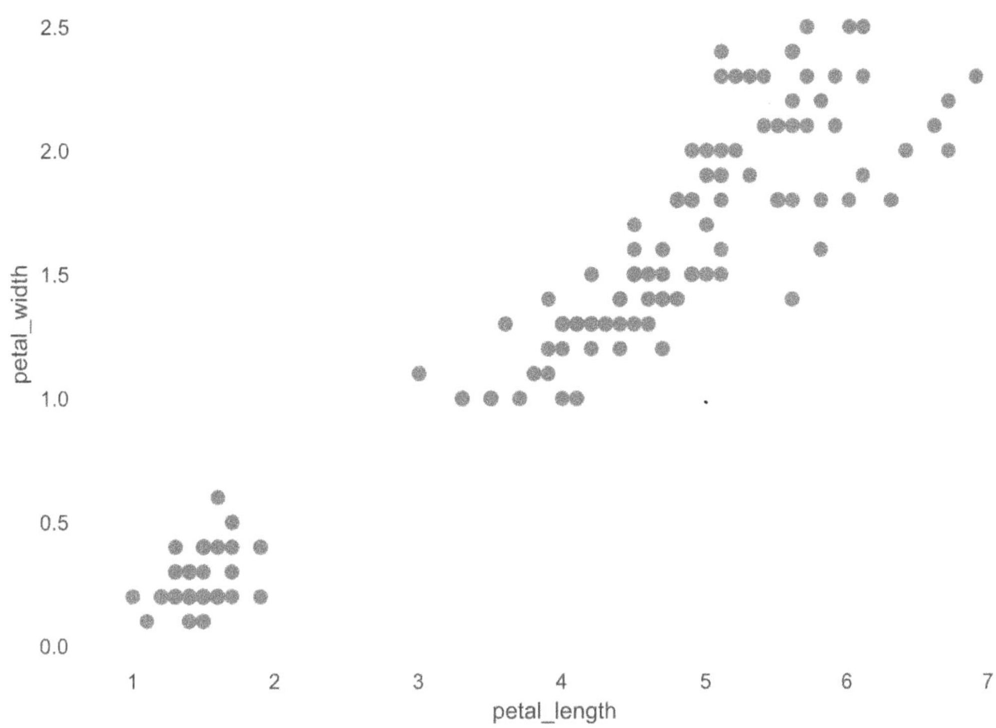

Figure 3.6 – Interaction chart between petal length and petal width

In the interaction chart, we can see how one variable influences the other. For example, as the petal length increases, the petal width increases as well. So, there is a linear relationship between the two. Since there is a strong interaction between these two variables, it is a good idea to deep dive into this interaction and examine in more detail the correlation plots for this pair.

Correlations

A **correlation matrix** also depicts the interactions between variables where each cell represents the relationship between two columns. The cells are color-coded based on the strength or type of interaction detected between the corresponding column pairs. This helps in identifying how strongly two variables are related. Positive correlations are typically shown in one color (e.g., blue), while negative correlations are shown in another (e.g., red), with the intensity of the color indicating the strength of the correlation.

For example, there might be a positive correlation between petal length and petal width, indicating that as the length of the petal increases, so does the width.

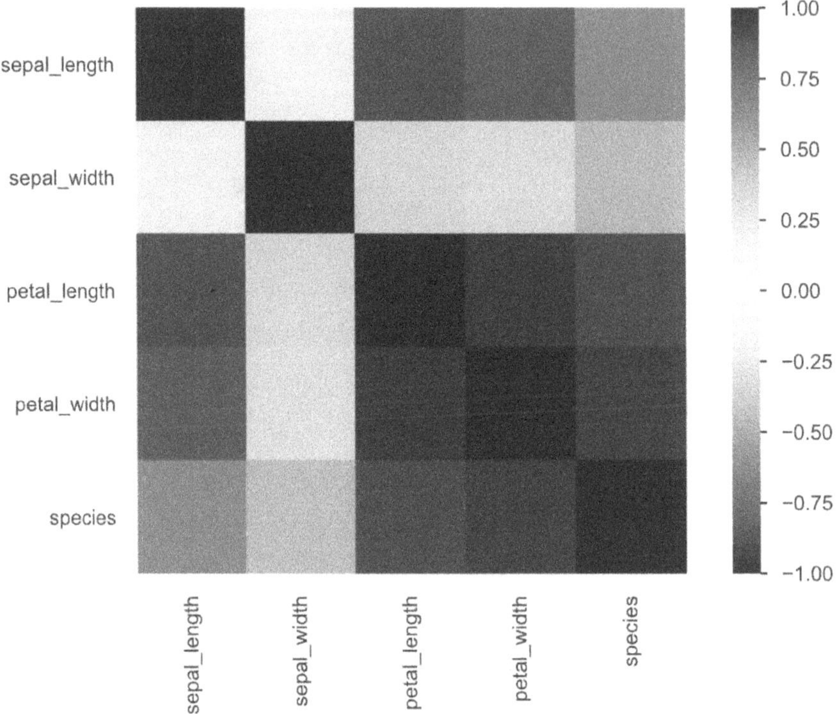

Figure 3.7 – Correlation chart between numeric variables

As we can see from the chart, the darker the color blue is, the stronger the correlation between the variables. Petal length and petal width have more than 0.75 positive correlation, showing that as one increases the other increases too. This is something we need to be aware of before proceeding to any modeling exercise as we may not need to keep both variables in the dataset, as having one can predict the other. For instance, if two variables are highly correlated, you might drop one of them or create a new feature that encapsulates the information from both. In some cases, removing highly correlated features can lead to faster training times for some machine learning algorithms, as the algorithm doesn't need to deal with redundant information; also, we can simplify models, making them easier to understand and less prone to overfitting.

> Note
>
> **High correlation threshold**: Set a threshold for high correlation (e.g., 0.8 or 0.9). Variables with correlation coefficients above this threshold are considered highly correlated.

Missing values

Another key aspect of data quality is **missing values**. It refers to the absence of data in specific entries or variables within a dataset. They can occur for various reasons, such as data entry errors, sensor malfunctions, or errors in the ingestion process. If ignored, it can lead to biased and inaccurate results.

The following figure shows the percentage of non-missing values for each column in the data:

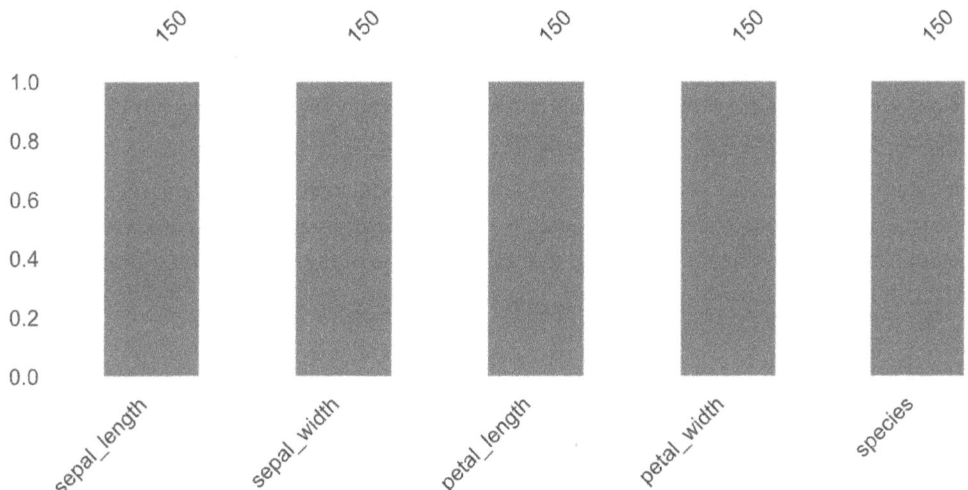

Figure 3.8 – Percentage of non-missing values in the data

In the present example, we can see that all the values in the dataset are complete and that all the features have 150 non-null values. So, that is good news for us and we can proceed to the next check.

Duplicate rows

A **duplicate row** in a dataset refers to a row that is identical to another row in every column. This means that for every column in the dataset, the values in the duplicate row are the same as those in the row it duplicates. Surfacing the presence and extent of duplicate rows helps us quickly identify potential data quality issues. As we've said, duplicate rows can arise due to various reasons, such as data integration problems, incorrect deduplication processes, or simply the nature of the data collection process.

In a profiling report, we can see the duplicate rows under **Most frequently occurring**, where a sample of the duplicates in the dataset is presented.

Most frequently occurring

	sepal_length	sepal_width	petal_length	petal_width	species	# duplicates
0	5.8	2.7	5.1	1.9	virginica	2

Figure 3.9 – Duplicate rows in the data

In general, to find the duplicate rows, you need to identify key columns or a combination of columns that should be unique. Typically, we identify duplicates based on all columns in the dataset. If there are duplicates, it indicates we have duplicate rows. There are only two duplicated rows in the dataset in our example, as shown in the preceding figure.

At this stage of the analysis, we are not handling duplicates since our focus is on understanding the data's structure and characteristics. However, we will need to investigate the nature and source of these duplicates. Given that they represent a small proportion of the data, we could simply drop one of each pair of identical rows.

Sample dataset

Sampling refers to the process of selecting a subset of data from a larger dataset instead of working with the entire dataset. In the EDA step, we usually work on a sample of data as it can provide initial insights and help in formulating hypotheses before committing resources to a full analysis.

	sepal_length	sepal_width	petal_length	petal_width	species
0	5.1	3.5	1.4	0.2	setosa
1	4.9	3.0	1.4	0.2	setosa
2	4.7	3.2	1.3	0.2	setosa
3	4.6	3.1	1.5	0.2	setosa
4	5.0	3.6	1.4	0.2	setosa
5	5.4	3.9	1.7	0.4	setosa
6	4.6	3.4	1.4	0.3	setosa
7	5.0	3.4	1.5	0.2	setosa
8	4.4	2.9	1.4	0.2	setosa
9	4.9	3.1	1.5	0.1	setosa

Figure 3.10 – Sample dataset

Now that we've understood how to build data profiles with the `ydata_profiling` library, let's have a closer look at a very popular but similar profiler called the **pandas data profiler**.

Profiling high volumes of data with the pandas data profiler

Pandas profiling is a powerful library for generating detailed reports on datasets. However, for large datasets, the profiling process can become time-consuming and memory-intensive. When dealing with large datasets, you may need to consider a few strategies to optimize the profiling process:

- **Sampling**: Instead of profiling the entire dataset, you can take a random sample of the data to generate the report. This can significantly reduce the computation time and memory requirements while still providing a representative overview of the dataset:

```
from ydata_profiling import ProfileReport
sample_df = iris_data.sample(n=1000)  # Adjust the sample size
as per your needs
report = ProfileReport(sample_df)
```

- **Subset selection**: If you're interested in specific columns or subsets of the dataset, you can select only those columns for profiling. This reduces the computational load and narrows down the focus to the variables of interest:

```
subset_df = iris_data [['sepal_length', 'sepal_width']]  #
Select columns to profile
report = ProfileReport(subset_df)
```

- **Configuring profiler options**: The pandas profiling library provides several configuration options that allow you to fine-tune the profiling process. You can adjust these options to limit the depth of analysis, reduce computations, or skip certain time-consuming tasks if they are not necessary for your analysis:

```
report = ProfileReport(df, minimal=True)  # Generate a minimal
report
```

- **Parallel processing**: If your system supports parallel processing, you can leverage it to speed up the profiling process. By distributing the workload across multiple cores or machines, you can potentially reduce the time required for profiling large datasets:

```
import multiprocessing
with multiprocessing.Pool() as pool:
    report = pool.map(ProfileReport, [df1, df2, df3])  #
Profiling multiple DataFrames in parallel
```

- **Incremental profiling**: If your dataset is too large to fit in memory, you can consider performing incremental profiling by splitting the data into smaller chunks and profiling them individually. You can then combine the profiling results to get an overview of the entire dataset:

```
chunk_size = 10000
chunks = [df[i:i + chunk_size] for i in range(0, len(df), chunk_
size)]
reports = [ProfileReport(chunk) for chunk in chunks]
combined_report = ProfileReport(pd.concat(reports))
```

> **Note**
>
> Some of these strategies aim to optimize the profiling process for large datasets, but they may result in some loss of granularity and detail compared to profiling the entire dataset. It's essential to strike a balance between computational efficiency and the level of insight required for your analysis.

The next tool we are going to review is usually used in data-engineering heavy workflows as it provides a lot of flexibility, automation, and easy integration with other tools.

Data validation with the Great Expectations library

Great Expectations is an open source Python library that facilitates data validation and documentation. It provides a framework for defining, managing, and executing data quality checks, making it easier to ensure data integrity and reliability throughout the data pipeline. Quality checks can be executed at different stages of the data life cycle, as shown in the following diagram:

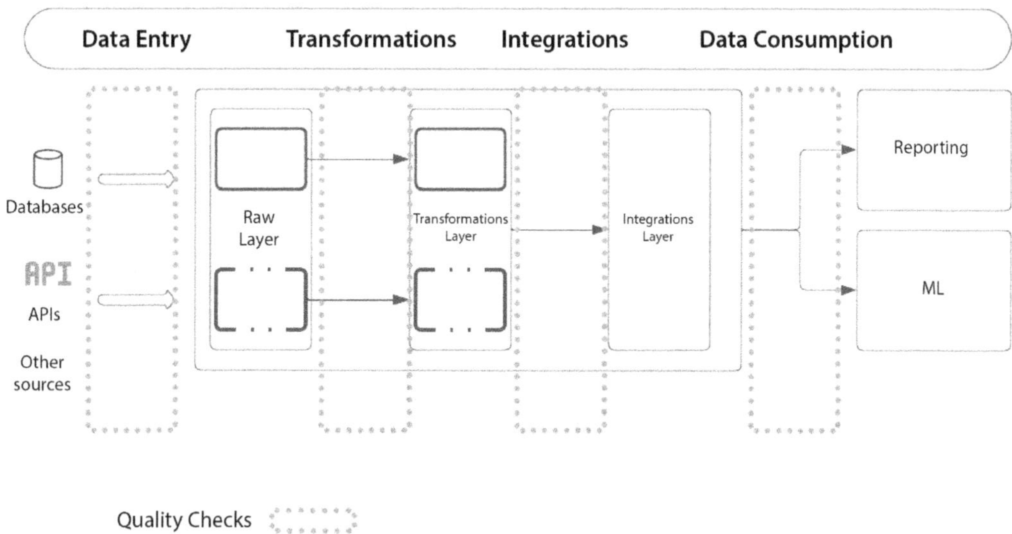

Figure 3.11 – Quality checks at different stages of the data life cycle

Let's discuss each of the touch points in the data life cycle where quality checks can be applied, as illustrated in the preceding figure:

- **Data entry**: During data entry or data collection, checks are conducted to ensure that the data is accurately captured and recorded. This can involve verifying the format, range, and type of data, as well as performing validation checks against predefined rules or standards.
- **Data transformation**: If data undergoes any transformations or conversions, such as data cleansing or data normalization, quality checks are performed to validate the accuracy of the transformed data. This helps ensure that the data retains its integrity throughout the process.
- **Data integration**: When combining data from different sources or systems, data quality checks are necessary to identify any inconsistencies or discrepancies. This may involve checking for duplicate records, resolving missing or mismatched data, and reconciling any conflicting information.
- **Data consumption**: Prior to performing any data analysis or generating reports, it is essential to run data quality checks to ensure the integrity of the data. This involves validating the data against predefined criteria, checking for outliers or anomalies, and verifying the overall quality of the dataset.

Great Expectations allows you to set Expectations or rules for your data and then validate your data against these Expectations at any point in the data life cycle. *Figure 3.12* illustrates the features of this library in more detail.

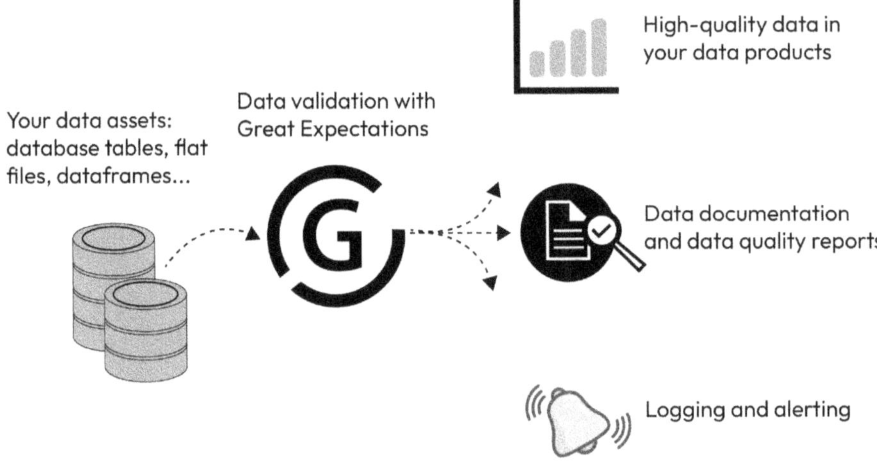

Figure 3.12 – Great Expectations process from data collection to data quality results

As you can see, there are three main steps to be aware of when working with Great Expectations:

- Bringing/collecting all the data you want to apply Expectations on
- Writing the Expectations and applying them to the different data
- Enjoying the benefits of clean, high-quality, and trustworthy data coming to life

In the next section, we will go through how to configure Great Expectations to validate the dataset.

Configuring Great Expectations for your project

You can validate your data against the defined Expectations using Great Expectations. The library provides functions to execute these validations and identify any inconsistencies or issues in the data.

You will need to install the `great-expectations` library for data profiling. You can use the following command to install the library in any IDE or a terminal:

```
pip install great-expectations==0.18.16
```

This should install the library. We are going to use the same dataset as before so that we can showcase the difference between the tools:

1. Let's start by setting up your project. Open your terminal, navigate to the desired location where you want to set up your new project, and then set up a new folder by running the following command:

   ```
   mkdir great_expectations
   ```

2. Then, we will go into the newly created folder by typing the following:

   ```
   cd great_expectations
   ```

   The first command will create a folder named `great_expectations` in your project directory where we'll store all the Expectations we are going to build. The second command will navigate you inside the `great_expectations` folder we just created.

3. Next, we will create some folders for our data and the code we will need to use to run our example. Make sure you are in the `great_expectations` directory and run the following:

   ```
   mkdir code
   mkdir data
   ```

   You should have created the following project structure:

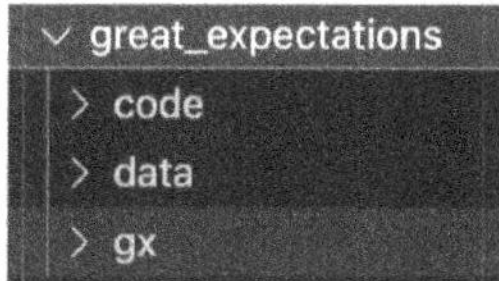

Figure 3.13 – Great Expectations project initialization

4. Next, we'll run the following command to initialize a new Great Expectations project. Make sure you are in the `great_expectations` folder:

   ```
   great_expectations init
   ```

   The preceding command sets up the basic project structure, configuration files, and directories needed to start using Great Expectations. The initialization process involves creating the necessary project files, such as the `great_expectations.yml` configuration file, the Expectations and data directories, and other project-specific files. The initialization step is a one-time setup that allows you to define and manage your data Expectations, create validation checkpoints, and generate documentation using Great Expectations. It helps you establish a project-specific configuration and directory structure that enables you to organize your Expectations and maintain consistency across your data pipelines. Once you have initialized the project, you can define Expectations, validate data, and generate reports based on those Expectations using Great Expectations.

The preceding code will display the following output:

```
             ~ Always know what to expect from your data ~

Let's create a new Data Context to hold your project configuration.

Great Expectations will create a new directory with the following structure:

    great_expectations
    |-- great_expectations.yml
    |-- expectations
    |-- checkpoints
    |-- plugins
    |-- .gitignore
    |-- uncommitted
        |-- config_variables.yml
        |-- data_docs
        |-- validations

OK to proceed? [Y/n]:
```

Figure 3.14 – Great Expectations project initialization

5. Hit *Y* when prompted and Great Expectations will go ahead and build a project structure for us in the `great_expectations` folder. The folder structure will look like the following:

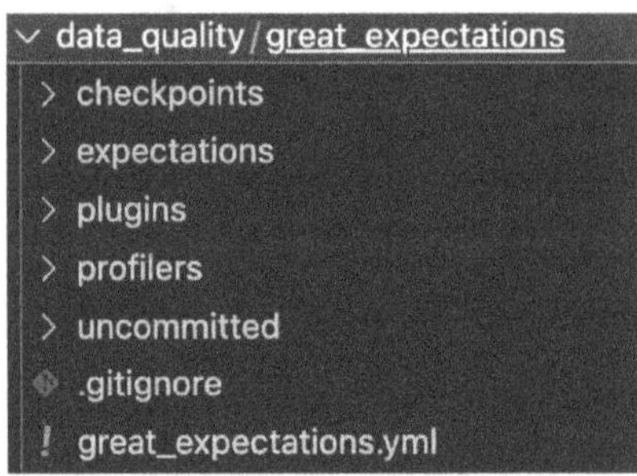

Figure 3.15 – Great Expectations folder structure

The folder structure of Great Expectations follows a specific convention to organize the configuration, Expectations, and data documentation related to your data pipeline. Let's learn a little more about the structure:

- `/uncommitted/`: This directory contains all the uncommitted configuration and validation files. It is where you define and modify Expectations, validations, and data documentation.
- `/checkpoints/`: This directory stores checkpoint files, which hold the sets of Expectations to be validated against specific data batches. Checkpoints are useful for running validations on specific portions or subsets of your data.
- `/expectations/`: This directory holds the Expectation Suites and Expectation files. An Expectation Suite is a collection of related Expectations, while Expectation files contain individual Expectations. You can create subdirectories within this folder to organize your Expectations based on a data source or data asset.
- `/plugins/`: This folder is used to store custom plugins and extensions that you may develop to extend the functionality of Great Expectations.
- `great_expectations.yml`: This configuration file stores the deployment settings for Great Expectations. It contains essential information and parameters that define how Great Expectations operates within your deployment environment.

Now that we've set up and initialized a Great Expectations project, let's create our first data source using Great Expectations.

Create your first Great Expectations data source

So far, we have created the project structure to create our Expectations. The next step is to get some data to build Expectations on. In order to retrieve the dataset, go to the repository at `https://github.com/PacktPublishing/Python-Data-Cleaning-and-Preparation-Best-Practices/tree/main/chapter03/great_expectations/code`, get the `1.data_set_up.py` script, and save it under the `great_expectations/code/` folder. Now, let's write some test data to our folder by running the following Python script: `great_expectations/code/1.data_set_up.py`. Here's what the script looks like:

```
import numpy as np
import pandas as pd
```

```
# Load the 'iris' dataset from seaborn library
iris_data = pd.read_csv('https://raw.githubusercontent.com/mwaskom/
seaborn-data/master/iris.csv')

iris_data.to_csv('../data/iris_data.csv', index=False)
print("File written! :)")
```

In your terminal, in the `great_expectations/code/` directory, execute the following command:

```
python 1.data_set_up.py
```

This script performs a simple task of loading the `iris` dataset from a remote source, from the `seaborn` library's GitHub repository, using the pandas library. It then saves this dataset to a local file named `iris_data.csv` in the `great_expectations/data` directory. Finally, it prints a confirmation message to indicate that the file has been successfully saved.

Now, we need to tell Great Expectations which data we want to use to build Great Expectations and where to find this data. In your terminal, execute the following command:

```
great_expectations datasource new
```

This will display the following prompt:

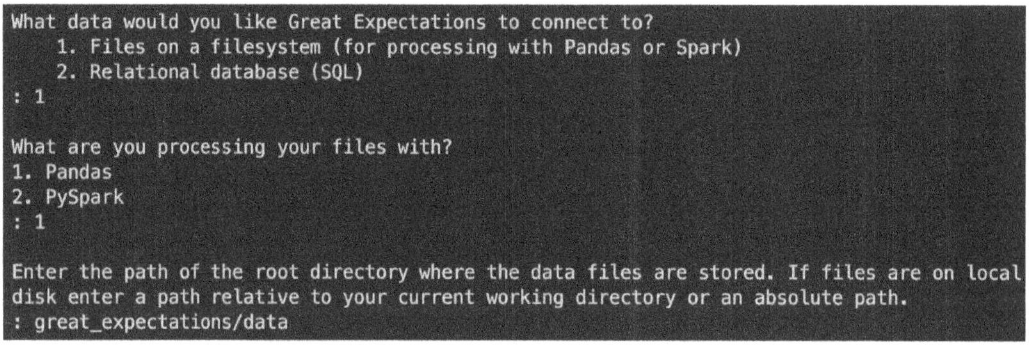

```
What data would you like Great Expectations to connect to?
    1. Files on a filesystem (for processing with Pandas or Spark)
    2. Relational database (SQL)
: 1

What are you processing your files with?
1. Pandas
2. PySpark
: 1

Enter the path of the root directory where the data files are stored. If files are on local
disk enter a path relative to your current working directory or an absolute path.
: great_expectations/data
```

Figure 3.16 – Great Expectations file configuration

Follow the steps in the terminal, as shown in *Figure 3.16*, making sure you choose option `1` as we are going to work with files and not SQL databases. Since our datasets are small enough to fit in memory, we can manipulate them with pandas. So, we'll choose option `1` again. It will then prompt you to enter the path to the dataset file and since we saved our dataset in the `data` folder, enter `../data`.

After this step, Great Expectations automatically creates a Jupyter Notebook for us to explore! This notebook is stored at the `great_expectations/gx/uncommitted/datasource_new.ipynb` path and after the execution, you can just delete it if you don't want to maintain unnecessary code. The purpose of this notebook is to help you create a pandas data source configuration and avoid any manual mistakes.

Let's open the notebook, update `datasource_name`, as shown in the following screenshot, and execute all the cells in the notebook.

Customize Your Datasource Configuration

If you are new to Great Expectations Datasources, you should check out our how-to documentation

My configuration is not so simple - are there more advanced options? Glad you asked! Datasources are versatile. Please see our How To Guides!

Give your datasource a unique name:

```
In [2]: datasource_name = "iris_data"
```

For files based Datasources:

Here we are creating an example configuration. The configuration contains an **InferredAssetFilesystemDataConnector** which will add a Data Asset for each file in the base directory you provided. It also contains a **RuntimeDataConnector** which can accept filepaths. This is just an example, and you may customize this as you wish!

Also, if you would like to learn more about the **DataConnectors** used in this configuration, including other methods to organize assets, handle multi-file assets, name assets based on parts of a filename, please see our docs on InferredAssetDataConnectors and RuntimeDataConnectors.

```
In [3]: example_yaml = f"""
name: {datasource_name}
class_name: Datasource
execution_engine:
  class_name: PandasExecutionEngine
data_connectors:
  default_inferred_data_connector_name:
    class_name: InferredAssetFilesystemDataConnector
    base_directory: ..\data
    default_regex:
      group_names:
        - data_asset_name
      pattern: (.*)
  default_runtime_data_connector_name:
    class_name: RuntimeDataConnector
    assets:
      my_runtime_asset_name:
        batch_identifiers:
          - runtime_batch_identifier_name
"""
print(example_yaml)
```

Figure 3.17 – Great Expectations – customizing data source name

We can give it any name we want at this point, but to be consistent with the incoming data, let's name it `iris_data`. From now on, when we refer to `iris_data`, we know we are working on the Expectations for the `iris` data source we created in the previous step.

> **Note**
>
> Maintaining consistent and clear naming between Expectation validations and data sources enhances readability, reduces errors, and simplifies maintenance and debugging!

Creating your first Great Expectations suite

Now that we have declared which data source we want to build an Expectation for, let's go ahead and build the first suite for our `iris` dataset.

Open your terminal and execute the following command:

```
great_expectations suite new
```

There are multiple ways to create your Expectation Suite, as you can see from the following figure.

```
How would you like to create your Expectation Suite?
    1. Manually, without interacting with a sample Batch of data (default)
    2. Interactively, with a sample Batch of data
    3. Automatically, using a Data Assistant
```

Figure 3.18 – Great Expectations – options for creating your suite

Let's explore each of the options:

- `Manually, without interacting with a sample batch of data (default)`: This approach involves manually defining Expectations and configuring the suite without directly interacting with a sample batch of data. Expectations are typically based on your knowledge of the data and the specific requirements of your project. You define Expectations by specifying conditions, ranges, patterns, and other criteria that you expect the data to meet. This approach requires a thorough understanding of the data and domain knowledge to define accurate Expectations.
- `Interactively, with a sample batch of data`: In this approach, you load a small representative batch of data into Great Expectations and use it to interactively define Expectations. This allows you to visually inspect the data, identify patterns, and explore various data statistics. You can iteratively build and refine Expectations based on your observations and understanding of the data.

- `Automatically, using a Data Assistant`: Great Expectations provides a Data Assistant feature that automatically suggests Expectations based on the data. The Data Assistant analyzes the data and generates a set of suggested Expectations, which you can review and customize. This approach can be helpful when you have limited knowledge about the data or want to quickly generate a starting point for your Expectations. You can leverage the suggested Expectations as a foundation and further refine them based on your domain knowledge and specific requirements. The Data Assistant accelerates the process of building a suite by automating the initial Expectation generation.

In this example, we will use the third option to build the suite automatically. This functionality is similar to the one that pandas profiling offers and we have explored it before in the *Profiling data with pandas' ydata_profiling* section. So, go ahead and choose option `3` in the terminal, as shown in *Figure 3.19*.

```
How would you like to create your Expectation Suite?
    1. Manually, without interacting with a sample Batch of data (default)
    2. Interactively, with a sample Batch of data
    3. Automatically, using a Data Assistant
: 3

A batch of data is required to edit the suite - let's help you to specify it.


Which data asset (accessible by data connector "default_inferred_data_connector_name") would
 you like to use?
    1. iris_data.csv
```

Figure 3.19 – Great Expectations – the Data Assistant option

As a next step, you will be asked to choose which data source you want to create the suite for, which is the output from the previous step. Type `1` for the `iris_data` source we built before and then input the name of the new Expectation Suite: `expect_iris`.

After executing the preceding command, a new notebook will be created automatically at `great_expectations/gx/uncommitted/edit_expect_iris.ipynb`. Open and read the notebook to understand the logic of the code; in summary, this notebook helps you choose columns and other factors from the data that you care about and lets the profiler create some Expectations for you that you can adjust later.

You have the option to create Expectations for all the columns in your dataset or a subset of them, as shown in *Figure 3.20*.

Initialize a new Expectation Suite by profiling a batch of your data.

This process helps you avoid writing lots of boilerplate when authoring suites by allowing you to select columns and other factors that you care about and letting a profiler write some candidate expectations for you to adjust.

Expectation Suite Name: expect_iris

```
In [ ]: import datetime

        import pandas as pd

        import great_expectations as gx
        import great_expectations.jupyter_ux
        from great_expectations.core.batch import BatchRequest
        from great_expectations.checkpoint import SimpleCheckpoint
        from great_expectations.exceptions import DataContextError

        context = gx.get_context()

        batch_request = {'datasource_name': 'iris_data.csv', 'data_connector_name': 'default_inferred_data_connector_name', 'da

        expectation_suite_name = "expect_iris"

        validator = context.get_validator(
            batch_request=BatchRequest(**batch_request),
            expectation_suite_name=expectation_suite_name
        )
        column_names = [f'"{column_name}"' for column_name in validator.columns()]
        print(f"Columns: {', '.join(column_names)}.")
        validator.head(n_rows=5, fetch_all=False)
```

Select columns

Select the columns on which you would like to set expectations and those which you would like to ignore.

Great Expectations will choose which expectations might make sense for a column based on the **data type** and **cardinality** of the data in each selected column.

Simply comment out columns that are important and should be included. You can select multiple lines and use a Jupyter keyboard shortcut to toggle each line: **Linux/Windows**: `Ctrl-/` , **macOS**: `Cmd-/`

Other directives are shown (commented out) as examples of the depth of control possible (see documentation for details).

```
In [ ]: exclude_column_names = [
            "sepal_length",
            "sepal_width",
            "petal_length",
            "petal_width",
            "species",
        ]
```

Figure 3.20 – Great Expectations – columns included in the Suite

You can add all the column names for which you do *not* want to create Expectations in the `exclude_column_name` list. For any columns not added to the list, `great_expectations` will build Expectations for you. In our case, we want to create Expectations for all the columns, so we will leave the list empty, as shown in *Figure 3.21*.

Select columns

Select the columns on which you would like to set expectations and those which you would like to ignore.

Great Expectations will choose which expectations might make sense for a column based on the **data type** and **cardinality** of the data in each selected column.

Simply comment out columns that are important and should be included. You can select multiple lines and use a Jupyter keyboard shortcut to toggle each line: **Linux/Windows**: `Ctrl-/` , **macOS**: `Cmd-/`

Other directives are shown (commented out) as examples of the depth of control possible (see documentation for details).

```
In [5]: exclude_column_names = [
        ]
```

Figure 3.21 – Great Expectations – excluding columns from the Suite

Remember to execute all the cells in the notebook and let's have a look at all the different Expectations built automatically by `great_expectations` for us.

Great Expectations Suite report

Let's have a look at the profiling created by `great_expectations`. As you can see from *Figure 3.22*, 52 Expectations were created and all have successfully passed. We can monitor the success percentage in the **Overview** tab to get a quick idea of how many Expectations pass every time a new data feed is coming to your pipeline.

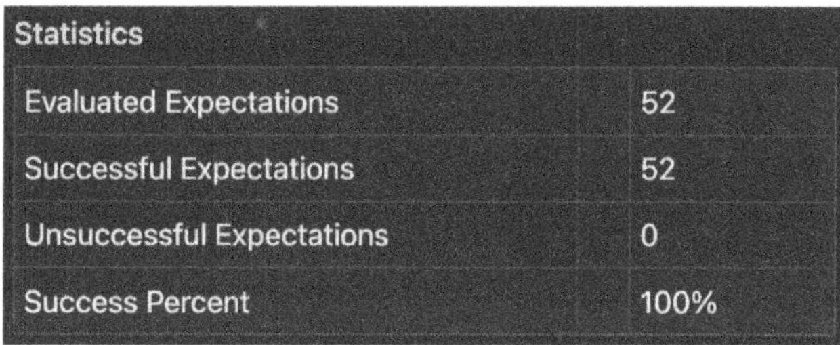

Statistics	
Evaluated Expectations	52
Successful Expectations	52
Unsuccessful Expectations	0
Success Percent	100%

Figure 3.22 – Report overview statistics

Let's have a closer look at what Expectations we are validating our data against. The first thing to consider is across-the-table or table-level Expectations, as shown in the following screenshot:

Table-Level Expectations

- Must have greater than or equal to 150 and less than or equal to 150 rows.
- Must have at least these columns (in any order): species, sepal_length, petal_length, petal_width, sepal_width

Notes

This Expectation suite currently contains 52 total Expectations across 5 columns.

Figure 3.23 – Table-level expectations

These Expectations check if the columns in the dataset match a given set of column names and if the dataset has the expected number of columns. It can be useful for ensuring all expected columns are present in the incoming data. If the incoming data does not contain all the columns shown in the Expectations, then the process will fail.

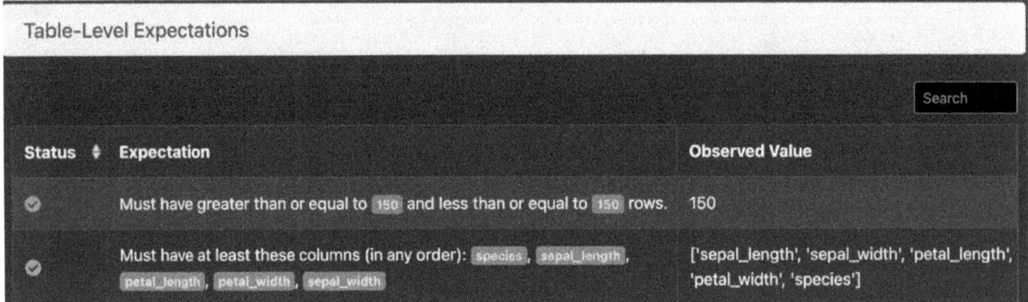

Table-Level Expectations

Search

Status	Expectation	Observed Value
✓	Must have greater than or equal to 150 and less than or equal to 150 rows.	150
✓	Must have at least these columns (in any order): species, sepal_length, petal_length, petal_width, sepal_width	['sepal_length', 'sepal_width', 'petal_length', 'petal_width', 'species']

Figure 3.24 – Column-level Expectations

The next set of Expectations is created for each of the columns in the table and we will refer to them as feature Expectations.

Status	Expectation	Observed Value
✓	values must never be null.	100% not null
✓	minimum value must be greater than or equal to 1.0 and less than or equal to 1.0.	1
✓	maximum value must be greater than or equal to 6.9 and less than or equal to 6.9.	6.9
✓	values must be greater than or equal to 1.0 and less than or equal to 6.9.	0% unexpected
✓	quantiles must be within the following value ranges. Quantile / Min Value / Max Value Q1 / 1.6 / 1.6 Median / 4.3 / 4.3 Q3 / 5.1 / 5.1	Quantile Value Q1 1.6 Median 4.3 Q3 5.1
✓	median must be greater than or equal to 4.35 and less than or equal to 4.35.	4.35
✓	mean must be greater than or equal to 3.7580000000000005 and less than or equal to 3.7580000000000005.	≈3.758
✓	standard deviation must be greater than or equal to 1.7652982332594662 and less than or equal to 1.7652982332594662.	≈1.765298233
✓	values must belong to this set: 1.2 1.0 1.4 1.3 1.1 1.5 1.6 1.7 1.9 3.0 3.3 3.5 4.0 5.0 6.0 4.1 4.4 4.8 4.6 4.9 5.1 5.4 5.5 5.6 5.9 6.1 6.4 6.6 6.9 3.6 3.7 3.8 4.2 4.3 4.7 4.5 5.2 5.3 5.7 5.8 6.3 6.7 3.9.	0% unexpected
✓	must have greater than or equal to 43 and less than or equal to 43 unique values.	43
✓	fraction of unique values must be exactly 0.28666666666666667.	≈0.2866666667

Figure 3.25 – Feature-level Expectations

These Expectations are checked separately for each column, and they can contain min and max values for the feature, whether we accept null values in the column or not, and many others. Remember, up to this point, all the Expectations were built automatically by us using a tool that does not understand the business context of the data. So, remember to check the Expectations and update them based on the business understanding of the data, as we will show in the next part.

Manually edit Great Expectations

While automatically generated Expectations provide a good starting point, they may not be sufficient for production-ready data validation. At this stage, it is important to further refine and customize the suite. You have the option to edit the suite manually or interactively. In general, manual editing is preferred when you have a clear understanding of the expected data properties and want to define Expectations efficiently and precisely. Since we've already done a basic automatic profiling of the data, we will proceed with the manual editing approach.

Open the terminal and execute the following command:

```
great_expectations suite edit expect_iris
```

You will be prompted to choose how you want to update the suite, either manually or interactively. We will proceed with it manually.

Upon providing the necessary input, Great Expectations opens the Jupyter Notebook available at the following location: `great_expectations/gx/uncommitted/edit_expect_iris.ipynb`. The notebook includes a comprehensive display of all the Expectations that were automatically generated. This allows you to review and examine the Expectations in detail, providing you with a clear overview of the validation rules that Great Expectations has inferred from the data. Have a look at all the Expectations we created and update them as necessary. In case you don't want to use notebooks, you can go open the `great_expectations/gx/expectations/expect_iris.json` file and update it there.

Checkpoints

So far, we have established a connection to our training dataset and defined our Expectations based on the training data. The next step is to apply these Expectations to our new stream of data in order to validate the new dataset and make sure it passes the checks. So, we need to create the connection between the Great Expectation suite and the new data to validate. We can do this with a checkpoint. To achieve this, we will first mock some test data to apply the Expectations. You can find the script at the following location: `https://github.com/PacktPublishing/Python-Data-Cleaning-and-Preparation-Best-Practices/blob/main/chapter03/great_expectations/code/2.mock_test_dataset.py`.

Save it under the `great_expectations/code/` folder. The script takes care of saving the test file in the required location, which is `great_expectations/data/`.

In your terminal, in the `great_expectations/code/` directory, execute the following:

```
python 2.mock_test_dataset.py
```

Let's have a closer look at the code we just executed, starting with the import statements:

```
import numpy as np
import pandas as pd
```

Load the `iris` dataset from the `seaborn` library:

```
iris_data = pd.read_csv('https://raw.githubusercontent.com/mwaskom/
seaborn-data/master/iris.csv')
```

We will do some transformations that will fail the Expectations and, in this case, we will update the `sepal_length` values to `60`, which will break our Expectations:

```
iris_data['sepal_length'] = 60
```

We will also rename columns to showcase the change in column names and by extension to the expected schema of the data:

```
iris_data.rename(columns={'petal_width': 'petal_w'}, inplace=True)
```

We will write DataFrame that will work as a new data feed to test our Expectations against:

```
iris_data.to_csv('../data/iris_data_test.csv', index=False)
print("File written! :)")
```

Then, we need to create a checkpoint that will execute the Great Expectation Suite we created on the test dataset. To initiate the checkpoint, you can run the following command in your terminal:

```
great_expectations checkpoint new expect_iris_ckpnt
```

Upon execution, Great Expectations automatically generates a Jupyter Notebook that provides valuable information about the checkpoint here: `/great_expectations/gx/uncommitted/edit_checkpoint_expect_iris_ckpnt.ipynb`. This includes details about the data to which the checkpoint will be applied. Before executing the notebook, we need to update the file name and point it to the test file, as shown here:

```
my_checkpoint_name = "expect_iris_ckpnt" # This was populated from
your CLI command.

yaml_config = f"""
name: {my_checkpoint_name}
config_version: 1.0
class_name: SimpleCheckpoint
run_name_template: "%Y%m%d-%H%M%S-my-run-name-template"
validations:
    - batch_request:
        datasource_name: iris_data.csv
```

```
        data_connector_name: default_inferred_data_connector_name
        data_asset_name: iris_data_test.csv
        data_connector_query:
          index: -1
    expectation_suite_name: expect_iris
"""
print(yaml_config)
```

Uncomment the last two lines and then execute all the cells of the notebook:

```
context.run_checkpoint(checkpoint_name=my_checkpoint_name)
context.open_data_docs()
```

The preceding notebook will apply the checkpoint to the new dataset and create a report of all the Expectations that have passed or failed. Let's see the results!

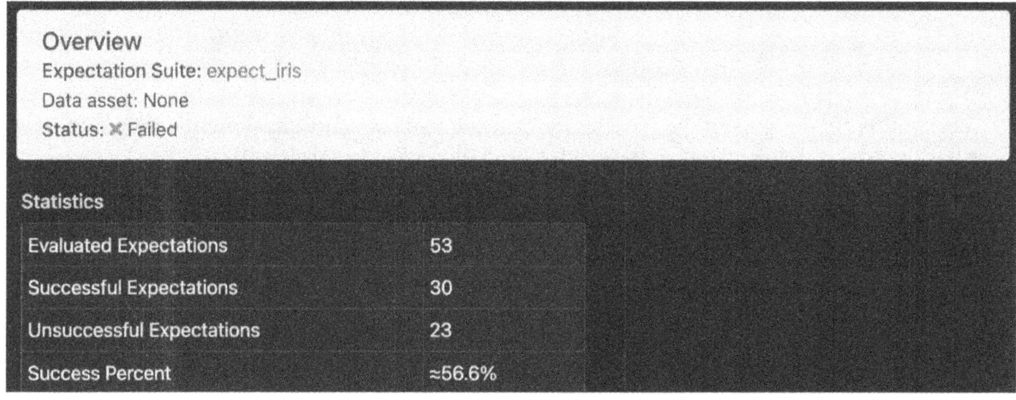

Figure 3.26 – Expectations results

As expected, our Expectations failed on the column names and on the petal width as it cannot find the right column names because of schema changes.

values must be greater than or equal to 0.1 and less than or equal to 2.5.

expect_column_values_to_be_between raised an exception:

Error: The column "petal_width" in BatchData does not exist.

Show exception traceback...

Figure 3.27 – Expectations failures because of schema changes

It also alerted us about the `sepal_length` variable as all the values are unexpected and outside of the accepted range of values it has seen so far!

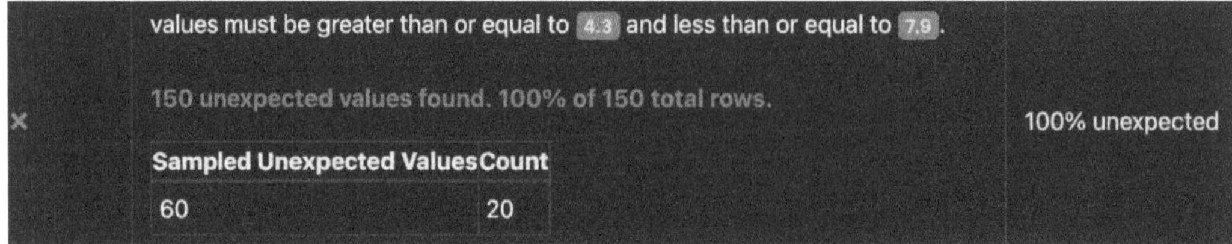

Figure 3.28 – Expectations failures because of out-of-range values

Can you see how many problems it could save us from? If this data was not checked and had been ingested, the subsequent processes and integration pipelines would fail, and a lot of work would be needed to try and identify which process failed and why. In our case, we know exactly where the problem started, and we have a clear idea of what we need to do to fix it.

> **Note**
>
> Checkpoints are designed to be reusable, so you can run the same checkpoint configuration against multiple batches of data as they arrive. This allows you to consistently validate incoming data against the same set of Expectations. Additionally, checkpoints can be enhanced with various actions, such as sending notifications, updating data documentation (Data Docs), or triggering downstream processes based on the validation results.

Now, if you are impressed by the automation that Great Expectations provides and you wish to see how you can migrate all the pandas profiling you've been doing so far to Great Expectations Suites, then we've got you covered. Just keep reading.

Using pandas profiler to build your Great Expectations Suite

The pandas profiler has a functionality that allows you to build Expectation Suites out of a pandas profiling exercise. Let's look at the following example `great_expectations/code/3.with_pandas_profiler.py`:

```
import pandas as pd
from ydata_profiling import ProfileReport

# Load the 'iris' dataset from seaborn library
iris_data = pd.read_csv('https://raw.githubusercontent.com/mwaskom/seaborn-data/master/iris.csv')

# run Pandas Profiling
```

```
profile = ProfileReport(iris_data, title="Pandas Profiling Report",
explorative=True)

# obtain an Expectation Suite from the profiling
suite = profile.to_expectation_suite(suite_name="my_pandas_profiling_
suite")
```

In this code snippet, we have taken our data and have created a pandas profiling. Then, we obtained an Expectation Suite from the report created previously. We can use this suite to further validate and check another batch of data.

So far, we have reviewed different profiling tools and how they work. The next step is to get a better understanding of when to use which tool and where to start.

Comparing Great Expectations and pandas profiler – when to use what

Pandas profiling and Great Expectations are both valuable tools for data profiling and analysis, but they have different strengths and use cases. Here's a comparison between the two tools.

	Pandas Profiler	**Great Expectations**
Data Exploration	Provides quick insights and exploratory data summaries	Focuses on data validation and documentation
Data Validation	Limited data validation capabilities	Advanced data validation with explicit Expectations and rules
Customization	Limited customization options	Extensive customization for defining Expectations and rules
Learning Curve	Relatively easy to use	A steeper learning curve for defining Expectations and configuration
Scalability	Suitable for small- to medium-scale data	Scalable for big data environments with distributed processing
Visualizations	Generates interactive visualizations	Focuses more on validating and documenting data rather than visuals
Use Case	Quick data exploration and initial insights	Data quality control and enforcing data consistency

Table 3.2 – Great Expectations and pandas profiler comparison

Pandas profiling is well suited for quick data exploration and initial insights, while Great Expectations excels in data validation, documentation, and enforcing data quality rules. Pandas profiling is more beginner-friendly and provides immediate insights, while Great Expectations offers more advanced customization options and scalability for larger datasets. The choice between the two depends on the specific requirements of the project and the level of data quality control needed.

As the volume of data increases, we need to make sure that the choice of tools we've made can scale as well. Let's have a look at how we can do this with Great Expectations.

Great Expectations and big data

While Great Expectations can be used effectively with smaller datasets, it also provides mechanisms to address the challenges associated with scaling data validation and documentation for big data environments. Here are some considerations for scaling Great Expectations as data size increases:

- **Distributed processing frameworks**: Great Expectations integrates seamlessly with popular distributed processing frameworks, such as Apache Spark. By leveraging the parallel processing capabilities of these frameworks, Great Expectations can distribute the data validation workload across a cluster, allowing for efficient processing and scalability.
- **Partitioning and sampling**: Great Expectations simplifies the process of partitioning and sampling large datasets and enhancing performances and scalability. Unlike the manual partitioning required in tools such as pandas profiling, Great Expectations automates the creation of data subsets or partitions for profiling and validation. This feature allows you to validate specific subsets or partitions of the data, rather than processing the entire dataset at once. By automating the partitioning process, Great Expectations streamlines the profiling workflow and eliminates the need for manual chunk creation, saving time and effort.
- **Incremental validation**: Instead of revalidating the entire big dataset every time, Great Expectations supports incremental validation. This means that as new data is ingested or processed, only the relevant portions or changes need to be validated, reducing the overall validation time and effort. This is a great trick to reduce the time it takes to check the whole data and optimize for cost!
- **Caching and memoization**: Great Expectations incorporates caching and memoization techniques to optimize performance when repeatedly executing the same validations. This can be particularly beneficial when working with large datasets, as previously computed results can be stored and reused, minimizing redundant computations.
- **Cloud-based infrastructure**: Leveraging cloud-based infrastructure and services can enhance scalability for Great Expectations. By leveraging cloud computing platforms, such as AWS or Azure, you can dynamically scale resources to handle increased data volumes and processing demands

- **Efficient data storage**: Choosing appropriate data storage technologies optimized for big data, such as distributed file systems or columnar databases, can improve the performance and scalability of Great Expectations. These technologies are designed to handle large-scale data efficiently and provide faster access for validation and processing tasks.

> **Note**
>
> While Great Expectations offers scalability options, the specific scalability measures may depend on the underlying infrastructure, data storage systems, and distributed processing frameworks employed in your big data environment.

Summary

This chapter detailed how data profiling is crucial for ensuring the quality, integrity, and reliability of datasets. The process involves in-depth analysis to understand the structure, patterns, and potential issues within the data. For effective profiling, tools such as pandas profiling and Great Expectations offer powerful solutions. Pandas profiling automates the generation of comprehensive reports, providing valuable insights into data characteristics. Great Expectations, on the other hand, facilitates the creation of data quality Expectations and allows for systematic validation. While these tools excel in smaller datasets, scaling profiling to big data requires specialized approaches. Learning the tips and tricks, such as data sampling and parallel processing, enables efficient and scalable profiling on large datasets.

In the next chapter, we will focus on how to clean and manipulate data to make sure it is in the right format to pass Expectations and be successfully ingested.

4

Cleaning Messy Data and Data Manipulation

In this chapter, we'll dive into the strategies of **data manipulation**, focusing on efficient techniques to clean and fix messy datasets. We'll remove irrelevant columns, systematically address inconsistent data types, and fix dates and times.

In this chapter, we'll cover the following topics:

- Renaming columns
- Removing irrelevant or redundant columns
- Fixing data types
- Working with dates and times

Technical requirements

You can find all the code for this chapter in the following GitHub link: `https://github.com/PacktPublishing/Python-Data-Cleaning-and-Preparation-Best-Practices/tree/main/chapter04`.

Each file is named according to the respective sections covered in this chapter.

Renaming columns

Renaming columns with more descriptive and meaningful names makes it easier to understand the content and purpose of each column. Clear and intuitive column names enhance the interpretability of the dataset, especially when sharing or collaborating with others.

To better understand all the concepts introduced in this chapter, we will use a scenario across the chapter. Let's consider an e-commerce company that wants to analyze customer purchase data to optimize its marketing strategies. The dataset includes information about customer transactions, such as purchase amount, payment method, and timestamp of the transactions. However, the dataset is messy and requires cleaning and manipulation to derive meaningful insights.

The distribution of the features is presented in the following figure. To build the following statistic charts, execute the file at `https://github.com/PacktPublishing/Python-Data-Cleaning-and-Preparation-Best-Practices/blob/main/chapter04/1.descriptive_stats.py`. The data and the following charts are created automatically once you run this script.

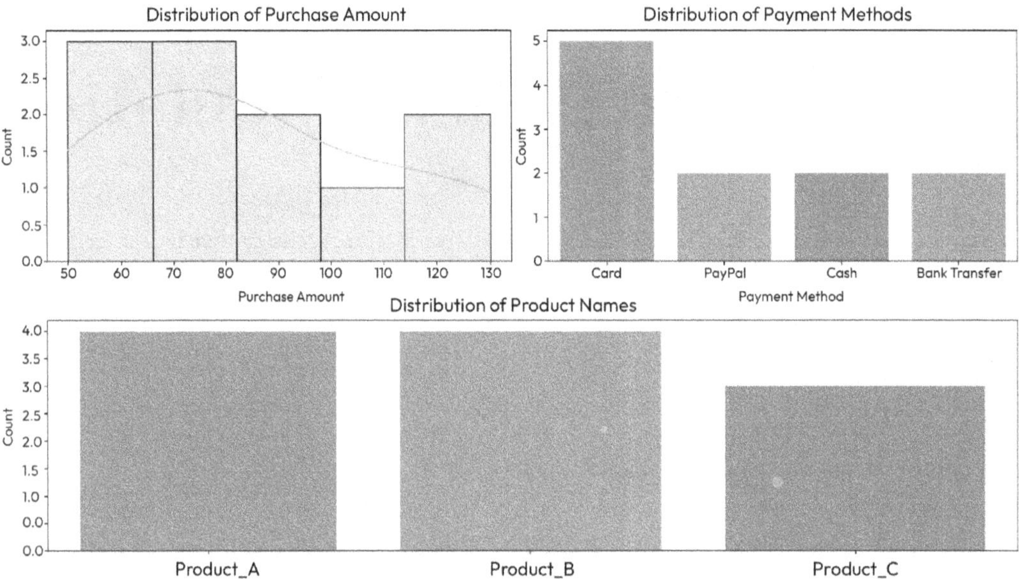

Figure 4.1 – Distribution of features before any data transformation

We have five columns in the dataset:

- `CustomerID`: A unique identifier for each customer. In this example, customer IDs range from `1` to `11`.
- `ProductName`: This represents the name of the purchased product. In the dataset, three products are considered: `Product_A`, `Product_B`, and `Product_C`.
- `PurchaseAmount`: This indicates the amount spent by the customer on a particular product. The amounts are in an arbitrary currency.
- `PaymentMethod`: This describes the method used by the customer to make the purchase. Payment methods include `Card`, `PayPal`, `Cash`, and `Bank Transfer`.
- `Timestamp`: This represents the date and time when the purchase occurred. It is formatted as a `datetime` object.

The first thing we are going to check and update are the column names. Let's start with this in the following section.

Renaming a single column

Now, the e-commerce company has decided to rebrand its products, necessitating a change in the column names related to product information. We'll start by renaming a single column, and then we'll further rename multiple columns to align with the rebranding initiative. For the renaming example, go to the file at `https://github.com/PacktPublishing/Python-Data-Cleaning-and-Preparation-Best-Practices/blob/main/chapter04/2.rename_columns.py`.

Let's have a look at how we can rename one column in the dataset:

```
df.rename(columns={'ProductName': 'OldProductName'}, inplace=True)
```

The `inplace=True` argument is an optional parameter in pandas DataFrame methods that allows you to modify the DataFrame directly without creating a new DataFrame object.

When `inplace` is set to `True`, the DataFrame is modified in place, meaning the changes are applied to the original DataFrame object. This can be useful when you want to update or modify the DataFrame without assigning the modified DataFrame to a new variable.

If `inplace=True` is not specified or set to `False` (which is the default behavior), the DataFrame methods return a new modified DataFrame object, leaving the original DataFrame unchanged. In such cases, you need to assign the modified DataFrame to a new variable to store the changes.

> **Note**
>
> It's important to note that using `inplace=True` can be a destructive operation since it modifies the original DataFrame directly. Therefore, it's recommended to use it with caution and ensure that you have a backup of the original DataFrame if needed. If you have a large dataset, modifying it in place can help conserve memory.

In the next section, we will rename multiple columns to align with the rebranding initiative.

Renaming all columns

Following a rebranding initiative, the company decided to rename `OldProductName` as `NewProductName` and `PurchaseAmount` as `NewPurchaseAmount` to align with the updated product names. This code demonstrates how to rename multiple columns at once:

```
df.rename(columns={'OldProductName': 'NewProductName',
'PurchaseAmount': 'NewPurchaseAmount'}, inplace=True)
```

If you want to rename columns in a DataFrame and need to ensure a smooth and error-free process, we can add error handling. For example, ensure that the columns you intend to rename actually exist in the DataFrame. If a column is misspelled or does not exist, the renaming operation will raise an error:

```
if 'OldProductName' in df.columns:
try:
# Attempt to rename multiple columns
  df.rename(columns={'OldProductName': 'NewProductName',
'PurchaseAmount': 'NewPurchaseAmount'}, inplace=True)
except ValueError as ve:
  print(f"Error: {ve}")
else:
  print("Error: Column 'OldProductName' does not exist in the
DataFrame.")
```

> **Note**
>
> Ensure that the new column names do not already exist in the DataFrame to avoid overwriting existing columns.

Renaming a column is one of the simplest things we can do to make our data cleaner and easier to understand. The next thing we usually do is keep only the columns we need or care about, as we will discuss in the next section.

Removing irrelevant or redundant columns

Large datasets often contain numerous columns, some of which may be irrelevant to the specific analyses or tasks at hand. By eliminating these columns, we can get some significant benefits. Firstly, storage requirements are dramatically reduced, leading to cost savings and more efficient use of resources. Additionally, the streamlined dataset results in faster query performance, optimized memory usage, and expedited processing times for complex analyses. This not only improves the overall efficiency of data processing tasks but also facilitates easier management and maintenance of large datasets. Furthermore, in cloud-based environments, where storage costs are a factor, the removal of unnecessary columns directly contributes to cost efficiency. So, let's have a look at how we can drop columns in an efficient way.

In the e-commerce dataset we presented earlier, we have collected information about customer purchases. However, as your analysis focuses on product-related metrics and customer behavior, certain columns, such as `CustomerID` and `Timestamp`, may be considered irrelevant to the current analysis. The goal is to streamline the dataset by dropping these columns. You can follow along with this example using this Python script `https://github.com/PacktPublishing/Python-Data-Cleaning-and-Preparation-Best-Practices/blob/main/chapter04/3.dropping_columns.py`:

```
columns_to_drop = ['CustomerID', 'Timestamp'] # Replace with the names
of the columns you want to drop
```

```
try:
# Drop columns considered irrelevant for the current analysis
    df.drop(columns=columns_to_drop, inplace=True)
except KeyError as ke:
    print(f"Error: {ke}")
```

Now, if you have a look at the dataset, the column before the deletion was this:

```
Index(['CustomerID', 'NewProductName', 'NewPurchaseAmount',
'PaymentMethod','Timestamp'],dtype='object')
```

After dropping the two columns, we have the following:

```
Index(['NewProductName', 'NewPurchaseAmount', 'PaymentMethod'],
dtype='object')
```

> **Note**
>
> Python, by default, is case-sensitive. This means that `ColumnName` and `columnname` are considered different.

We successfully removed the unnecessary columns, as demonstrated earlier. To further assess memory efficiency, we can calculate the memory consumption of the DataFrame both before and after the column deletion. The following code provides a Python example of how to calculate the memory used by the DataFrame before and after the drop of columns:

```
print("Initial Memory Usage:")
print(df.memory_usage().sum() / (1024 ** 2), "MB") # Convert bytes to
megabytes

print("\nMemory Usage After Dropping Columns:")
print(df.memory_usage().sum() / (1024 ** 2), "MB") # Convert bytes to
megabytes
```

The initial memory usage of the DataFrame was approximately 0.00054 megabytes, and after dropping columns, it reduced to around 0.00037 megabytes. The achieved reduction in memory usage showcases an optimization of nearly 31%.

While this example involves a small dataset, the principles of memory efficiency hold significant implications when extrapolated to big data scenarios. In large-scale datasets, the impact of removing unnecessary columns becomes even more pronounced.

To underline the significance of the operation, consider a scenario with a substantial dataset. Initially, the dataset size was 100,000 megabytes, and after the removal of unnecessary columns, it was reduced to 69,000 megabytes. To execute the same workload, the initial option would be utilizing an AWS EC2 instance of type `r7g.4xlarge` with an hourly rate of $1.0064 and memory of 128 GiB, as we need 100 gigabytes of memory to load just the dataset. However, by reducing the dataset size to 61 gigabytes, an alternative, more cost-effective option is available, employing an `r7g.2xlarge` instance at $0.5032 per hour and memory of 64 GiB. In the context of a five-minute workload runtime, the cost associated with the operation before dropping the data was as follows:

```
Cost_before = (Hourly Rate/60) * Runtime(in minutes) = (1.0064/60) * 5
= 0.0838$
Cost_after = (Hourly Rate/60) * Runtime(in minutes) = (0.5032/60) * 5
= 0.041$
```

The solution became approximately 50% more cost-effective after dropping unnecessary columns. This represents the cost savings achieved by optimizing the dataset and utilizing a more suitable AWS instance type.

The simplicity of this example underscores a crucial message:

Streamline your data operations by focusing on what is truly essential and let this simplicity drive cost-effectiveness.

Transitioning from dropping columns to fixing inconsistent data types involves addressing the quality and integrity of the remaining columns in your dataset.

Dealing with inconsistent and incorrect data types

When working with a DataFrame, it's important to ensure that each column has the correct data type. Inconsistent or incorrect data types can lead to errors in analysis, unexpected behavior, and difficulties in performing operations. Let's review how you can handle such situations. You can find the code for this example here: `https://github.com/PacktPublishing/Python-Data-Cleaning-and-Preparation-Best-Practices/blob/main/chapter04/4.data_types.py`.

Inspecting columns

Inspecting the data types of each column in the data is an essential step in identifying any inconsistencies or incorrect data types. The `dtypes` attribute of a DataFrame provides information about the data types of each column. Let's check the data types of the columns in our dataset:

```
print("\nUpdated Data Types of Columns:")
print(df.dtypes)
```

The types are presented here:

```
CustomerID          int64
ProductName        object
PurchaseAmount      int64
PaymentMethod      object
Timestamp          object
```

Inspecting the data types allows you to understand the current representation of the data and determine if any data type conversions or transformations are needed for further analysis or data cleaning tasks. In the following sections, we will perform different type transformations.

Columnar type transformations

In the data world, various **type transformations** can be applied to manipulate and convert data into different formats. In those situations, the `astype` method is your friend. The most common type transformations that you should be comfortable with are presented in the next sections.

Converting to numeric types

In pandas, the `astype()` function is used to convert a column to a specified numeric data type. For example, to convert a column named `PurchaseAmount` to an integer type, you can use the following:

```
df['PurchaseAmount'] = pd.to_numeric(df['PurchaseAmount'],
errors='coerce')
```

Now, let's see how we can turn columns into strings.

Converting to string types

You can use the `astype()` function to convert a column into a string type:

```
df['ProductName'] = df['ProductName'].astype('str')
```

Now, let's see how we can turn columns into categorical types.

Converting to categorical types

A **categorical type** refers to a data type that represents categorical or discrete variables. Categorical variables can take on a limited, and usually fixed, number of distinct categories or levels. These variables often represent qualitative data or attributes with no inherent order:

```
df['PaymentMethod'] = df['PaymentMethod'].astype('category')
```

The last transformation we will discuss is the Boolean.

Converting to Boolean types

A **Boolean transformation** refers to converting values in a dataset or a specific column to Boolean (`True`/`False`) values based on certain conditions or criteria. This transformation is often used to create binary indicators or flags, making it easier to work with and analyze data:

```
df['HasDive'] = df['ProductName'].str.contains('Dive', case=False)
```

The preceding code part checks whether each element in the `ProductName` column contains the substring `Dive`. It returns a Boolean Series where each element is `True` if the condition is met and `False` otherwise:

```
df['HasDive'] = df['HasDive'].astype('bool')
```

The `astype('bool')` method is used to explicitly cast the data type of the `HasDive` column to Boolean.

> **Things to be aware of when using astype(bool)**
>
> If you're experiencing a situation where all values are being converted to `True`, it could be due to one of the following reasons:
>
> 1. **Non-zero values**: Since non-zero values are considered `True` in a Boolean context, `.astype(bool)` will convert all non-zero values to `True`. In such cases, consider if the column contains unexpected or unintended non-zero values.
>
> 2. **Non-missing values**: If your column has missing values represented as **Not a Number** (**NaN**), they are treated as non-zero and are converted to `True` when using `.astype(bool)`. Check if there are missing values present in the column and consider how you want to handle them. You may need to fill in or drop missing values before the conversion.

In the last section of this chapter, we'll discuss how to handle dates and times.

Working with dates and times

Imagine you have data that includes information about when things happened – being able to understand and handle that time-related data is key for making sense of patterns and trends. It's not just about understanding when things happened, but also about making it easier to visualize and tell stories with your data. Whether you're diving into trends over time, filtering data for specific periods, or making predictions with machine learning, being good with dates and times is key to unlocking valuable insights from datasets that involve the dimension of time.

Now that we understand why dealing with dates and time is so important, the next step is learning how to grab that time-related info and make it work for us.

Importing and parsing date and time data

Python provides several main functions to parse dates, depending on the format of the input date string and the desired output. Let's discuss the commonly used functions for parsing dates.

pd.to_datetime() from the pandas library

This function is specifically designed for parsing date strings within pandas DataFrames or Series, but it can also be used independently. It is suitable when working with tabular data and allows handling multiple date formats simultaneously:

```
df['Timestamp3'] = pd.to_datetime(df['Timestamp'], format='%Y-%m-%d
%H:%M:%S')
```

The `format` parameter specifies the expected format of the input string. In this example, `%Y` represents the four-digit year, `%m` represents the month, `%d` represents the day, `%H` represents the hour, `%M` represents the minute, and `%S` represents the second.

Considerations

If your dataset contains missing or inconsistent timestamp values, consider using the `errors` parameter. For example, `errors='coerce'` will replace parsing errors with **Not a Time (NaT)** values.

While `pd.to_datetime` is efficient, it may have performance implications for large datasets. For improved performance, consider using the `infer_datetime_format=True` parameter to automatically infer the format (works well for standard formats). When `infer_datetime_format` is set to `True`, and `parse_dates` is enabled, Pandas will try to automatically deduce the format of datetime strings in the columns. If successful, it switches to a more efficient parsing method, potentially boosting parsing speed by 5-10 times in certain scenarios.

If your data involves different time zones, consider using the `utc` and `tz` parameters to handle **Coordinated Universal Time** (**UTC**) conversion and time zone localization.

In the next section, we will introduce another method, `strftime`. This method allows for the customization of datetime values, enabling the creation of specific and readable representations of time.

strftime() from the datetime module

This function is used to parse a date string into a datetime object based on *a specified format string*. It is suitable when you have a known date format and want precise control over the parsing process:

```
df['FormattedTimestamp'] = df['Timestamp'].dt.strftime('%b %d, %Y
%I:%M %p')
```

The resulting DataFrame is as follows:

```
            Timestamp     FormattedTimestamp
0  2022-01-01 08:30:45  Jan 01, 2022 08:30 AM
1  2022-01-02 14:20:30  Jan 02, 2022 02:20 PM
```

The format is controlled by format specifiers, each starting with a percent (`%`) character, representing different components of the date and time (e.g., `%Y` for the year, `%m` for the month, `%d` for the day, `%H` for the hour, `%M` for the minute, `%S` for the second, etc.). A full list of format specifiers can be found in the Python documentation: `https://strftime.org/`.

Unlike the rigid structure required by `strftime`, `dateutil.parser.parse()` excels in interpreting a wide range of date and time representations, offering a dynamic solution for parsing diverse datetime strings, as we will see in the next section.

dateutil.parser.parse() from the dateutil library

This function provides a flexible approach to parse date strings, *automatically inferring* the format based on the input. It is useful when dealing with a variety of date formats or when the format is unknown:

```
df['Timestamp2'] = df['Timestamp'].apply(parser.parse)
```

One thing to note about this method is that the parser can infer and *handle time zone information*, making it convenient for working with data originating from different time zones.

In the next section, instead of treating dates and times, we shift our approach to splitting them into individual parts, such as days, months, and years.

Extracting components from dates and times

You can extract specific components of a datetime object, such as year, month, day, hour, minute, or second, using the attributes provided by the datetime module:

```
df['Day'] = df['Timestamp'].dt.day
df['Month'] = df['Timestamp'].dt.month
df['Year'] = df['Timestamp'].dt.year
```

Using the `.dt` accessor, we can extract the day, month, and year components from the `Timestamp` column and create new columns, `Day`, `Month`, and `Year`, as presented here:

```
            Timestamp  Day  Month  Year
0 2022-01-01 08:30:45    1      1  2022
1 2022-01-02 14:20:30    2      1  2022
```

Extracting components is useful in the following cases:

- **Temporal analysis**: If your analysis involves patterns or trends that vary across days, months, or years, extracting these components facilitates a more focused exploration.
- **Grouping and aggregation**: When grouping data based on temporal patterns, extracting components allows for easy aggregation and summarization.

- **Time series analysis**: For time series analysis, breaking down datetime values into components is essential for understanding seasonality and trends.

Moving on to calculate time differences and durations elevates our exploration of temporal data by introducing a dynamic dimension.

Calculating time differences and durations

When calculating the time difference between two datetime objects using subtraction, you harness the inherent capability of Python's `datetime` library to produce a `timedelta` object. This object encapsulates the duration between the two timestamps, providing a comprehensive representation of the temporal gap in terms of days, hours, minutes, and seconds. The code for this section can be found here: `https://github.com/PacktPublishing/Python-Data-Cleaning-and-Preparation-Best-Practices/blob/main/chapter04/8.time_deltas.py`:

```
df['TimeSincePreviousPurchase'] = df['Timestamp'].diff()
```

This pandas function, `.diff()`, calculates the difference between each element and the previous element in the `Timestamp` column. It effectively computes the time elapsed since the previous timestamp for each row.

```
df['TimeUntilNextPurchase'] = -df['Timestamp'].diff(-1)
```

Similar to the first line, this computes the difference between each element and the following element in the `Timestamp` column. It calculates the time duration until the next timestamp. The negative sign is applied to reverse the sign of the time differences. This is done to get a positive representation of the time until the next purchase.

Let's see how the time delta is depicted in the data:

```
            Timestamp TimeSincePreviousPurchase TimeUntilNextPurchase
0 2022-01-01 08:30:45                       NaT       1 days 05:49:45
1 2022-01-02 14:20:30           1 days 05:49:45       1 days 05:54:40
```

If you are wondering when it is a good idea to consider adding some time differences in your data workflow, then read the following:

- **Time-based analysis**: Calculating time differences allows for analyzing the duration between events or timestamps. It helps quantify the time taken for different processes, activities, or intervals.
- **Performance measurement**: By measuring the duration of tasks or events, you can evaluate performance metrics, such as response time, processing time, or time taken to complete an operation. This information can guide optimization efforts and identify areas for improvement.

- **Event sequencing**: By comparing timestamps, you can determine the chronological order in which events occurred. This sequencing helps you understand the relationships between events and their dependencies.
- **Service-level agreement (SLA) monitoring**: Time differences are useful for SLA monitoring. By comparing timestamps related to SLA metrics, such as response time or resolution time, you can ensure compliance with agreed-upon service levels. Monitoring time differences helps identify SLA breaches and take appropriate actions.

The `.diff()` method in pandas is primarily used to compute the difference between *consecutive* elements in a Series or DataFrame. While it's straightforward to compute first-order differences (i.e., differences between adjacent elements), there are additional considerations and variations to explore.

Specifying time intervals

You can customize `.diff()` to compute the difference between elements at a specific *time interval*. This is achieved by passing the `periods` parameter to specify the number of elements to shift:

```
df['TimeDifference'] = df['Timestamp'].diff(periods=2)
```

Let's observe the following results:

```
            Timestamp TimeSincePreviousPurchase TimeDifference2periods
0 2022-01-01 08:30:45                       NaT                    NaT
1 2022-01-02 14:20:30           1 days 05:49:45                    NaT
2 2022-01-03 20:15:10           1 days 05:54:40        2 days 11:44:25
3 2022-01-04 12:45:30           0 days 16:30:20        1 days 22:25:00
```

As you can see, `.diff(periods=2)` calculated the difference between each timestamp and the two positions before it. The `periods` parameter allows you to specify the number of elements to shift when computing the difference. In this case, it is `periods=2`, but you can assign to it any value that makes sense for your use case.

Handling missing values

The `.diff()` method introduces a NaN for the first 2 elements when used with `diff(periods=2)` since there are no previous elements to calculate the difference from. You can handle or fill in these missing values based on your specific use case:

```
df['TimeDifference'] = df['Timestamp']. diff(periods=2).fillna(0)
```

Let's observe the results:

```
            Timestamp TimeDiff2periods_nonulls TimeDifference2periods
0  2022-01-01 08:30:45                        0                    NaT
1  2022-01-02 14:20:30                        0                    NaT
2  2022-01-03 20:15:10         2 days 11:44:25        2 days 11:44:25
3  2022-01-04 12:45:30         1 days 22:25:00        1 days 22:25:00
```

As you can see, `fillna(0)` replaced the NaN values with `0`.

Moving on from time differences and durations to time zones and daylight saving time, we'll now address the nuances of handling temporal data across different regions.

Handling time zones and daylight saving time

Handling time zones is key when dealing with data that spans multiple geographical regions or when accurate time representation is crucial. Time zones help standardize time across different locations, considering the offset from UTC due to geographic boundaries and daylight-saving time adjustments. In our example dataset, we'll demonstrate how to handle time zones using pandas:

```
df['Timestamp_UTC'] = df['Timestamp'].dt.tz_localize('UTC')
```

We localized the timestamps to a specific time zone, in this case, `'UTC'`.

```
df['Timestamp_NY'] = df['Timestamp_UTC'].dt.tz_convert('America/New_
York')
```

We then converted the localized timestamps to a different time zone, in this case, `'America/New_York'`. Let's observe the following results:

```
            Timestamp             Timestamp_UTC              Timestamp_NY
0  2022-01-01 08:30:45 2022-01-01 08:30:45+00:00 2022-01-01 03:30:45-05:00
1  2022-01-02 14:20:30 2022-01-02 14:20:30+00:00 2022-01-02 09:20:30-05:00
```

Curious about the significance of managing time zones? Let's understand why it matters:

- When working with data collected from different time zones, it is essential to handle time zones to ensure accurate analysis and interpretation. Without proper time zone handling, the analysis might be skewed due to inconsistencies in time representation.
- For applications that require precise time representation, such as financial transactions, log entries, or event tracking, handling time zones becomes crucial.

- When integrating data from various sources or merging datasets, handling time zones becomes necessary to align timestamps accurately. This ensures the correct chronological ordering of events and prevents inconsistencies in time-based analysis.
- If you are developing applications or services that serve users across different time zones, handling time zones is crucial for providing accurate and relevant information to users based on their local time.

> **Considerations**
>
> Time zone handling should be implemented consistently throughout the data processing pipeline to avoid inconsistencies or errors.

Let's summarize the learnings from this chapter.

Summary

This chapter was about the techniques for cleaning and manipulating data. Beginning with the challenges of messy data, we covered the removal of irrelevant columns and the handling of inconsistent data types. Practical use cases were demonstrated with an e-commerce dataset, showcasing Python code for effective data transformations. The importance of dropping unnecessary columns was emphasized, highlighting potential cost reductions and memory efficiency gains, particularly for big data. Data type transformations, including numeric, string, categorical, and Boolean conversions, were illustrated with practical examples. The chapter then explored intricate aspects of working with dates and times, showcasing methods such as `pd.to_datetime()`, `strftime`, and `dateutil.parser.parse()`.

As we wrap up this chapter, it lays a solid foundation for the upcoming one in which data merging and transformations will be discussed.

5

Data Transformation – Merging and Concatenating

Understanding how to transform and manipulate data is crucial for unlocking valuable insights. Techniques such as joining, merging, and appending allow us to blend information from various sources and organize and analyze subsets of data. In this chapter, we'll learn how to merge multiple datasets into a single dataset and explore the various techniques that we can use. We'll understand how to avoid duplicate values while merging datasets and some tricks to improve the process of merging datasets.

In this chapter, we'll cover the following topics:

- Joining datasets
- Handling duplicates when merging datasets
- Performance tricks for merging
- Concatenating DataFrames

Technical requirements

You can find all the code for the chapter at the following link: `https://github.com/PacktPublishing/Python-Data-Cleaning-and-Preparation-Best-Practices/tree/main/chapter05`.

Each section is followed by a script with a similar naming convention, so feel free to execute the scripts and/or follow along by reading the chapter.

Joining datasets

In data analysis projects, it is common to encounter data that is spread across multiple sources or datasets. Each dataset may contain different pieces of information related to a common entity or subject. **Data merging**, also known as data joining or data concatenation, is the process of combining these separate datasets into a single cohesive dataset. In data analysis projects, it's common to encounter situations where information about a particular subject or entity is spread across multiple datasets. For instance, imagine you're analyzing customer data for a retail business. You might have one dataset containing customer demographics, such as names, ages, and addresses, and another dataset with their purchase history, such as transaction dates, items bought, and total spending. Each of these datasets provides valuable insights but, individually, they don't give a complete picture of customer behavior. To gain a comprehensive understanding, you need to combine these datasets. By merging the customer demographics with their purchase history based on a common identifier, such as a customer ID, you create a single dataset that allows for richer analysis. For example, you could identify patterns such as which age groups are buying specific products or how spending habits vary by location.

Choosing the correct merge strategy

Choosing the correct join type is crucial as it determines which rows from the input DataFrames are included in the joined output. Python's pandas library provides several join types, each with different behaviors. Let's introduce the use case example we are going to work on in this chapter and then expand on the different types of joins.

In this chapter, our use case involves employee data and project assignments for a company managing its workforce and projects. You can execute the following script to see the DataFrames in more detail: `https://github.com/PacktPublishing/Python-Data-Cleaning-and-Preparation-Best-Practices/blob/main/chapter05/1.use_case.py`.

The `employee_data` DataFrame represents employee details, such as their names and departments, as presented here:

```
   employee_id    name department
0            1   Alice         HR
1            2     Bob         IT
```

The `project_data` DataFrame contains information about project assignments, including the project names:

```
   employee_id project_name
0            2     ProjectA
1            3     ProjectB
```

In the following sections, we will discuss the different DataFrame merging options, starting with the inner join.

Inner merge

The inner merge returns only the rows that have matching values in both DataFrames for the specified join columns. It's very important to note the following:

- Rows with non-matching keys in either DataFrame will be excluded from the merged output
- Rows with missing values in the key columns will be excluded from the merged result

The result of an inner merge is presented in the following figure:

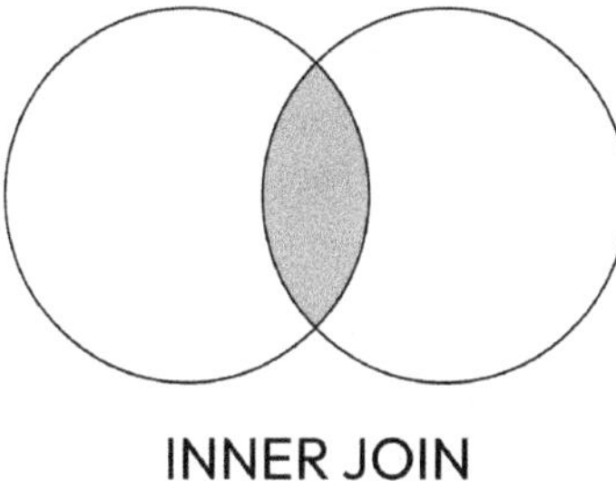

INNER JOIN

Figure 5.1 – Inner merge

Let's have a look at how we can achieve the preceding result using the pandas `merge` function, using the example presented in the previous section:

```
merged_data = pd.merge(employee_data, project_data, on='employee_id',
how='inner')
```

As we see in the preceding code snippet, the `pd.merge()` function is used to merge the two DataFrames. The `on='employee_id'` argument specifies that the `employee_id` column should be used as the key on which to join the DataFrames. The `how='inner'` argument specifies that an inner join should be performed. This type of join returns only the rows that have matching values in both DataFrames, which, in this case, are the rows where `employee_id` matches in both `employee_data` and `project_data`. In the following table, you can see the output of the inner join of the two DataFrames:

```
   employee_id     name department project_name
0            2      Bob         IT     ProjectA
1            3  Charlie  Marketing     ProjectB
2            4    David    Finance     ProjectC
3            5      Eva         IT     ProjectD
```

This approach ensures that the data from both DataFrames is combined based on a **common key**, with rows included only when there is a match across both DataFrames, adhering to the principles of an inner join.

If this is still not clear, in the following list, we present specific examples from the data world where an inner merge is crucial:

- **Match tables**: Inner joins are ideal when you need to match data from different tables. For example, if you have a table of employees and another table of department names, you can use an inner join to match each employee with their respective department.
- **Data filtering**: Inner joins can act as a filter to exclude rows that do not have corresponding entries in both tables. This is useful in scenarios where you only want to consider records that have complete data across multiple tables. For instance, matching customer orders with product details only where both records exist.
- **Efficiency in query execution**: Since inner joins only return rows with matching values in both tables, they can be more efficient in terms of query execution time compared to outer joins, which need to check for and handle non-matching entries as well.
- **Reducing data duplication**: Inner joins help in reducing data duplication by only returning matched rows, thus ensuring that the data in the result set is relevant and not redundant.
- **Simplifying complex queries**: When dealing with multiple tables, inner joins can be used to simplify queries by reducing the number of rows to be examined and processed in subsequent query operations. This is particularly useful in complex database schemas where multiple tables are interrelated.

Moving from an inner join to an outer join expands the scope of the merged data, incorporating all available rows from both datasets, even if they don't have corresponding matches.

Outer merge

The outer merge (also known as full outer join) returns all the rows from both DataFrames, combining the matching rows as well as the non-matching rows. The full outer join ensures that no data is lost from either DataFrame, but it can introduce NaN values where there are unmatched rows in either DataFrame.

The result of an outer merge is presented in the following figure:

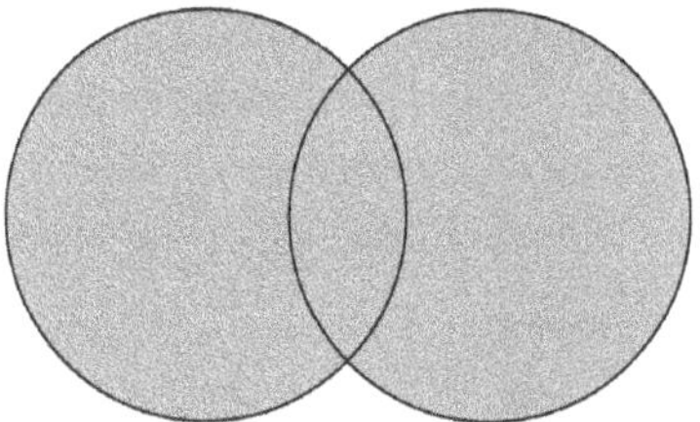

FULL OUTER JOIN

Figure 5.2 – Outer merge

Let's have a look at how we can achieve the preceding result using the pandas `merge` function, using the example presented in the previous section:

```
full_outer_merged_data = pd.merge(employee_data, project_data,
on='employee_id', how='outer')
```

As we see in the preceding code snippet, the `pd.merge()` function is used to merge the two DataFrames. The `on='employee_id'` argument specifies that the `employee_id` column should be used as the key on which to merge the DataFrames. The `how='outer'` argument specifies that a full outer join should be performed. This type of join returns all rows from both DataFrames, filling in `NaN` where there is no match. In the following table, you can see the output of the outer join of the two DataFrames:

```
   employee_id     name  department  project_name
0            1    Alice          HR           NaN
1            2      Bob          IT      ProjectA
2            3  Charlie   Marketing      ProjectB
3            4    David     Finance      ProjectC
4            5      Eva          IT      ProjectD
5            6      NaN         NaN      ProjectE
```

This approach ensures that the data from both DataFrames is combined, allowing for a comprehensive view of all available data, even if some of it is incomplete due to mismatches between the DataFrames.

In the following list, we present specific examples from the data world where an outer merge is crucial:

- **Including optional data**: Outer joins are ideal when you want to include rows that have optional data in another table. For instance, if you have a table of users and a separate table of addresses, not all users might have an address. An outer join allows you to list all users and show addresses for those who have them, without excluding users without addresses.
- **Data integrity and completeness**: In scenarios where you need a comprehensive dataset that includes records from both tables, regardless of whether there's a matching record in the joined table or not, outer joins are essential. This ensures that you have a complete view of the data, which is particularly important in reports that need to show all entities, such as a report listing all customers and their purchases, including those who have not made any purchases.
- **Mismatched data analysis**: Outer joins can be used to identify discrepancies or mismatches between tables. For example, if you are comparing a list of registered users against a list of participants in an event, an outer join can help identify users who did not participate and participants who are not registered.
- **Complex data merging**: When merging data from multiple sources that do not perfectly align, outer joins can be used to ensure that no data is lost during the merging process. This is particularly useful in complex data environments where data integrity is critical.

Transitioning from an outer join to a right join narrows the focus of the merged data, emphasizing the inclusion of all rows from the right DataFrame while maintaining matches from the left DataFrame.

Right merge

The right merge (also known as right outer join) returns all the rows from the right DataFrame and the matching rows from the left DataFrame. The result of a right merge is presented in the following figure:

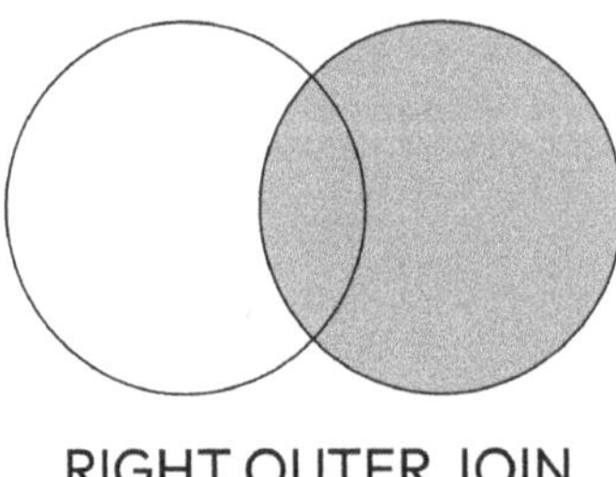

Figure 5.3 – Right merge

Let's have a look at how we can achieve the preceding result using the pandas `merge` function, using the example presented in the previous section:

```
right_merged_data = pd.merge(employee_data, project_data,
on='employee_id', how='right')
```

The `how='right'` argument specifies that a right outer join should be performed. This type of join returns all rows from the right DataFrame (`project_data`), and the matched rows from the left DataFrame (`employee_data`). Where there is no match, the result will have `NaN` in the columns of the left DataFrame. In the following table, you can see the output of the preceding join of the two DataFrames:

```
   employee_id     name department project_name
0            2      Bob         IT     ProjectA
1            3  Charlie  Marketing     ProjectB
2            4    David    Finance     ProjectC
3            5      Eva         IT     ProjectD
4            6      NaN        NaN     ProjectE
```

In the following list, we present specific examples from the data world where a right merge is crucial:

- **Completing data**: A right merge is useful when you need to ensure that all entries from the right DataFrame are retained in the result, which is important when the right DataFrame contains essential data that must not be lost
- **Data enrichment**: This type of join can be used to enrich a dataset (right DataFrame) with additional attributes from another dataset (left DataFrame) while ensuring that all records from the primary dataset are preserved
- **Mismatched data analysis**: Like outer joins, right merges can help identify which entries in the right DataFrame do not have corresponding entries in the left DataFrame, which can be critical for data cleaning and validation processes

Transitioning from a right to a left merge shifts the perspective of the merged data, prioritizing the inclusion of all rows from the left DataFrame while maintaining matches from the right DataFrame.

Left merge

The left merge (also known as left outer join) returns all the rows from the left DataFrame and the matching rows from the right DataFrame. The result of a left merge is presented in the following figure:

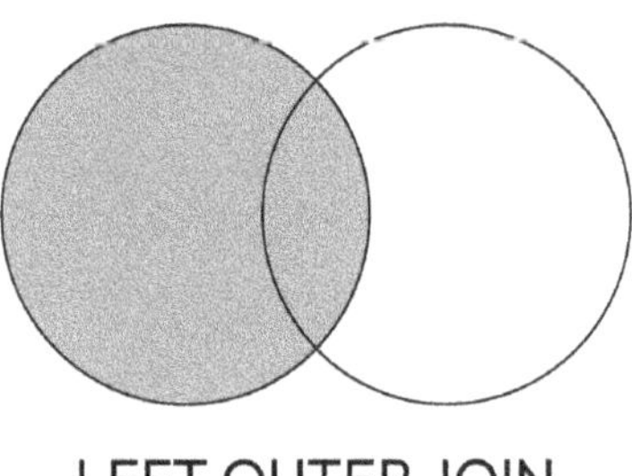

Figure 5.4 – Left merge

Let's have a look at how we can achieve the preceding result using the pandas `merge` function, using the example presented in the previous section:

```
left_merged_data = pd.merge(employee_data, project_data, on='employee_
id', how='left')
```

The `how='left'` argument specifies that a left outer join should be performed. This type of join returns all rows from the left DataFrame (`employee_data`), and the matched rows from the right DataFrame (`project_data`). Where there is no match, the result will have `NaN` in the columns of the right DataFrame. In the following table, you can see the output of the preceding join of the two DataFrames:

```
   employee_id     name department project_name
0            1    Alice         HR          NaN
1            2      Bob         IT     ProjectA
2            3  Charlie  Marketing     ProjectB
3            4    David    Finance     ProjectC
4            5      Eva         IT     ProjectD
```

If you are wondering when the left merge should be used, then the considerations presented in the previous section about the right merge apply in this case too. Now that we have discussed merge operations, let's move on to handling duplicates that may arise during the merging process.

Handling duplicates when merging datasets

Handling duplicate keys before performing merge operations is crucial because duplicates can lead to unexpected results, such as Cartesian products, where rows are multiplied by the number of matching entries. This can not only distort the data analysis but also significantly impact performance due to the increased size of the resulting DataFrame.

Why handle duplication in rows and columns?

Duplicate keys can lead to a range of problems that may compromise the accuracy of your results and the efficiency of your data processing. Let's explore why it's a good idea to handle duplicate keys prior to merging data:

- If there are duplicate keys in either table, merging these tables can result in a **Cartesian product**, where each duplicate key in one table matches with each occurrence of the same key in the other table, leading to an exponential increase in the number of rows
- Duplicate keys might represent data errors or inconsistencies, which can lead to incorrect analysis or conclusions
- Reducing the dataset size by removing duplicates can lead to faster processing times during the merge operation

Having understood the importance of handling duplicate keys, let's now examine various strategies to effectively manage these duplicates before proceeding with merge operations.

Dropping duplicate rows

Dropping duplicate entries in your datasets involves identifying and removing any duplicate rows based on specific key columns, which ensures that each entry is unique. This step not only simplifies subsequent data merging but also enhances the reliability of the analysis by eliminating potential sources of error caused by duplicate data. To showcase the dropping of duplicates, we will expand the example we have been using to add more duplicated rows in each of the DataFrames. As always, you can follow the full code here: `https://github.com/PacktPublishing/Python-Data-Cleaning-and-Preparation-Best-Practices/blob/main/chapter05/6a.manage_duplicates.py`.

Let's first create the sample employee data with some duplicate keys in the `employee_id` column:

```
employee_data = pd.DataFrame({
    'employee_id': [1, 2, 2, 3, 4, 5, 5],
    'name': ['Alice', 'Bob', 'Bob', 'Charlie', 'David', 'Eva', 'Eva'],
    'department': ['HR', 'IT', 'IT', 'Marketing', 'Finance', 'IT',
'IT']
})
```

Let's also create the sample project data with some duplicate keys in the `employee_id` column:

```
project_data = pd.DataFrame({
    'employee_id': [2, 3, 4, 5, 5, 6],
    'project_name': ['ProjectA', 'ProjectB', 'ProjectC', 'ProjectD',
'ProjectD', 'ProjectE']
})
```

Now, we want to merge these datasets. But first, we'll drop any duplicates so that we can make the merge operation as lightweight as possible. Dropping the duplicates before the merge is shown in the following code snippet:

```
employee_data = employee_data.drop_duplicates(subset='employee_id',
keep='first')
project_data = project_data.drop_duplicates(subset='employee_id',
keep='first')
```

As shown in the code, `drop_duplicates()` is used to remove duplicate rows based on `employee_id`. The `keep='first'` parameter ensures that only the first occurrence is kept while others are removed.

After dropping the duplicates, you can proceed with the merge operation, as shown in the following code:

```
merged_data = pd.merge(employee_data, project_data, on='employee_id',
how='inner')
```

The merged dataset can be seen here:

```
   employee_id     name department project_name
0            2      Bob         IT     ProjectA
1            3  Charlie  Marketing     ProjectB
2            4    David    Finance     ProjectC
3            5      Eva         IT     ProjectD
```

The `merged_data` DataFrame includes columns from both the `employee_data` and `project_data` DataFrames, showing the `employee_id`, `name`, `department`, and `project_name` values for each employee that exists in both datasets. The duplicates are removed, ensuring each employee appears only once in the final merged dataset. The `drop_duplicates` operation is crucial for avoiding data redundancy and potential conflicts during the merge operation. Next, we will discuss how we can guarantee that the merge operation respects the uniqueness of the keys and adheres to specific constraints.

Validating data before merging

When merging datasets, especially large and complex ones, ensuring the integrity and validity of the merge operation is crucial. pandas provides the `validate` parameter in the `merge()` function to enforce specific conditions and relationships between the keys used in the merge. This helps in identifying and preventing unintended duplications or data mismatches that could compromise the analysis.

The following code demonstrates how to use the `validate` parameter to enforce `merge()` constraints and handle exceptions when these constraints are not met. You can see the full code at `https://github.com/PacktPublishing/Python-Data-Cleaning-and-Preparation-Best-Practices/blob/main/chapter05/6b.manage_duplicates_validate.py`:

```
try:
    merged_data = pd.merge(employee_data, project_data, on='employee_
id', how='inner', validate='one_to_many')
    print("Merged Data Result:")
    print(merged_data)
except ValueError as e:
    print("Merge failed:", e)
```

In the preceding code snippet, the merge operation is wrapped in a `try-except` block. This is a way to handle exceptions, which are errors occurring during a program's execution. The `try` block contains the code that might raise an exception, in this case, the merge operation. If an exception occurs, the code execution moves to the `except` block.

If the merge operation fails the validation check (in our case, if there are duplicate keys in the left DataFrame when they are expected to be unique), `ValueError` exception will be raised, and the `except` block will be executed. The `except` block catches the `ValueError` exception and prints a `Merge failed:` message, followed by the error message provided by pandas.

After executing the preceding code, you will see the following error message:

```
Merge failed: Merge keys are not unique in left dataset; not a one-to-
many merge
```

The `validate='one_to_many'` parameter is included in the merge operation. This parameter tells pandas to check that the merge operation is of the specified type. In this case, `one_to_many` means that the merge keys should be unique in the left DataFrame (`employee_data`) but can have duplicates in the right DataFrame (`project_data`). If the validation check fails, pandas will raise a `ValueError` exception.

When to use which approach

Use **manual duplicate removal** when you need fine control over how duplicates are identified and handled, or when duplicates require special processing (e.g., aggregation or transformation based on other column values).

Use **merge validation** when you want to ensure the structural integrity of your data model directly within the merge operation, especially in straightforward cases where the relationship between the tables is well-defined and should not include duplicate keys according to the business logic or data model.

If there is a good reason for the existence of duplicates in the data, we can consider employing aggregation methods during the merge to consolidate redundant information.

Aggregation

Aggregation is a powerful technique for managing duplicates in datasets, particularly when dealing with key columns that should be unique but contain multiple entries. By grouping data on these key columns and applying aggregation functions, we can consolidate duplicate entries into a single, summarized record. Aggregation functions such as sum, average, or maximum can be used to combine or summarize the data in a way that aligns with the analytical goals.

Let's see how aggregation can be employed to effectively deal with duplicates before merging data. We will extend the dataset a little bit to help us with this example, as shown here. You can see the full example at `https://github.com/PacktPublishing/Python-Data-Cleaning-and-Preparation-Best-Practices/blob/main/chapter05/6c.merge_and_aggregate.py`:

```
employee_data = pd.DataFrame({
    'employee_id': [1, 2, 2, 3, 4, 5, 5],
    'name': ['Alice', 'Bob', 'Bob', 'Charlie', 'David', 'Eva', 'Eva'],
    'department': ['HR', 'IT', 'IT', 'Marketing', 'Finance', 'IT',
'IT'],
    'salary': [50000, 60000, 60000, 55000, 65000, 70000, 70000]
})

# Sample project assignment data with potential duplicate keys
project_data = pd.DataFrame({
    'employee_id': [2, 3, 4, 5, 7, 6],
    'project_name': ['ProjectA', 'ProjectB', 'ProjectC', 'ProjectD',
'ProjectD', 'ProjectE']
})
```

Now, let's perform the aggregation step:

```
aggregated_employee_data = employee_data.groupby('employee_id').agg({
    'name': 'first', # Keep the first name encountered
    'department': 'first', # Keep the first department encountered
    'salary': 'sum' # Sum the salaries in case of duplicates
}).reset_index()
```

The `groupby()` method is used on `employee_data` with `employee_id` as the key. This groups the DataFrame by `employee_id`, which is necessary because of the duplicate `employee_id` values.

The `agg()` method is then applied to perform specific aggregations on different columns:

- `'name': 'first'` and `'department': 'first'` ensure that the first encountered values for these columns are retained in the grouped data
- `'salary': 'sum'` sums up the salaries for each `employee_id` value, which is useful if the duplicates represent split records of cumulative data

In the final step, the `pd.merge()` function is used to combine `aggregated_employee_data` with `project_data` using an inner join on the `employee_id` column:

```
merged_data = pd.merge(aggregated_employee_data, project_data,
on='employee_id', how='inner')
```

This ensures that only employees with project assignments are included in the result. The result after the merge is as follows:

```
   employee_id     name department  salary project_name
0            2      Bob         IT  120000     ProjectA
1            3  Charlie  Marketing   55000     ProjectB
2            4    David    Finance   65000     ProjectC
3            5      Eva         IT  140000     ProjectD
```

The `agg()` method in pandas is highly versatile, offering numerous options beyond the simple "keep first" approach demonstrated in the previous example. This method can apply a wide range of aggregation functions to consolidate data, such as summing numerical values, finding averages, or selecting maximum or minimum entries. We will dive deeper into the various capabilities of the `agg()` method in the next chapter, exploring how these different options can be applied to enhance data preparation and analysis.

Let's move on from using aggregation as a way to handle duplicates to concatenating duplicated rows when dealing with text or categorical data.

Concatenation

The concatenation of values from duplicate rows into a single row can be a useful technique, especially when dealing with categorical or textual data that may have multiple valid entries for the same key. This approach allows you to preserve all the information across duplicates without losing data.

Let's see how concatenation of rows can be employed to effectively deal with duplicates before merging data. To showcase this method, we will use the following DataFrame:

```
employee_data = pd.DataFrame({
    'employee_id': [1, 2, 2, 3, 4, 5, 5],
    'name': ['Alice', 'Bob', 'Bob', 'Charlie', 'David', 'Eva', 'Eva'],
```

```
    'department': ['HR', 'IT', 'Marketing', 'Marketing', 'Finance',
'IT', 'HR']
})
```

Now, let's perform the concatenation step, as shown in the following code snippet:

```
employee_data['department'] = employee_data.groupby('employee_id')
['department'].transform(lambda x: ', '.join(x))
```

In the concatenation step, the `groupby('employee_id')` method groups the data by `employee_id`. The `transform(lambda x: ', '.join(x))` method is then applied to the `department` column. The `transform` function is used here with a `lambda` function that joins all entries of the column department for each group (i.e., `employee_id`) into a single string separated by commas.

The result of this operation replaces the original `department` column in `employee_data`, where each `employee_id` now has a single entry for `department` that includes all original department data concatenated into one string, as shown in the following table:

```
   employee_id     name      department
0            1    Alice              HR
1            2      Bob   Marketing, IT
3            3  Charlie       Marketing
4            4    David         Finance
5            5      Eva          IT, HR
```

Use concatenation when you need to preserve all categorical or textual data across duplicate entries without preferring one entry over another.

This method is useful for summarizing textual data in a way that is still readable and informative, especially when dealing with attributes that can have multiple valid values (e.g., an employee belonging to multiple departments).

Once duplicate rows in each DataFrame are resolved, attention shifts to identifying and resolving duplicate columns across the DataFrames to be merged.

Handling duplication in columns

When merging data from different sources, it's not uncommon to encounter DataFrames with overlapping column names. This challenge often arises when combining data from similar datasets.

Expanding the example data we have been using so far, we will adjust the DataFrames to help us showcase the options we have when dealing with common columns across DataFrames. The data can be seen here:

```
employee_data_1 = pd.DataFrame({
    'employee_id': [1, 2, 3, 4, 5],
```

```
    'name': ['Alice', 'Bob', 'Charlie', 'David', 'Eva'],
    'department': ['HR', 'IT', 'Marketing', 'Finance', 'IT']
})

employee_data_2 = pd.DataFrame({
    'employee_id': [6, 7, 8, 9, 10],
    'name': ['Frank', 'Grace', 'Hannah', 'Ian', 'Jill'],
    'department': ['Logistics', 'Marketing', 'IT', 'Marketing',
'Finance']
})
```

Let's see how we can merge these datasets by applying different techniques without breaking the merge operation.

Handling duplicate columns while merging

The columns in the two DataFrames presented previously share identical names and may represent the same data. However, we have decided to retain both sets of columns in the merged DataFrame. This decision is based on the suspicion that, despite having the same column names, the entries are not entirely identical, indicating that they may be different representations of the same data. This is an issue that we can address later, following the merge operation.

The best approach to keep both sets of columns is to use the `suffixes` parameter in the `merge()` function. This will allow you to differentiate between the columns from each DataFrame without losing any data. Here's how you can implement this in Python using pandas:

```
merged_data = pd.merge(employee_data_1, employee_data_2, on='employee_
id', how='outer', suffixes=('_1', '_2'))
```

The `pd.merge()` function is used to merge the two DataFrames on `employee_id`. The `how='outer'` parameter is used to ensure all records from both DataFrames are included in the merged DataFrame, even if there are no matching `employee_id` values. The `suffixes=('_1', '_2')` parameter adds suffixes to the columns from each DataFrame to differentiate them in the merged DataFrame. This is crucial when columns have the same names but come from different sources. Let's review the output DataFrame:

```
   employee_id   name_1 department_1  name_2 department_2
0            1    Alice           HR     NaN          NaN
1            2      Bob           IT     NaN          NaN
2            3  Charlie    Marketing     NaN          NaN
3            4    David      Finance     NaN          NaN
4            5      Eva           IT     NaN          NaN
5            6      NaN          NaN   Frank    Logistics
6            7      NaN          NaN   Grace    Marketing
7            8      NaN          NaN  Hannah           IT
```

```
8         9         NaN         NaN     Ian   Marketing
9        10         NaN         NaN    Jill     Finance
```

This approach is particularly useful in scenarios where merging data from different sources involves overlapping column names but where it's also important to retain and clearly distinguish these columns in the resulting DataFrame. Another point to consider is that suffixes allow for identifying which DataFrame the data originated from, which is useful in analyses involving data from multiple sources.

In the next section, we will explain how to deal with duplicate columns by dropping them *before* the merge.

Dropping duplicate columns before the merge

If we find that we have copies of the same column in both DataFrames we want to merge and that the column in one of the DataFrames is sufficient or more reliable than the other, then it may be more practical to drop one of the duplicate columns before a merge operation instead of keeping both. This decision can be driven by the need to simplify the dataset, reduce redundancy, or when one of the columns does not provide additional value to the analysis. Let's have a look at the data for this example:

```
employee_data_1 = pd.DataFrame({
    'employee_id': [1, 2, 3, 4, 5],
    'name': ['Alice', 'Bob', 'Charlie', 'David', 'Eva'],
    'department': ['HR', 'IT', 'Marketing', 'Finance', 'IT']
})

employee_data_2 = pd.DataFrame({
    'employee_id': [1, 2, 3, 4, 5],
    'name': ['Alice', 'Bob', 'Charlie', 'David', 'Eva'],
    'department': ['Human Resources', 'Information Technology',
'Sales', 'Financial', 'Technical']
})
```

If we take a closer look at this data, we can see that the `department` column in the two DataFrames captures the same information but in different formats. For the sake of our example, let's assume that we know the HR system tracks the department of each employee in the format presented in the first DataFrame. That's why we will trust this column more than the one in the second DataFrame. Therefore, we will drop the second one before the merge operation. Here's how you can drop the column before the merge:

```
employee_data_2.drop(columns=['department'], inplace=True)
```

Before merging, the `department` column from `employee_data_2` is dropped because it's deemed less reliable. This is done using the `drop(columns=['department'], inplace=True)` method. Having dropped the required columns, we can proceed with the merge:

```
merged_data = pd.merge(employee_data_1, employee_data_2,
on=['employee_id', 'name'], how='inner')
```

The DataFrames are merged using the `employee_id` and `name` columns as keys with the `pd.merge()` function. The `how='inner'` parameter is used to perform an inner join, which includes only rows that have matching values in both DataFrames.

To optimize the merging process and improve performance, it's often beneficial to drop unnecessary columns before performing the merge operation for the following reasons:

- It leads to improved performance by significantly reducing the memory footprint during the merge operation, as it minimizes the amount of data to be processed and combined, thereby expediting the process.
- The resulting DataFrame becomes simpler and cleaner, facilitating easier data management and subsequent analysis. This reduction in complexity not only streamlines the merge operation but also reduces the likelihood of errors.
- In resource-constrained environments, such as those with limited computing resources, minimizing the dataset before intensive operations such as merging enhances resource efficiency and ensures smoother execution.

In the case that we have identical columns across the DataFrames, another option is to consider whether we can use them as keys in the merge operation.

Duplicate keys

When encountering identical keys across DataFrames, a smart approach is to merge based on these common columns. Let's revisit the example presented in the previous section:

```
merged_data = pd.merge(employee_data_1, employee_data_2,
on=['employee_id', 'name'], how='inner')
```

We can see here that we used `['employee_id', 'name']` as keys in the merge. If `employee_id` and `name` are reliable identifiers that ensure accurate matching of records across DataFrames, they should be used as keys. This ensures that the merged data accurately represents combined records from both sources.

As the volume and complexity of data continues to grow, it is crucial to efficiently combine datasets, as we will learn in the following section.

Performance tricks for merging

When working with large datasets, the performance of merge operations can significantly impact the overall efficiency of data processing tasks. Merging is a common and often necessary step in data analysis, but it can be computationally intensive, especially when dealing with big data. Therefore, employing performance optimization techniques is crucial to ensure that merges are executed as quickly and efficiently as possible.

Optimizing merge operations can lead to reduced execution time, lower memory consumption, and an overall smoother data-handling experience. In the following sections, we will explore various performance tricks that can be applied to merge operations in pandas, such as utilizing indexes, sorting indexes, choosing the right merge method, and reducing memory usage.

Set indexes

Utilizing indexes in pandas is a critical aspect of data manipulation and analysis, particularly when dealing with large datasets or performing frequent data retrieval operations. Indexes serve as a tool for both identification and efficient data access, providing several benefits that can significantly enhance performance. Specifically, when merging DataFrames, utilizing indexes can lead to performance improvements. Merging on indexes, rather than on columns, is generally faster because pandas can perform the merge operation using optimized index-based joining methods, which is more efficient than column-based merging. Let's revisit the employee example to prove this concept. The full code for this example can be found at `https://github.com/PacktPublishing/Python-Data-Cleaning-and-Preparation-Best-Practices/blob/main/chapter05/8a.perfomance_benchmark_set_index.py`.

First, let's import the necessary libraries:

```
import pandas as pd
import numpy as np
from time import time
```

Select the number of rows for the benchmarking example for each DataFrame:

```
num_rows = 5
```

Let's create the DataFrames for the example, which will have the number of rows as defined in the `num_rows` variable. The first employee DataFrame is defined here:

```
employee_data_1 = pd.DataFrame({
  'employee_id': np.arange(num_rows),
  'name': ['Alice', 'Bob', 'Charlie', 'David', 'Eva'],
  'department': ['HR', 'IT', 'Marketing', 'Finance', 'IT'],
  'salary': [50000, 60000, 70000, 80000, 90000]
})
```

The second DataFrame is as follows:

```
employee_data_2 = pd.DataFrame({
  'employee_id': np.arange(num_rows),
  'name': ['Alice', 'Bob', 'Charlie', 'David', 'Eva'],
  'department': ['HR', 'IT', 'Sales', 'Finance', 'Operations'],
```

```
    'bonus': [3000, 4000, 5000, 6000, 7000]
})
```

To demonstrate the effectiveness of the performance tricks we applied, we will initially perform the merge *without utilizing the index*. We'll calculate the time taken for this operation. Subsequently, we'll set the index in both DataFrames and repeat the merge operation, recalculating the time. Finally, we will present the results. Let's hope this approach yields the desired outcome! Let's start the clock:

```
start_time = time()
```

Let's perform the merge operation without using indexes, just by inner joining on `['employee_id', 'name']`:

```
merged_data = pd.merge(employee_data_1, employee_data_2,
on=['employee_id', 'name'], how='inner', suffixes=('_1', '_2'))
```

Let's calculate the time it took to perform the merge:

```
end_time = time()
merge_time = end_time - start_time
Merge operation took around 0.00289 seconds
```

> **Note**
>
> The timings may vary depending on the computer used to execute the program. The idea is that the optimized version takes less time than the original merge operation.

By setting `employee_id` as the index for both DataFrames (`employee_data_1` and `employee_data_2`), we allow pandas to use optimized index-based joining methods. This is particularly effective because indexes in pandas are implemented via hash tables or B-trees, depending on the data type and the sortedness of the index, which facilitates faster lookups:

```
employee_data_1.set_index('employee_id', inplace=True)
employee_data_2.set_index('employee_id', inplace=True)
```

Let's repeat the merge operation after setting the indexes and calculate the time once more:

```
start_time = time()
merged_data_reduced = pd.merge(employee_data_1, employee_data_2, left_
index=True, right_index=True, suffixes=('_1', '_2'))
end_time = time()
merge_reduced_time = end_time - start_time

Merge operation with reduced memory took around 0.00036 seconds
```

Now, if we calculate the percentage difference from the initial time to the final one, we see that we managed to drop the time by around 88.5%, just by setting the index. This seems impressive but let's also discuss some considerations when setting indexes.

Index considerations

It's important to choose the right columns for indexing based on the query patterns. Over-indexing can lead to unnecessary use of disk space and can degrade write performance due to the overhead of maintaining the indexes.

Rebuilding or **reorganizing indexes** is essential for optimal performance. These tasks address index fragmentation and ensure consistent performance over time.

While indexes can significantly improve read performance, they can also impact write performance. It's crucial to find a balance between optimizing for read operations (such as searches and joins) and maintaining efficient write operations (such as inserts and updates).

Multi-column indexes, or concatenated indexes, can be beneficial when multiple fields are often used together in queries. However, the order of the fields in the index definition is important and should reflect the most common query patterns.

Having proved the importance of setting indexes, let's go a step further and discuss the option of sorting the index before merging.

Sorting indexes

Sorting the index in pandas can be particularly beneficial in scenarios where you are frequently merging or performing join operations on large DataFrames. When indexes are sorted, pandas can take advantage of more efficient algorithms to align and join data, which can lead to significant performance improvements. Let's deep dive into this before proceeding to the code example:

- When indexes are sorted, pandas can use binary search algorithms to locate the matching rows between DataFrames. Binary search has a time complexity of *O(log n)*, which is much faster than the linear search required for unsorted indexes, especially as the size of the DataFrame grows.
- Sorted indexes facilitate quicker data alignment. This is because pandas can make certain assumptions about the order of the data, which streamlines the process of finding corresponding rows in each DataFrame during a merge.
- With sorted indexes, pandas can avoid unnecessary comparisons that would be required if the indexes were unsorted. This reduces the computational overhead and speeds up the merging process.

Let's go back to the code example by adding the sorting of the index. The original data remains the same; however, in this experiment, we are comparing the time it takes to perform the merge operation after the setting of the index versus the time it takes to perform the merge operation after the setting and sorting of indexes. The following code shows the main code components but, as always, you can follow the full example at `https://github.com/PacktPublishing/Python-Data-Cleaning-and-Preparation-Best-Practices/blob/main/chapter05/8b.performance_benchmark_sort_indexes.py`:

```
employee_data_1.set_index('employee_id', inplace=True)
employee_data_2.set_index('employee_id', inplace=True)
```

Let's perform the merge operation without sorting indexes:

```
merged_data = pd.merge(employee_data_1, employee_data_2, left_
index=True, right_index=True, suffixes=('_1', '_2'))
```

```
Merge operation with setting index took around 0.00036 seconds
```

Let's repeat the merge operation after sorting the indexes and calculate the time once more:

```
employee_data_1.sort_index(inplace=True)
employee_data_2.sort_index(inplace=True)

merged_data_reduced = pd.merge(employee_data_1, employee_data_2, left_
index=True, right_index=True, suffixes=('_1', '_2'))
```

```
Merge operation after sorting took around 0.00028 seconds.
```

Now, if we calculate the percentage difference from the initial time to the final one, we see that we managed to drop the time by an extra ~22%, by sorting the index. This seems impressive but let's also discuss some considerations when setting indexes.

Sorting index considerations

Sorting a DataFrame's index is not free of computational cost. The initial sorting operation itself takes time, so it's most beneficial when the sorted DataFrame will be used in multiple merge or join operations, amortizing the cost of sorting over these operations.

Sorting can sometimes increase the memory overhead, as pandas may create a sorted copy of the DataFrame's index. This should be considered when working with very large datasets where memory is a constraint.

Sorting the index is most beneficial when the key used for merging is not only unique but also has some logical order that can be leveraged, such as time-series data or ordered categorical data.

Index management and maintenance are crucial aspects you should consider when working with pandas DataFrames, especially when dealing with large datasets. Maintaining a well-managed index requires careful consideration. For example, regularly updating or reindexing a DataFrame can introduce computational costs, similar to sorting operations. Each time you modify the index—by sorting, reindexing, or resetting—it can result in additional memory usage and processing time, particularly with large datasets.

Indexes need to be maintained in a way that balances performance and resource usage. For instance, if you frequently merge or join DataFrames, ensuring that the index is properly sorted and unique can significantly speed up these operations. However, continuously maintaining a sorted index can be resource-intensive, so it's most beneficial when the DataFrame will be involved in multiple operations that can leverage the sorted index.

Additionally, choosing the right index type—whether it's a simple integer-based index, a datetime index for time-series data, or a multi-level index for hierarchical data—can influence how efficiently pandas handles your data. The choice of index should align with the structure and access patterns of your data to minimize unnecessary overhead.

In the next section, we will discuss how using the `join` function instead of `merge` can impact performance.

Merge versus join

While merging is a commonly used method to combine datasets based on specific conditions or keys, there is another approach: the `join` function. This function provides a streamlined way to perform merges primarily based on indexes, offering a simpler alternative to the more general merge function. The `join` method in pandas is particularly useful when the DataFrames involved have their indexes set up as the keys for joining, allowing for efficient and straightforward data combinations without the need for specifying complex join conditions.

Using the `join` function instead of `merge` can impact performance in several ways, primarily due to the underlying mechanisms and default behaviors of these two functions:

- The `join` function in pandas is optimized for index-based joining, meaning it's designed to be efficient when joining DataFrames on their indexes. If your DataFrames are already indexed by the keys you want to join on, using `join` can be more performance-efficient because it leverages the optimized index structures [2][6][7].
- Join is a simplified version of merge that defaults to joining on indexes. This simplicity can translate into performance benefits, especially for straightforward joining tasks where the complexity of merge is unnecessary. By avoiding the overhead of aligning non-index columns, join can execute more quickly in these scenarios [2][6].
- Under the hood, join uses merge [2][6].
- When joining large DataFrames, the way join and merge handle memory can impact performance. A join, by focusing on index-based joining, might manage memory usage more efficiently in certain scenarios, especially when the DataFrames have indexes that pandas can optimize on [1][3][4].

- While merge offers greater flexibility by allowing joins on arbitrary columns, this flexibility comes with a performance cost, especially for complex joins involving multiple columns or non-index joins. Join offers a performance advantage in simpler, index-based joins due to its more specific use case [2][6].

In summary, choosing between `join` and `merge` depends on the specific requirements of your task. If your joining operation is primarily based on indexes, join can offer performance benefits due to its optimization for index-based joining and its simpler interface. However, for more complex joining needs that involve specific columns or multiple keys, merge provides the necessary flexibility, albeit with potential impacts on performance.

Concatenating DataFrames

When you have datasets spread across multiple DataFrames with similar structures (same columns or same rows) and you want to combine them into a single DataFrame, this is where concatenating shines. The concatenation process can be along a particular axis, either row-wise (`axis=0`) or column-wise (`axis=1`).

Let's deep dive into the row-wise concatenation, also known as append.

Row-wise concatenation

The row-wise concatenation is used to concatenate one DataFrame to another along `axis=0`. To showcase this operation, two DataFrames, `employee_data_1` and `employee_data_2`, created with the same structure but different data can be seen here:

```
employee_data_1 = pd.DataFrame({
  'employee_id': np.arange(1, 6),
  'name': ['Alice', 'Bob', 'Charlie', 'David', 'Eva'],
  'department': ['HR', 'IT', 'Marketing', 'Finance', 'IT']
})

employee_data_2 = pd.DataFrame({
  'employee_id': np.arange(6, 11),
  'name': ['Frank', 'Grace', 'Hannah', 'Ian', 'Jill'],
  'department': ['Logistics', 'HR', 'IT', 'Marketing', 'Finance']
})
```

Let's perform the row-wise concatenation, as shown in the following code snippet:

```
concatenated_data = pd.concat([employee_data_1, employee_data_2],
axis=0)
```

The `pd.concat()` function is used to concatenate the two DataFrames. The first argument is a list of DataFrames to concatenate, and the `axis=0` parameter specifies that the concatenation should be row-wise, stacking the DataFrames on top of each other.

The result can be seen here:

```
   employee_id     name department
0            1    Alice         HR
1            2      Bob         IT
2            3  Charlie  Marketing
3            4    David    Finance
4            5      Eva         IT
0            6    Frank  Logistics
1            7    Grace         HR
2            8   Hannah         IT
3            9      Ian  Marketing
4           10     Jill    Finance
```

Some things you need to consider when performing row-wise concatenation are as follows:

- Ensure that the columns you want to concatenate are aligned correctly. pandas will automatically align columns by name and fill any missing columns with `NaN` values.
- After concatenation, you may want to reset the index of the resulting DataFrame to avoid duplicate index values, especially if the original DataFrames had their own range of indices. Observe the index in the following example before performing a `reset` operation:

```
   employee_id     name department
0            1    Alice         HR
1            2      Bob         IT
2            3  Charlie  Marketing
3            4    David    Finance
4            5      Eva         IT
0            6    Frank  Logistics
1            7    Grace         HR
2            8   Hannah         IT
3            9      Ian  Marketing
4           10     Jill    Finance
```

Let's now perform the `reset` operation on the index:

```
concatenated_data_reset = concatenated_data.reset_
index(drop=True)
```

Let's see the output once more:

```
   employee_id     name  department
0            1    Alice          HR
1            2      Bob          IT
2            3  Charlie   Marketing
3            4    David     Finance
4            5      Eva          IT
5            6    Frank   Logistics
6            7    Grace          HR
7            8   Hannah          IT
8            9      Ian   Marketing
9           10     Jill     Finance
```

Resetting the index creates a new, continuous index for the concatenated DataFrame. The `drop=True` parameter is used to avoid adding the old index as a column in the new DataFrame. This step is crucial for maintaining a clean DataFrame, especially when the index itself does not carry meaningful data. A continuous index is often easier to work with, particularly for indexing, slicing, and potential future merges or joins.

- Concatenation can increase the memory usage of your program, especially when working with large DataFrames. Be mindful of the available memory resources.

In the next section, we will discuss the column-wise concatenation.

Column-wise concatenation

Concatenating DataFrames column-wise in pandas involves combining two or more DataFrames side by side, aligning them by their index. To showcase this operation, the two DataFrames we have been using so far, `employee_data_1` and `employee_data_2`, will be used, and the operation can be done as shown here:

```
concatenated_data = pd.concat([employee_data_1, employee_performance],
axis=1)
```

The `pd.concat()` function is used with the `axis=1` parameter to concatenate the DataFrames side by side. This aligns the DataFrames by their index, effectively adding new columns from `employee_performance` to `employee_data_1`. This will display the following output:

```
   employee_id     name department employee_id performance_rating
0            1    Alice         HR           1                  3
1            2      Bob         IT           2                  4
2            3  Charlie  Marketing           3                  5
3            4    David    Finance           4                  3
4            5      Eva         IT           5                  4
```

Some things you need to consider when performing column-wise concatenation are as follows:

- The indices of the DataFrames to be concatenated are aligned properly. When concatenating DataFrames column-wise, each row in the resulting DataFrame should ideally represent data from the same entity (e.g., the same employee). Misaligned indexes can lead to a scenario where data from different entities is erroneously combined, leading to inaccurate and misleading results. For example, if the index represents employee IDs, misalignment could result in an employee's details being incorrectly paired with another employee's performance data.
- If the DataFrames contain columns with the same name but are intended to be distinct, consider renaming these columns before concatenation to avoid confusion or errors in the resulting DataFrame.
- While column-wise concatenation typically does not increase memory usage as significantly as row-wise concatenation, it is still important to monitor memory usage, especially with large DataFrames.

> **Join versus concatenation**
>
> **Concatenation** is primarily used for combining DataFrames along an axis (either rows or columns) without considering the values within. It's ideal for situations where you simply want to stack DataFrames together based on their order or extend them with additional columns.
>
> **Joins** are used to combine DataFrames based on one or more keys (a common identifier in each DataFrame). This is more about merging datasets based on shared data points, which allows for more complex and conditional combinations of data.

Having explored the nuances of concatenation in pandas, including its importance for aligning indexes and how it contrasts with join operations, let's now summarize the key points discussed to encapsulate our understanding and highlight the critical takeaways from our exploration of DataFrame manipulations in pandas.

Summary

In this chapter, we explored various aspects of DataFrame operations in pandas, focusing on concatenation, merging, and the importance of managing indexes.

We discussed merging, which is suited for complex combinations based on shared keys, offering flexibility through various join types such as inner, outer, left, and right joins. We also discussed how concatenation is used to combine DataFrames along a specific axis (either row-wise or column-wise) and is particularly useful for appending datasets or adding new dimensions to data. The performance implications of these operations were discussed, highlighting that proper index management can significantly enhance the efficiency of these operations, especially in large datasets.

In the upcoming chapter, we will deep dive into how the `groupby` function can be leveraged alongside various aggregation functions to extract meaningful insights from complex data structures.

References

1. `https://github.com/pandas-dev/pandas/issues/38418`
2. `https://realpython.com/pandas-merge-join-and-concat/`
3. `https://datascience.stackexchange.com/questions/44476/merging-dataframes-in-pandas-is-taking-a-surprisingly-long-time`
4. `https://stackoverflow.com/questions/40860457/improve-pandas-merge-performance`
5. `https://www.youtube.com/watch?v=P6hSBrxs0Eg`
6. `https://pandas.pydata.org/pandas-docs/version/1.5.1/user_guide/merging.html`
7. `https://pandas.pydata.org/pandas-docs/version/0.20/merging.html`

6

Data Grouping, Aggregation, Filtering, and Applying Functions

Data grouping and **aggregation** are fundamental techniques in data cleaning and preprocessing, serving several critical purposes. Firstly, they enable the summarization of large datasets, transforming extensive raw data into concise, meaningful summaries that facilitate analysis and insight derivation. Additionally, aggregation helps manage missing or noisy data by smoothing out inconsistencies and filling gaps with combined data points. These techniques also contribute to reducing data volume, enhancing processing efficiency, and creating valuable features for further analysis or machine learning models.

The main components of data grouping and aggregation include group keys, which define how data is segmented; aggregation functions, which perform operations such as summing, averaging, counting, and more; and output columns, which display the group keys and aggregated values.

In this chapter, we'll cover the following main points:

- Grouping data using one or multiple keys
- Applying aggregate functions on grouped data
- Applying functions on grouped data
- Data filtering

Technical requirements

You can find the code for the chapter in the following GitHub repository: `https://github.com/PacktPublishing/Python-Data-Cleaning-and-Preparation-Best-Practices/tree/main/chapter06`.

Grouping data using one or multiple keys

In pandas, grouping data is a fundamental operation that involves splitting data into groups based on one or more keys and then performing operations within each group. Grouping is often used in data analysis to gain insights and perform aggregate calculations on subsets of data. Let's dive deeper into grouping data and provide examples to illustrate their usage. The code for this section can be found here: `https://github.com/PacktPublishing/Python-Data-Cleaning-and-Preparation-Best-Practices/blob/main/chapter06/2.groupby_full_example.py`.

Grouping data using one key

Grouping data with pandas using one key is a common operation for data analysis.

To group data using one key, we use the `groupby()` method of a DataFrame and specify the column that we want to use as the key for grouping:

```
grouped = df.groupby('column_name')
```

After grouping, you typically want to perform some aggregation. Common aggregation functions include the following:

- `grouped.sum()`: This calculates the sum of all numeric columns
- `grouped.mean()`: This calculates the average (arithmetic mean)
- `grouped.count()`: This counts the number of non-null values
- `grouped.agg(['sum', 'mean', 'count'])`: This applies multiple aggregation functions at once: `sum`, `mean`, and `count`

Let's present a common use case on which to apply our learnings. Let's pretend we are working for an electronics retail company and we need to analyze the sales data for the different products. A sample of the data is presented here:

```
      Category Sub-Category Region  Sales        Date
0  Electronics       Mobile  North    200  2023-01-01
1  Electronics       Laptop  South    300  2023-01-02
2  Electronics       Tablet   East    250  2023-01-03
3  Electronics       Laptop   West    400  2023-01-04
4    Furniture        Chair  North    150  2023-01-05
5    Furniture        Table  South    350  2023-01-06
```

In data analysis, certain columns are typically candidates for grouping due to their *categorical* nature. These columns often represent categories, classifications, or time-related segments that make sense to aggregate data around:

- **Category columns**: Columns that represent distinct groups or types within the data. Examples include product categories, user types, or service types. These columns help in understanding the performance or behavior of each group.

- **Geographical columns**: Columns that denote geographical divisions, such as country, region, city, or store location. These are useful for regional performance analysis.
- **Temporal columns**: Columns representing time-related information, such as year, quarter, month, week, or day. Grouping by these columns helps in trend analysis over time.
- **Demographic columns**: Columns that describe demographic attributes, such as age group, gender, or income level. These are useful for segmenting data based on population characteristics.
- **Transaction-related columns**: Columns related to the nature of transactions, such as transaction type, payment method, or order status. These help in understanding different aspects of transactional data.

Given the data we have in our example, the candidate columns for grouping are `Category`, `Subcategory`, and `Region`. `Date` could also be a candidate if we had multiple records per day and we wanted to calculate the number of sales per day. In our case, our manager asked us to report on the total number of sales (sales volume) for each category. Let's see how to calculate this:

```
category_sales = df.groupby('Category')['Sales'].sum().reset_index()
```

In this code example, we group the data by the `Category` column, sum the `Sales` column for each category, and reset the index. The resulting DataFrame is as follows:

```
      Category  Sales
0     Clothing   1070
1  Electronics   1370
2    Furniture   1270
```

Now that we have seen how to group data by a single key, let's add more complexity by grouping the data by `Category` and `Region`.

Grouping data using multiple keys

Grouping by multiple keys allows for a more granular and detailed examination of the data. This approach helps uncover insights that may be hidden when only using a single key, offering a deeper understanding of relationships and patterns within the dataset. In our example, grouping by both `Region` and `Category` allows the company to see not only the overall sales performance but also how different categories perform in each region. This helps in identifying which products are popular in specific regions and tailoring marketing strategies accordingly.

To group data using multiple keys, we pass a list of column names to the `groupby()` method. Pandas will create groups based on *unique combinations* of values from these columns:

```
category_region_sales = df.groupby(['Category', 'Region'])['Sales'].
sum().reset_index()
```

In this piece of code, we group the data by both the `Category` and `Region` columns, and then we perform the aggregation by summing the `Sales` column for each group. Finally, we reset the index. Let's see the output from this operation:

```
       Category Region  Sales
0      Clothing   East    420
1      Clothing  North    100
2      Clothing  South    250
3      Clothing   West    300
4   Electronics   East    250
5   Electronics  North    420
6   Electronics  South    300
7   Electronics   West    400
8     Furniture   East    200
9     Furniture  North    150
10    Furniture  South    350
11    Furniture   West    570
```

With just a line of code, we have managed to summarize and present all the sales for each `Category` and `Region` value, making our manager very happy. Now, let's have a look at some best practices when working with groupby statements.

Best practices for grouping

When grouping data in pandas, there are several things to consider to ensure accurate results:

- **Missing data**: Be aware of missing data in the columns used for grouping. Pandas will *exclude* rows with missing data from the grouped result, which can affect the final calculations.
- `MultiIndex`: When grouping by multiple columns, pandas returns a hierarchical index (`MultiIndex`). Be familiar when working with `MultiIndex` and consider resetting the index if needed as we have been doing for simplicity.
- **Order of operations**: The order in which you perform groupings and aggregations *can affect the results*. Be mindful of the sequence in which you apply grouping and aggregation functions.
- **Grouping large datasets**: For large datasets, grouping can be memory intensive. Consider using techniques such as chunking or parallel processing to manage memory usage and computation time.

Our managing team saw the efficiency of the groupby operations we performed, and they asked us for a more detailed summary of sales! With multiple keys in place, we can further enhance our analysis by applying multiple aggregation functions to the `Sales` column. This will give us a more detailed summary of the data.

Applying aggregate functions on grouped data

In pandas, after grouping data using the `groupby()` method, you can apply aggregate functions to perform calculations on the grouped data. **Aggregate functions** are used to summarize or compute statistics for each group, resulting in a new DataFrame or Series. Let's dive deeper into applying aggregate functions on grouped data and provide examples to illustrate their usage.

Basic aggregate functions

We have touched base on the basic aggregation function in the first section as you cannot perform groupby without an aggregation function. In this section, we will expand a bit more on what each function does and when should we use each one, starting by presenting all the available functions in the following table:

Aggregation function	Description	When to use	Code example
`sum`	Adds up all values in a group	When you need the total value for each group. Example: Total sales per category.	`df.groupby('Category')['Sales'].sum()`
`mean`	Calculates the average of values in a group	When you need the average value for each group. Example: Average sales per region.	`df.groupby('Category')['Sales'].mean()`
`count`	Counts the number of non-null values in a group	When you need to know the number of occurrences in each group. Example: Number of sales transactions per sub-category.	`df.groupby('Category')['Sales'].count()`
`min`	Finds the minimum value in a group	When you need the smallest value in each group. Example: Minimum sales value per region.	`df.groupby('Category')['Sales'].min()`

Aggregation function	Description	When to use	Code example
`max`	Finds the maximum value in a group	When you need the largest value in each group. Example: Maximum sales value per category.	`df.groupby('Category')['Sales'].max()`
`median`	Finds the median value in a group	When you need the middle value in a sorted list of numbers. Example: Median sales value per category.	`df.groupby('Category')['Sales'].median()`
`std` (Standard Deviation)	Measures the spread of values in a group	When you need to understand the variation in values. Example: Standard deviation of sales per region.	`df.groupby('Category')['Sales'].std()`

Table 6.1 – Summary table of the basic aggregation functions

You can call each of these functions one by one or all together, for example:

```
total_sales = df.groupby('Category')['Sales'].sum().reset_index()
```

This calculates the number of sales per category, as we've learned, and it's sufficient if this is the only aggregate information you want to extract from the dataset. However, if you find yourself being asked to produce multiple sale aggregates for the different product categories, a more efficient way is to perform all the aggregates at once:

```
category_region_sales_agg = df.groupby(['Category', 'Region'])
['Sales'].agg(['sum', 'mean']).reset_index()
```

In this code, we apply multiple aggregation functions (`sum` and `mean`) to the `Sales` column. The result is as follows:

```
      Category Region  sum   mean
0     Clothing   East  420  210.0
1     Clothing  North  100  100.0
2     Clothing  South  250  250.0
3     Clothing   West  300  300.0
4  Electronics   East  250  250.0
5  Electronics  North  420  210.0
```

```
6   Electronics  South  300  300.0
7   Electronics   West  400  400.0
8     Furniture   East  200  200.0
9     Furniture  North  150  150.0
10    Furniture  South  350  350.0
11    Furniture   West  570  285.0
```

> **Note**
>
> We can add as many aggregations as we want in the group by clause.

We've been really efficient with calculating all the different metrics our managing team asked for, and as a result, they are now keen to understand the sales metrics and the number of unique sub-category sales per region and category. Let's do that next.

Advanced aggregation with multiple columns

To understand the sales metrics and the number of unique sub-category sales per region and category, we can group additional columns and apply multiple aggregations to both the `Sales` and `Subcategory` columns:

```
advanced_agg = df.groupby(['Category', 'Region']).agg({
  'Sales': ['sum', 'mean', 'count'],
  'Sub-Category': 'nunique' # Unique count of Sub-Category
}).reset_index()
```

In this code, we group the DataFrame by `Category` and `Region` and we perform a couple of aggregations:

- `'Sales': ['sum', 'mean', 'count']` calculates the total sales, average sales, and number of transactions (count of rows) for each group
- `'Sub-Category': 'nunique'` calculates the number of unique sub-categories within each group of `Category` and `Region`

The summarized results are presented here:

```
     Category Region Sales                Sub-Category
                       sum   mean count        nunique
0    Clothing   East   420  210.0     2              2
1    Clothing  North   100  100.0     1              1
2    Clothing  South   250  250.0     1              1
3    Clothing   West   300  300.0     1              1
4 Electronics   East   250  250.0     1              1
5 Electronics  North   420  210.0     2              1
6 Electronics  South   300  300.0     1              1
```

Now, you may be wondering, what have we learned by making these calculations? Let me answer that! Total sales, average sales, and transaction count were calculated to understand the financial performance across different category-region combinations. Additionally, the `Sub-Category` unique count revealed crucial aspects of our product distribution strategy. This analysis serves multiple purposes: it provides insights into the diversity of products within each category-region segment, for example, in the context of our data, knowing the number of unique products (sub-categories) sold in each region under different categories provides insights into the market segmentation and product assortment strategies. It also aids in assessing market penetration by highlighting regions with a broader product offering and supports strategic decisions in product portfolio management, including expansion and inventory strategies tailored to regional preferences.

Standard aggregation functions, such as sum, mean, and count, provide fundamental statistics. However, custom functions allow you to calculate metrics that are specific to your business needs or analysis goals. For example, calculating the range or coefficient of variation in sales data can reveal insights into the distribution and variability of sales within different groups. As you can imagine, we were asked to implement these custom metrics, which we'll do next.

Applying custom aggregate functions

Custom functions are valuable when the aggregation requires complex calculations that go beyond simple statistics. You can use them when you need to calculate metrics that are unique to your analysis objectives or business context. For example, in sales analysis, you might want to compute profit margins, customer lifetime value, or churn rates, which are not typically available through standard aggregation functions.

Let's go back to our example and build the metrics we were asked about: For each region, we want to calculate the sales range and the variability of sales. Let's have a look at the following code:

1. We create a function that calculates the range (difference between max and min) of sales:

   ```
   def range_sales(series):
     return series.max() - series.min()
   ```

2. Then, we create a function that computes the coefficient of variation of sales, which measures the relative variability in relation to the mean:

   ```
   def coefficient_of_variation(series):
     return series.std() / series.mean()
   ```

3. The `df` DataFrame is then grouped by `Region`:

   ```
   advanced_agg_custom = df.groupby('Region').agg({
     'Sales': ['sum', 'mean', 'count', range_sales, coefficient_of_
   variation],
   ```

```
    'Sub-Category': 'nunique'
}).reset_index()
```

`Sales: ['sum', 'mean', 'count', range_sales, coefficient_of_variation]` calculates the total sales, average sales, transaction count, sales range, and coefficient of variation using the custom functions. `'Sub-Category':'nunique'` counts the number of unique sub-categories within each group. Then, we reset the index to flatten the `df` DataFrame and make it easier to work with.

4. Finally, we rename the aggregated columns for clarity and better readability of the output:

```
advanced_agg_custom.columns = [
   'Region', 'Total Sales', 'Average Sales', 'Number of
Transactions',
   'Sales Range', 'Coefficient of Variation', 'Unique
Sub-Categories'
]
```

5. Let's print the final DataFrame:

```
print(advanced_agg_custom)
```

The final DataFrame is presented here:

```
  Region  TotalSales SalesRange  Coef Unique Sub-Categories
0   East         870        120  0.24                     4
1  North         670        120  0.32                     3
2  South         900        100  0.16                     3
3   West        1270        230  0.34                     4
```

Let's spend some time understanding the sales variability for each region. The range of sales within each region can reveal the **spread** or **difference** between the highest and lowest sales figures. For instance, a wide range may indicate significant variability in consumer demand or sales performance across different regions. The coefficient of variation helps to standardize the variability of sales relative to their average. A higher coefficient suggests greater relative variability, which may prompt further investigation into factors influencing sales fluctuations.

> **Note**
>
> I hope it's clear to you that you can build any function that you want as a custom aggregate function as long as it computes a *single* aggregation result from an input series of values. The function should also return a single scalar value, which is the result of the aggregation for that group.

Now, let's have a look at some best practices when working with aggregate functions.

Best practices for aggregate functions

When working with aggregate functions in pandas, there are several things to consider to ensure accurate results:

- Write efficient custom functions that minimize computational overhead, especially when working with large datasets. Avoid unnecessary loops or operations that can slow down processing time.
- Clearly document the logic and purpose of your custom aggregation functions. This helps in maintaining and sharing code within your team or organization, ensuring transparency and reproducibility of analysis.
- Validate the accuracy of custom aggregation functions by comparing results with known benchmarks or manual calculations. This step is crucial to ensure that your custom metrics are reliable and correctly implemented.

In pandas, when using the `.agg()` method with `groupby`, the aggregation function you define should *ideally* return a single scalar value for each column it operates on within each group. However, there are scenarios where you might want to return multiple values or perform more complex operations. While the pandas `.agg()` method expects scalar values, you can achieve more complex aggregations by using custom functions that return tuples or lists. However, this requires careful handling and often isn't straightforward within pandas' native aggregation framework. For more complex scenarios where you need to return multiple values or perform intricate calculations, we can use `apply()` instead of `agg()`, which is more flexible, as we will see in the next section.

Using the apply function on grouped data

The `apply()` function in Pandas is a powerful method used to apply a custom function along an axis of a DataFrame or Series. It is highly versatile and can be used in various scenarios to manipulate data, compute complex aggregations, or transform data based on custom logic. The `apply()` function can be used to do the following:

- Apply functions row-wise or column-wise
- Apply functions to groups of data when used in conjunction with `groupby()`

In the next section, we will focus on using the `apply` function on groups of data by first grouping on the column we want and then performing the `apply` operation.

> **Note**
>
> Using the `apply` function without `groupby` allows you to apply a function across either rows or columns of a DataFrame directly. This is useful when you need to perform row-wise or column-wise operations that don't require grouping the data. Apply the same learnings and just skip the group by clause.

When using the `apply` function in pandas, `axis=0` (default) applies the function to each column, while `axis=1` applies it to each row. Let's go a little bit deeper on this.

`axis=0` applies the function along the *rows*. In other words, it processes each *column* independently. This is typically used when you want to aggregate data column-wise (e.g., summing up values in each column), as shown in the following figure:

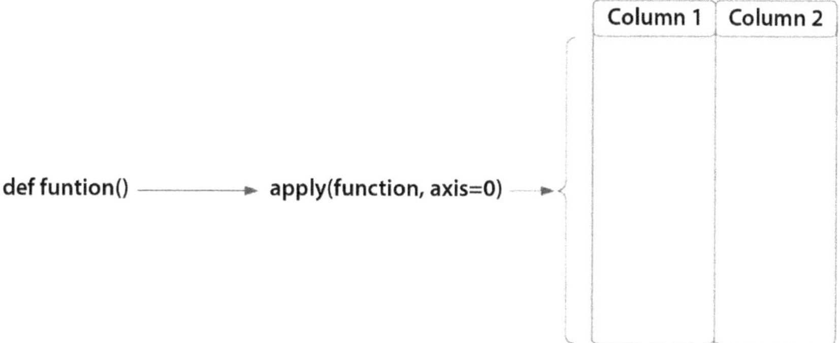

Figure 6.1 – Apply() with axis=0

If we go back to our use case, the managing team wants to understand more about the actual quantity sold for the products per category apart from the sum of dollars in sales we achieved. Our example gets more and more complex; so, it's a good idea to implement that with `apply()`. Let's see this with a code example that can also be found here: `https://github.com/PacktPublishing/Python-Data-Cleaning-and-Preparation-Best-Practices/blob/main/chapter06/3.apply_axis0.py`.

1. Let's extend our DataFrame to add the `Quantity` column:

```
data = {
  'Category': ['Electronics', 'Electronics', 'Furniture',
'Furniture', 'Clothing', 'Clothing'],
  'Sub-Category': ['Mobile', 'Laptop', 'Chair', 'Table', 'Men',
'Women'],
  'Sales': [100, 200, 150, 300, 120, 180],
  'Quantity': [10, 5, 8, 3, 15, 12],
  'Date': ['2023-01-01', '2023-01-02', '2023-01-03', '2023-01-
04', '2023-01-05', '2023-01-06']
}
df = pd.DataFrame(data)
```

2. We then convert the `Date` column to datetime format:

```
df['Date'] = pd.to_datetime(df['Date'])
```

3. Now, let's define a custom function to compute multiple statistics for `Sales` and `Quantity`:

```
def compute_statistics(series):
    sum_sales = series['Sales'].sum()
    mean_sales = series['Sales'].mean()
    std_sales = series['Sales'].std()
    cv_sales = std_sales / mean_sales
    sum_quantity = series['Quantity'].sum()
    mean_quantity = series['Quantity'].mean()
    std_quantity = series['Quantity'].std()
    cv_quantity = std_quantity / mean_quantity
    return pd.Series([sum_sales, mean_sales, std_sales, cv_
sales, sum_quantity, mean_quantity, std_quantity, cv_quantity],
index=['Sum_Sales', 'Mean_Sales', 'Std_Sales', 'CV_Sales',
'Sum_Quantity', 'Mean_Quantity', 'Std_Quantity', 'CV_Quantity'])
```

The custom function (`compute_statistics`) now computes multiple statistics (sum, mean, std, CV) for both the `Sales` and `Quantity` columns within each group defined by `Category`. For each category group (series), it calculates the following:

- `Sum_Sales`: The sum of sales
- `Mean_Sales`: The mean of sales
- `Std_Sales`: The standard deviation of sales
- `CV_Sales`: The **coefficient of variation** (**CV**) of sales
- `Sum_Quantity`: Sum of quantities
- `Mean_Quantity`: Mean of quantities
- `Std_Quantity`: Standard deviation of quantities
- `CV_Quantity`: The CV of quantities

In the end, it returns a pandas Series with these computed statistics, indexed appropriately.

4. Next, we'll perform a groupby operation on `Category` and apply our custom function to compute statistics of `Sales` and `Quantity`:

```
result_complex = df.groupby('Category').apply(compute_
statistics).reset_index()
```

We use `apply()` in conjunction with `groupby('Category')` to apply the `compute_statistics` function to each group of sales data defined by the `Category` column. The function operates on the entire group (series), allowing the computation of statistics for both the `Sales` and `Quantity` columns simultaneously. Finally, `reset_index()` is used to flatten the resulting DataFrame, providing a structured output with category-wise statistics for both columns. Let's have a look at the final DataFrame:

	Category	Sum_Sales	Mean_Sales	Std_Sales	CV_Sales	Sum_Quantity	Mean_Quantity	Std_Quantity	CV_Quantity
0	Clothing	300.0	150.0	42.426407	0.282843	27.0	13.5	2.121320	0.157135
1	Electronics	300.0	150.0	70.710678	0.471405	15.0	7.5	3.535534	0.471405
2	Furniture	450.0	225.0	106.066017	0.471405	11.0	5.5	3.535534	0.642824

By grouping the data by `Category`, we can analyze sales and quantity metrics at the category level, which helps us understand how different types of products (electronics, furniture, clothing) perform in terms of sales and quantity. As we can see from the presented results, `Furniture` is the key income generator as it has the highest `Sum_Sales` and `Mean_Sales`, indicating a category with popular or high-value products. Categories with lower `CV_Sales` and `CV_Quantity` values, such as `Clothing`, are more consistent in sales and quantity, suggesting stable demand or predictable sales patterns whereas categories with higher variability (`Std_Sales` and `Std_Quantity`) may indicate products with fluctuating sales or seasonal demand.

This is great in terms of data analysis, but now, we get asked to come up with some strategic decisions related to product assortment, pricing strategies, and marketing initiatives. Let's be a little bit more creative at this point:

- Categories with high `Sum_Sales` values and stable metrics (`CV_Sales`, `CV_Quantity`) are prime candidates for expanding product lines or investing in marketing efforts
- Categories with high variability (`Std_Sales`, `Std_Quantity`) may require dynamic pricing strategies or seasonal promotions to optimize sales
- We can use the `Mean_Sales` and `Mean_Quantity` values to identify categories with potential for growth

When using the `apply()` function in pandas without specifying the axis parameter, the default behavior is `axis=0`. This means that the function will be applied to each column (i.e., it will process each column independently). This is what we have applied in the example code provided earlier. Depending on your specific use case, adjust `apply()` to operate row-wise (`axis=1`) or column-wise (`axis=0`) as needed. Next, let's focus on `axis=1`.

`axis=1` applies the function along the columns, so it processes each row independently. This is typically used when you want to perform row-wise operations (e.g., calculating a custom metric for each row).

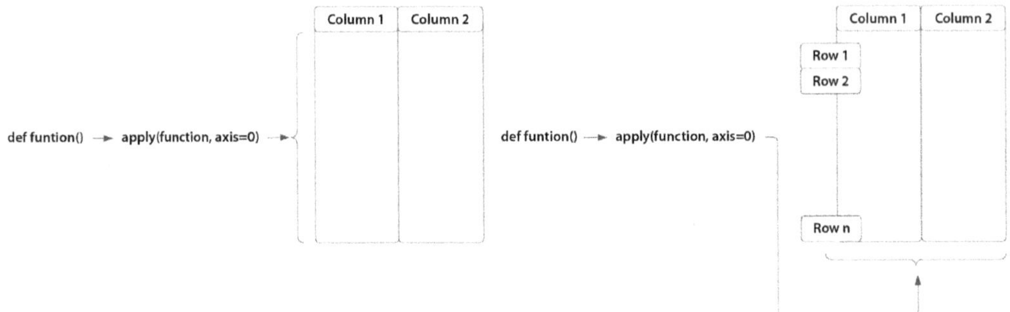

Figure 6.2 – Apply() with axis=1

Applying functions row-wise allows for row-level transformations and calculations. Let's see a code example with `axis=1`. Let's start by defining a function to be applied across columns (`axis=1`). The code can be found here: `https://github.com/PacktPublishing/Python-Data-Cleaning-and-Preparation-Best-Practices/blob/main/chapter06/4.apply_axis1.py`:

1. The `row_summary` function takes a single row of a DataFrame as input and returns a summary of that row's data. The input for the function is key to understanding that is a single row of the DataFrame, passed as a pandas Series:

```
def row_summary(row):
    total_sales_quantity = row['Sales'] + row['Quantity']
    sales_quantity_ratio = row['Sales'] / row['Quantity'] if
row['Quantity'] != 0 else np.nan
    return pd.Series(
        [total_sales_quantity,sales_quantity_ratio],
        index=['Total_Sales_Quantity',
        'Sales_Quantity_Ratio'])
```

   The `total_sales_quantity` variable will store the sum of `Sales` and `Quantity` for the row. The `sales_quantity_ratio` variable will store the ratio of `Sales` to `Quantity` for the row, or `np.nan` if the quantity is zero, providing insight into the sales efficiency.

2. We apply the function row-wise (`axis=1`):

```
df_row_summary = df.apply(row_summary, axis=1)
   Total_Sales_Quantity  Sales_Quantity_Ratio
0                 110.0                 10.00
1                 205.0                 40.00
2                 158.0                 18.75
3                 303.0                100.00
```

```
4            135.0              8.00
5            192.0             15.0
```

This will produce a new df_row_summary DataFrame where each row corresponds to the calculated values for total_sales_quantity and sales_quantity_ratio for the original rows in df.

3. Finally, we group by Category to calculate metrics per category:

```
category_metrics = df.groupby('Category')[['Total_Sales_
Quantity', 'Sales_Quantity_Ratio']].mean().reset_index()
```

Let's see the final result:

```
      Category  Total_Sales_Quantity  Sales_Quantity_Ratio
0     Clothing                 163.5                11.500
1  Electronics                 157.5                25.000
2    Furniture                 230.5                59.375
```

The total_sales_quantity metric provides a simple yet effective measure of overall sales performance per transaction, helping to understand the combined impact of the number of items sold (Quantity) and their sales value (Sales). By analyzing total_sales_quantity, we can identify transactions with high combined sales and quantity, which might indicate popular product categories or successful sales strategies. Conversely, it also helps recognize low-performing transactions, thereby guiding inventory management and promotional adjustments to improve sales efficiency and product performance. This dual insight aids in strategic decision-making to optimize sales and inventory management.

The sales_quantity_ratio metric provides valuable insights into the efficiency of sales per unit quantity, revealing how effectively products convert quantity into revenue. This metric is crucial for assessing the value derived from each unit sold. With this, we can identify products that generate high revenue per unit, indicating high-value items that may warrant prioritized marketing efforts. Conversely, it helps uncover products with a low revenue per unit, signaling potential areas for price adjustments, targeted promotions, or re-evaluation within the product portfolio to optimize profitability and sales performance.

> **Note**
>
> Whenever feasible, prefer using vectorized operations (built-in pandas methods or NumPy functions) over apply for performance reasons. Vectorized operations are generally faster because they leverage optimized C code under the hood.

The concepts and techniques we have explored so far directly lead to the importance of filtering in data cleaning. Once we have applied transformations or aggregated data, filtering allows us to focus on specific subsets of data that are relevant to our analysis or meet certain conditions. For example, after calculating sales performance metrics, such as `Total_Sales_Quantity` and `Sales_Quantity_Ratio`, across different product categories, filtering can help identify categories or products that require further investigation, such as those with unusually high or low performance metrics.

Data filtering

Data filtering is a fundamental operation in data manipulation that involves selecting a subset of data based on specified conditions or criteria. It is used to extract relevant information from a larger dataset, exclude unwanted data points, or focus on specific segments that are of interest for analysis or reporting.

In the following example, we filter the DataFrame to only include rows where the `Quantity` column is greater than `10`. This operation selects products that have sold more than 10 units, focusing our analysis on potentially high-performing products. `https://github.com/PacktPublishing/Python-Data-Cleaning-and-Preparation-Best-Practices/blob/main/chapter06/5.simple_filtering.py`:

```
filtered_data = df[df['Quantity'] > 10]
```

Let's have a look at the resulting DataFrame:

```
   Category Sub-Category  Sales  Quantity        Date
4  Clothing          Men    120        15  2023-01-05
5  Clothing        Women    180        12  2023-01-06
```

Moving beyond simple filters allows us to identify electronic products that satisfy more complex conditions, as we will see in the next section.

Multiple criteria for filtering

Filtering may involve complex conditions, such as combining logical AND and OR operations, or using nested conditions. Let's say that the management team wants us to identify high-value electronics products (`sales > 1000`) with relatively low sales quantities (`quantity < 30`).

Let's see how we can do this with multiple filtering criteria (the code can be found here: `https://github.com/PacktPublishing/Python-Data-Cleaning-and-Preparation-Best-Practices/blob/main/chapter06/6.advanced_filtering.py`):

```
filtered_data = df[(df['Sales'] > 1000) & (df['Quantity'] < 30)]
```

In this example, we define a filter condition that filters rows where the sales are greater than `1000` and the quantity is less than `30`. Let's have a look at the resulting DataFrame:

```
      Category Sub-Category  Sales  Quantity        Date
1  Electronics       Laptop   1500        25  2023-01-02
```

Filtering is a straightforward operation to implement, but let's explore some best practices to optimize its effectiveness.

Best practices for filtering

Let's explore the best practices that can enhance the effectiveness of filtering operations:

- Clearly define the filtering criteria based on the analysis goals. Use conditions that are specific and relevant to the insights you want to derive.
- Utilize built-in filter functions provided by data manipulation libraries, such as pandas in Python or SQL queries in databases. These functions are optimized for performance and ease of use.
- Ensure that the filtering criteria do not exclude important data points that might be valuable for analysis. Validate the results to confirm they align with the expected outcomes.
- Document the filtering criteria and steps applied to maintain transparency and facilitate reproducibility of the analysis.

As datasets grow, filtering becomes essential to manage and extract insights efficiently. Operations that involve processing large volumes of data can become prohibitively slow without effective filtering strategies. Filtering helps in optimizing resource utilization, such as memory and processing power, by reducing the amount of data that needs to be stored and processed at any given time.

Performance considerations as data grows

Let's have a look at things to keep in mind as data grows:

- Filtering operations optimize query execution by reducing the number of rows or columns that need to be processed, leading to faster response times for data queries and analyses.
- Large datasets consume significant memory and storage resources. Filtering reduces the amount of data held in memory or stored on disk, improving efficiency, and reducing operational costs associated with data storage.

Let's now summarize the learnings from this chapter.

Summary

In this chapter, we've explored some powerful techniques, such as grouping, aggregation, and applying custom functions. These methods are essential for summarizing and transforming data, enabling deeper insights into datasets. We've learned how to efficiently group data by categorical variables, such as `Category` and `Region`, and apply aggregate functions, such as sum, mean, and custom metrics to derive meaningful summaries.

Additionally, we deep-dived into the versatility of `apply` functions, which allow for row-wise or column-wise custom computations. Best practices, such as optimizing function efficiency, handling missing values, and understanding performance implications, were emphasized to ensure effective data processing. Finally, we discussed the strategic application of filters to refine datasets based on specific criteria, enhancing data analysis precision.

In the next chapter, we will discuss designing and optimizing data write operations to efficiently store the transformed and cleaned data.

7

Data Sinks

In the world of modern data processing, crucial decisions about data management, storage, and processing will determine successful outcomes. In this chapter, we will deep dive into three important pillars that underpin effective data processing pipelines: selecting the right **data sink**, choosing the optimal file type, and mastering partitioning strategies. By discussing these critical elements and their real-world applications, this chapter will equip you with the insights and strategies needed to architect data solutions that optimize efficiency, scalability, and performance within the complicated landscape of data processing technologies.

In this chapter, we will discuss the following topics:

- Choosing the right data sink for your use case
- Choosing the right file type for your use case
- Navigating partitioning
- Designing an online retail data platform

Technical requirements

For this chapter, we will need to install the following libraries:

```
pip install pymongo==4.8.0
pip install pyarrow
pip install confluent_kafka
pip install psycopg2-binary==2.9.9
```

As always, you can find all the code for this chapter in this book's GitHub repository: `https://github.com/PacktPublishing/Python-Data-Cleaning-and-Preparation-Best-Practices/tree/main/chapter07`.

Each section is followed by a script with a similar naming convention, so feel free to execute the scripts and/or follow along by reading this chapter.

Choosing the right data sink for your use case

A data sink refers to a destination or endpoint where data is directed or stored. The term "sink" is used metaphorically to convey the idea of data flowing into and being absorbed by a designated location. Data sinks are commonly used as storage locations where data can be permanently or temporarily stored. This storage can be in the form of databases, files, or other data structures.

Data engineers and data scientists often work with a variety of data sinks, depending on their specific tasks and use cases. Let's look at some common data sinks, along with code examples, while considering the pros and cons of each type.

Relational databases

Relational databases are a type of **database management system** (**DBMS**) that organizes data into tables with rows and columns, where each row represents a record and each column represents a field. The relationships between tables are established using keys. The primary key uniquely identifies each record in a table, and foreign keys create links between tables.

Overview of relational databases

The following is a quick overview of the key components of relational databases:

- **Tables**: Data is organized into tables, where each table represents a specific entity or concept. For example, in a database for a library, there might be tables for books, authors, and borrowers.
- **Rows and columns**: Each table consists of rows and columns. A row represents a specific record (e.g., a book), and each column represents a specific attribute or field of that record (e.g., title, author, and publication year).
- **Keys**: Keys are used to establish relationships between tables. The primary key uniquely identifies each record in a table, and foreign keys in related tables create links between them.
- **Structured Query Language** (**SQL**): Relational databases use SQL for querying and manipulating data. SQL allows users to retrieve, insert, update, and delete data, as well as define and modify the database's structure.

In the data field, we usually find relational databases in the following scenarios:

- **Structured data**: If your data has a well-defined structure with clear relationships between entities, a relational database is a suitable choice.
- **Data integrity requirements**: If maintaining data integrity is critical for your application (e.g., in financial systems or healthcare applications), a relational database provides mechanisms to enforce integrity constraints.

- **Atomicity, consistency, isolation, and durability (ACID) properties**: **Atomicity** ensures a transaction is an all-or-nothing operation: either all changes are committed, or none are. For instance, in transferring money between accounts, atomicity guarantees both balances are updated together or not at all. **Consistency** means transactions move the database from one valid state to another while adhering to integrity constraints. If a rule such as unique customer IDs is violated, the transaction is rolled back to maintain consistency. **Isolation** ensures transactions execute independently, preventing interference and visibility of uncommitted changes between concurrent transactions. This avoids issues such as dirty reads. Finally, **durability** guarantees that once committed, changes persist, even after system failures, ensuring the permanence of updates such as contact information in an online application. If your application demands adherence to ACID properties, relational databases are designed to meet these requirements.
- **Complex queries**: If your application involves complex queries and reporting needs, relational databases, with their SQL querying capabilities, are well-suited for such scenarios.

There are many different options for building relational databases, as we will see in the following section.

Different options for relational database management systems

There are many different **relational database management systems** (**RDBMSs**) out there. We've summarized the main ones in the following table:

Database	**Description**
MySQL	An open source RDBMS known for its speed, reliability, and wide usage in web development
PostgreSQL	An open source RDBMS with advanced features, extensibility, and support for complex queries
Oracle Database	A commercial RDBMS known for its scalability, security, and comprehensive set of data management features
Microsoft SQL Server	A commercial RDBMS by Microsoft that integrates with Microsoft technologies and has business intelligence support
SQLite	A lightweight, embedded, serverless RDBMS suitable for applications with low to moderate database requirements
MariaDB	An open source RDBMS forked from MySQL that aims for compatibility while introducing new features

Table 7.1 – Summary of RDBMSs

Now, let's see an example of how to quickly set up a local relational database, connect to it, and write a new table.

An example of a PostgreSQL database

First, we need to install and set up PostgreSQL. This differs depending on the **operating system** (**OS**), but the logic remains the same. The following script automates the process of installing and setting up PostgreSQL on macOS or Debian-based Linux systems: `https://github.com/PacktPublishing/Python-Data-Cleaning-and-Preparation-Best-Practices/blob/main/chapter07/setup/setup_postgres.sh`.

First, it detects the OS using the uname command:

```
OS=$(uname)
```

If macOS is detected, it uses Homebrew to update package lists, install PostgreSQL, and start the PostgreSQL service. If a Debian-based Linux OS is detected, it uses `apt-get` to update package lists, install PostgreSQL and its `contrib` package, and start the PostgreSQL service. Here's the code for installing macOS:

```
if [ "$OS" == "Darwin" ]; then
    echo "Detected macOS. Installing PostgreSQL via Homebrew..."
    brew update
    brew install postgresql
    brew services start postgresql
```

If your OS isn't supported by this script, then the following error message will be shown:

```
Unsupported OS. Please install PostgreSQL manually.
```

In this case, you will need to install PostgreSQL *manually and start the service*. Once you've done that, you can continue to the second part of the script. The script then switches to the default `postgres` user to execute SQL commands that create a new database user if it doesn't already exist, create a new database owned by this user, and grant all privileges on the database to the user, as shown here:

```
psql postgres << EOF
DO \$\$
BEGIN
    IF NOT EXISTS (
        SELECT FROM pg_catalog.pg_user
        WHERE usename = 'the_great_coder'
    ) THEN
        CREATE USER the_great_coder
        WITH PASSWORD 'the_great_coder_again';
    END IF;
END
\$\$;
EOF
```

```
psql postgres << EOF
CREATE DATABASE learn_sql2 OWNER the_great_coder;
EOF
psql postgres << EOF
-- Grant privileges to the user on the database
GRANT ALL PRIVILEGES ON DATABASE learn_sql2 TO the_great_coder;
EOF
```

To execute the preceding code, do the following:

1. Make sure you pull the repository to your local laptop.
2. Go to the folder where the repository is located.
3. Open a terminal in the repository folder location.
4. Execute the following commands to navigate to the right folder:

   ```
   cd chapter7
   cd setup
   ```

5. Validate that you're in the right location and that you can see the `setup_postgres.sh` script, as shown here:

   ```
   maria.zevrou@FVFGR3ANQ05P chapter7 % cd setup
   maria.zevrou@FVFGR3ANQ05P set up % ls
   setup_postgres.sh
   ```

6. Make the script executable by running the following command:

   ```
   chmod +x setup_postgres.sh
   ```

7. Finally, run the actual script using the following command:

   ```
   ./setup_postgres.sh
   ```

After executing the script, you should see a confirmation message that the PostgreSQL setup, including the database and user creation, has been completed:

```
PostgreSQL setup completed. Database and user created.
```

Now, we're ready to execute the script so that we can write our incoming data to the database we created in the previous step. You can find this script here: `https://github.com/PacktPublishing/Python-Data-Cleaning-and-Preparation-Best-Practices/blob/main/chapter07/1.postgressql.py`.

This script connects to the PostgreSQL database that we created previously and manages a table within it. Let's get started:

1. Let's start by importing the necessary libraries:

```
import pandas as pd
import psycopg2
from psycopg2 import sql
```

2. Then, we must define several functions, starting with `table_exists`. This function checks whether a specified table already exists in the database:

```
def table_exists(cursor, table_name):
    cursor.execute(
        sql.SQL("SELECT EXISTS ( \
                SELECT 1 FROM information_schema.tables \
                WHERE table_name = %s)"),
        [table_name]
    )
    return cursor.fetchone()[0]
```

3. The next function we need is the `create_table` function, which creates a new table if it doesn't already exist within a specific schema. In our case, it will have three columns: `id` as the primary key, then `name` and `age`:

```
def create_table(cursor, table_name):
    cursor.execute(
        sql.SQL("""
            CREATE TABLE {} (
                id SERIAL PRIMARY KEY,
                name VARCHAR(255),
                age INT
            )
        """).format(sql.Identifier(table_name))
    )
```

4. Then, we must define the `insert_data` function, which inserts rows of data into the table:

```
def insert_data(cursor, table_name, data):
    cursor.executemany(
        sql.SQL("INSERT INTO {} (name, age) \
                VALUES (%s, %s)"
        ).format(sql.Identifier(table_name)),
        data
    )
```

5. Finally, we must use the following function to display the retrieved data:

```
def print_table_data(cursor, table_name):
    cursor.execute(
        sql.SQL(
            "SELECT * FROM {}"
        ).format(sql.Identifier(table_name))
    )
    rows = cursor.fetchall()
    for row in rows:
        print(row)
```

At this point, the script will create a mock DataFrame containing sample data (names and ages):

```
data = {
    'name': ['Alice', 'Bob', 'Charlie'],
    'age': [25, 30, 22]
}
df = pd.DataFrame(data)
```

It establishes a connection to a PostgreSQL database using the specified connection parameters (database name, user, password, host, and port). These are the details we used in the previous step when we set up the database, so no change is required on your side:

```
db_params = {
    'dbname': 'learn_sql',
    'user': 'the_great_coder',
    'password': 'the_great_coder_again',
    'host': 'localhost',
    'port': '5432'
}
conn = psycopg2.connect(**db_params)
cursor = conn.cursor()
```

6. Finally, it checks whether a table named `example_table` exists, creates it if necessary, and then inserts the mock data into the table. After committing the changes to the database, the script fetches and prints the data from the table to confirm the successful insertion before finally closing the database connection:

```
table_name = 'example_table'

if not table_exists(cursor, table_name):
    create_table(cursor, table_name)

insert_data(cursor, table_name, df.values.tolist())
conn.commit()
```

```
    print_table_data(cursor, table_name)
    cursor.close()
    conn.close()
```

To execute the preceding script, just execute the following command in the `chapter7` folder:

```
python 1.postgressql.py
```

Important Note

Remember to always close the connections as it helps you avoid performance issues and ensures that new connections can be established when needed. It allows the database to free up resources associated with the connection, and it ensures that any uncommitted transactions are handled properly. Closing a connection returns it to the connection pool, making it available for reuse by other parts of your application.

To see the table that was created in the database, you can open the PSQL process in your terminal and connect to the `learn_sql` database by executing the following command:

```
psql -h localhost -U the_great_coder -d learn_sql
```

Then, run the following command to list all the available tables:

```
\dt
```

You should see something similar to the following:

```
learn_sql=> \dt
                 List of relations
 Schema |      Name      | Type  |      Owner
--------+----------------+-------+-----------------
 public | example_table  | table | the_great_coder
```

Figure 7.1 – List tables in the database

You can also interact now with the table by executing the following SQL commands:

```
learn_sql=> SELECT * FROM example_table;
 id |   name    | age
----+-----------+-----
  1 | Alice     |  25
  2 | Bob       |  30
  3 | Charlie   |  22
  4 | Alice     |  25
  5 | Bob       |  30
  6 | Charlie   |  22
```

Figure 7.2 – Showing all the rows in the table

If you rerun the same Python script without dropping the existing table first, you won't see a new table being created; instead, new rows will be added to the same table:

```
 id |   name    | age
----+-----------+-----
  1 | Alice     |  25
  2 | Bob       |  30
  3 | Charlie   |  22
  4 | Alice     |  25
  5 | Bob       |  30
  6 | Charlie   |  22
  7 | Alice     |  25
  8 | Bob       |  30
  9 | Charlie   |  22
```

Figure 7.3 – Showing all the rows in the table once the script has been rerun

Having understood how to set up a relational database and use it as a sink by writing new data in it, let's deep dive into the advantages and disadvantages of relational databases.

Advantages and disadvantages of relational databases

In this section, we'll summarize the advantages and disadvantages of using RDBMSs.

The advantages are as follows:

- RDBMS systems have ACID properties, providing a robust framework for reliable and secure transactions
- RDBMS technology has been around for decades, resulting in mature and well-established systems with extensive documentation and community support

However, they also have various disadvantages:

- The rigid schema of RDBMS can be a limitation when dealing with evolving or dynamic data structures as they require schema modifications. Schema changes might be required for new data.
- RDBMSs are primarily designed for structured data and may not be the best choice for handling unstructured or semi-structured data.

If you're wondering in what file type the data in the relational databases is written, then you'll find the following subsection fascinating.

Relational database file types

In relational databases, the choice of file types for storing data is typically *abstracted* from users and developers, and it is not common to interact with the underlying files directly. Relational databases manage data storage and retrieval through their internal mechanisms, which often involve *proprietary file formats.*

The process of storing and organizing data within relational databases is managed by the DBMS, and users interact with the data using SQL or other query languages. The DBMS abstracts the physical storage details from users, providing a logical layer that allows for data manipulation and retrieval without direct concern for the underlying file formats.

Let's discuss the key points regarding file types in relational databases:

- Relational database vendors often use proprietary file formats for their data storage. Each database management system may have *its own internal structure and mechanisms for managing data.*
- Relational databases typically organize data into *tablespaces*, which are logical storage containers. These tablespaces consist of pages or blocks where data is stored. The organization and structure of these pages are determined by the specific DBMS.
- Relational databases prioritize ACID properties to ensure data integrity and reliability. The internal file formats are designed to support these transactional guarantees.
- Relational databases use various indexing and optimization techniques to enhance query performance. The internal file structures, including B-trees or other indexing structures, are optimized for efficient data retrieval.
- Users interact with relational databases using SQL commands.

While users typically don't interact directly with the underlying file formats, understanding the concepts of tablespaces, pages, and how the DBMS manages data storage can be useful for database administrators and developers when they're optimizing performance or troubleshooting issues.

Transitioning from an RDBMS to a **not only SQL** (**NoSQL**) database involves a shift in data modeling, schema design, and querying approaches. We'll explore the differences in the following section.

NoSQL databases

NoSQL databases, also known as **not only SQL** or **non-SQL** databases, are a class of database systems that provide a flexible and scalable approach to handling and storing data. Unlike traditional relational databases, which enforce a structured schema with predefined tables, columns, and relationships, NoSQL databases are designed to handle various data models, accommodating different ways of structuring and organizing data and providing a more dynamic and adaptable approach to data modeling.

Overview of NoSQL databases

Here's a quick overview of the key components of NoSQL databases:

- NoSQL databases often use a schemaless or schema-flexible approach, allowing data to be stored without the need for a predefined schema. This flexibility is particularly useful in cases where the data structure is evolving or not well-known in advance.
- NoSQL databases come in different types, each with its own data model, such as document-oriented (such as MongoDB), key-value stores (such as Redis), column-family stores (such as Apache Cassandra), and graph databases (such as Neo4j). Data models vary to accommodate different types of data and use cases. The document-oriented data model stores data as JSON documents, allowing each document to have a different structure, which is ideal for semi-structured or unstructured data. The key-value data model stores data as key-value pairs, with the value being a simple type or a complex structure, offering fast data retrieval but limited query capabilities. The column-family data model organizes data into columns instead of rows, enabling efficient storage and retrieval of large datasets. Lastly, the graph data model represents data as nodes and edges, making it perfect for applications focused on relationships, such as social networks and network analysis.
- NoSQL databases are typically designed for horizontal scalability, meaning they can efficiently distribute data across multiple nodes.
- NoSQL databases often adhere to the **consistency, availability, and partition tolerance** (**CAP**) theorem, which states that a distributed system can provide – at most – two out of three of the guarantees. NoSQL databases may prioritize availability and partition tolerance over strict consistency.

In the data field, we usually find NoSQL databases as sinks in the following cases:

- When we're dealing with data models that may change frequently or aren't well-defined in advance.
- When an application anticipates or experiences rapid growth. In this case, horizontal scalability is essential.
- When the data can't be put into tables with fixed relationships. In this case, a more flexible storage model is needed.

- When rapid development and iteration are critical. In this case, we need to modify the data model on the fly.
- When the specific features and capabilities of a particular type of NoSQL database align with the requirements of the application (e.g., document-oriented for content-heavy applications, key-value stores for caching, and graph databases for relationship-centric data).

Let's see an example of how to connect to a NoSQL database and write a new table.

An example of a MongoDB database

Before diving into the code, we'll take some time to explain MongoDB and some important concepts related to it:

- **Document**: This is the basic unit of data in MongoDB and is represented as a **Binary JSON** (**BSON**) object. Documents are like rows in a relational database but can have varying structures. Documents are composed of fields (key-value pairs). Each field can contain different data types, such as strings, numbers, arrays, or nested documents.
- **Collection**: A grouping of MongoDB documents, analogous to a table in a relational database. Collections contain documents and serve as the primary method of organizing data. Collections don't require a predefined schema, allowing documents within the same collection to have different structures.
- **Database**: A container for collections. MongoDB databases hold collections and serve as the highest level of data organization. Each database is isolated from others, meaning operations in one database don't affect others.

Now that we have a better understanding, let's go through the code. To run this example, set up MongoDB locally by following the documentation for your OS. For mac instructions, go here: `https://www.mongodb.com/docs/manual/tutorial/install-mongodb-on-os-x/`. The following code example shows how to create a database and then write data in MongoDB using `pymongo`. Note that `pymongo` is the official Python driver for MongoDB and provides a Python interface for connecting to MongoDB databases, executing queries, and manipulating data using Python scripts.

Let's get started:

1. After installing MongoDB, open your Terminal and start the service. The commands presented here are for Mac; follow the commands in the documentation for your OS:

   ```
   brew services start mongodb-community@7.0
   ```

2. Validate whether the service is running by executing the following command:

   ```
   brew services list
   ```

You should see something similar to the following:

```
mongodb-community@7.0 started maria.zervou ~/Library/
LaunchAgents/h
```

Having finished with the installations, let's set up a MongoDB database.

3. In your Terminal, type the following command to go into the MongoDB editor:

```
Mongosh
```

4. Then, create a database called `no_sql_db`:

```
use no_sql_d
```

5. Next, create a collection called `best_collection_ever`:

```
db.createCollection("best_collection_ever")
```

You should see a response similar to the following:

```
{ ok: 1 }
```

At this point, we're ready to switch to Python and start adding data to this collection. You can find the code here: `https://github.com/PacktPublishing/Python-Data-Cleaning-and-Preparation-Best-Practices/blob/main/chapter07/2.pymongo.py`. Follow these steps:

1. First, let's import the required libraries:

```
from pymongo import MongoClient
```

2. Every time we connect to a NoSQL database, we need to provide the connection details. Update all the values for the parameters in the `mongo_params` dictionary, which contains the MongoDB server host, port, username, password, and authentication source:

```
mongo_params = {
    'host': 'localhost',
    'port': 27017,
    'username': 'your_mongo_username',
    'password': 'your_mongo_password',
    'authSource': 'your_auth_database'
}
```

3. Let's look at the different functions we will use in this example to insert the documents into the MongoDB database. The first function checks whether a collection exists in the database before creating a new one:

```
def collection_exists(db, collection_name):
    return collection_name in db.list_collection_names()
```

4. The following function takes a database and collection name as arguments and creates a collection with the name we pass (in our case, we provided `collection_name` as the name):

```
def create_collection(db, collection_name):
    db.create_collection(collection_name)
```

5. Finally, we will take the collection we created in the previous step, which is just a placeholder for now, and insert some data into it:

```
def insert_data(collection, data):
    collection.insert_many(data)
```

6. Let's create some data to be inserted into the collection:

```
documents = [
    {'name': 'Alice', 'age': 25},
    {'name': 'Bob', 'age': 30},
    {'name': 'Charlie', 'age': 22}
]
```

7. Let's specify the parameters that are required for the connection and create it:

```
db_name = ' no_sql_db'
collection_name = 'best_collection_ever'
client = MongoClient(**mongo_params)
db = client[db_name]
```

8. Now, let's check whether the collection with the provided name exists. If the collection exists, use the existing collection; if not, create a new collection with the provided name:

```
if not collection_exists(db, collection_name):
    create_collection(db, collection_name)
```

9. Then, take the collection and insert the provided data:

```
collection = db[collection_name]
insert_data(collection, documents)
```

10. Finally, close the MongoDB connection:

```
client.close()
```

After executing this script, you should be able to see the records being added to the collection:

```
{'_id': ObjectId('66d833ec27bc08e40e0537b4'), 'name': 'Alice', 'age':
25}
{'_id': ObjectId('66d833ec27bc08e40e0537b5'), 'name': 'Bob', 'age':
30}
```

```
{'_id': ObjectId('66d833ec27bc08e40e0537b6'), 'name': 'Charlie',
'age': 22}
```

This script demonstrates how to interact with MongoDB databases and collections. Unlike relational databases, MongoDB does not require *tables or schemas to be created upfront*. Instead, you work directly with databases and collections. As an exercise to understand more about this flexible data model, try inserting data into the collection of a different structure. This contrasts with relational databases, where you insert rows into a *table with a fixed schema*. You can find an example here: `https://github.com/PacktPublishing/Python-Data-Cleaning-and-Preparation-Best-Practices/blob/main/chapter07/3.pymongo_expand.py`.

Finally, the script uses the `find` method to query and retrieve documents from a collection. MongoDB queries are more flexible compared to SQL queries, especially for nested data.

> **Don't delete the MongoDB database**
>
> Please don't clean the Mongo resources we created as we will use them in the streaming sink example. We will clean up all the resources at the end of this chapter.

In the next section, we will discuss the advantages and disadvantages that NoSQL databases offer.

Advantages and disadvantages of NoSQL databases

Let's summarize the advantages and disadvantages of using NoSQL systems.

The advantages of NoSQL databases are as follows:

- **Scalability**: NoSQL databases are designed for horizontal scalability, allowing them to handle large volumes of data by distributing it across multiple servers. This makes them particularly suitable for big data applications and cloud environments.
- **Flexibility**: Unlike SQL databases, which require a fixed schema, NoSQL databases offer flexible schemas. This allows structured, semi-structured, and unstructured data to be stored, making it easier to adapt to changing data models without significant restructuring.
- **Performance**: NoSQL databases can offer superior performance for certain types of queries, especially when dealing with large datasets. They're often optimized for high-speed data retrieval and can handle large volumes of transactions per second.
- **Cost-effectiveness**: Many NoSQL databases are open source and can be scaled using commodity hardware, which reduces costs compared to the expensive hardware often required for scaling SQL databases.
- **Developer agility**: The flexibility in schema and data models allows developers to iterate quickly and adapt to new requirements without the need for extensive database administration

However, they also have their disadvantages:

- **Lack of standardization**: NoSQL databases don't have a standardized query language such as SQL. This can lead to a steeper learning curve and make it challenging to switch between different NoSQL systems.
- **Limited support for complex queries**: NoSQL databases generally lack the advanced querying capabilities of SQL databases, such as joins and complex transactions, which can limit their use in applications that require complex data relationships.
- **Data consistency**: Many NoSQL databases prioritize availability and partition tolerance over consistency (as per the CAP theorem). This can lead to eventual consistency models, which may not be suitable for applications that require strict data integrity.
- **Maturity and community support**: NoSQL databases are relatively newer compared to SQL databases, which means they may have less mature ecosystems and smaller communities. This can make finding support and resources more challenging.
- **Complex maintenance**: The distributed nature of NoSQL databases can lead to complex maintenance tasks, such as data distribution and load balancing, which require specialized knowledge

Now, let's discuss the file formats we may encounter when we're working with NoSQL databases.

NoSQL database file types

The most prevalent file formats are JSON and BSON. JSON is a lightweight, human-readable data interchange format that uses a key-value pair structure and supports nested data structures. It's widely adopted for web-based data exchange due to its simplicity and ease of parsing. JSON is language-agnostic, making it suitable for various programming languages. JSON's flexible and schemaless nature aligns well with the flexible schema approach of many NoSQL databases and it allows for easy handling of evolving data structures. NoSQL databases often deal with semi-structured or unstructured data, and JSON's hierarchical structure accommodates such data well. Here's an example of a JSON data file:

```
{
    "person": {
        "name": "John Doe",
        "age": 30,
        "address": {
            "city": "New York",
            "country": "USA"
        },
        "email": ["john.doe@email.com", "john@example.com"]
    }
}
```

BSON is a binary-encoded serialization for JSON-like documents and is designed to be efficient for storage and traversal. It adds additional data types not present in JSON, such as `date` and `binary`. BSON files are encoded before they're stored and decoded before they're displayed. BSON's binary format is more efficient for storage and serialization, making it suitable for scenarios where data needs to be represented compactly. BSON is the primary data format that's used in MongoDB. Let's have a look at the BSON representation of the file presented previously:

```
\x16\x00\x00\x00
{
    "person": {
        "name": "John Doe",
        "age": 30,
        "address": {
            "city": "New York",
            "country": "USA"
        },
        "email": ["john.doe@email.com", "john@example.com"]
    }
}\x00
```

In NoSQL databases, the choice between JSON and BSON often depends on the specific requirements of the database and the use case. While JSON is more human-readable and easy to work with in many scenarios, BSON's binary efficiency is beneficial in certain contexts, particularly where storage and serialization efficiency are critical.

In the next section, we will discuss data warehouses, what challenges they solve, and which use cases you should consider implementing when using them.

Data warehouses

Transitioning to a data warehouse becomes important when the volume and complexity of data, as well as the need for advanced analytics, surpass the capabilities of your existing relational or NoSQL databases. If your relational database struggles with large data volumes, complex queries, or performance issues during analytical processing, a data warehouse can offer optimized storage and query performance for such workloads. Similarly, NoSQL databases, while excellent for handling unstructured or semi-structured data and scaling horizontally, may lack the sophisticated query capabilities and performance needed for in-depth analytics and reporting. Data warehouses are designed to integrate data from multiple sources, including both relational and NoSQL databases, facilitating comprehensive analysis and reporting. They provide robust support for historical data analysis, complex queries, and data governance, making them an ideal solution when you need to enhance your data integration, analytics, and reporting capabilities beyond what traditional databases offer.

Overview of data warehouses

A data warehouse is a specialized database system designed for storing, organizing, and retrieving *large volumes of data* that are used for business intelligence and analytics efficiently. Unlike transactional databases, which are optimized for quick data updates and individual record retrievals, data warehouses are structured to support complex queries, aggregations, and reporting on historical and current data.

Here's a quick overview of the key components of data warehouses:

- Various data sources contribute to a data warehouse, including transactional databases, external files, logs, and more.
- **Extract, transform, and load** (ETL) processes are used to gather data from source systems, transform it into a consistent format, and load it into the data warehouse.
- Data warehouses employ optimized storage methods, such as columnar storage, to store large volumes of data efficiently.
- Indexes and pre-aggregated tables are used to optimize query performance. In a data warehouse, indexes play a crucial role in optimizing query performance. An index is a data structure that enhances the speed and efficiency of data retrieval from a table by creating a separate, organized subset of the data. Indexes are typically created on one or more columns to facilitate faster querying. Without indexes, the database must scan the entire table to locate relevant rows. Indexes help the database quickly find rows that meet query conditions. Common candidates for indexing include columns used in `WHERE` clauses, `JOIN` conditions, and `ORDER BY` clauses. However, over-indexing can lead to diminishing returns and increased maintenance overhead. While indexes improve query performance, they also consume additional storage and can slow down write operations due to the need to maintain the index.
- Techniques such as parallel processing and indexing are employed to enhance the speed of analytical queries.
- Integration with business intelligence tools allows users to create reports and dashboards and perform data analysis.
- Data is organized using multidimensional models, often in the form of star or snowflake schemas, to support analytics and reporting.

Let's expand on dimensional modeling. It's a design technique that's used in data warehousing to structure data so that it supports efficient retrieval for analytical queries and reporting. Unlike traditional relational models, dimensional models are optimized for query performance and ease of use. In the next section, we will present the main schema types in dimensional models.

Schema types in dimensional models

Dimensional modeling primarily involves two types of schemas: the star schema and the snowflake schema. The **star schema** is the simplest form of dimensional modeling and is where a central fact table is directly connected to multiple dimension tables, forming a star-like structure. This schema type is highly intuitive and easy to navigate, making it ideal for straightforward queries and reporting. Each dimension table in a star schema contains a primary key that maps to a foreign key in the fact table, providing descriptive context to the quantitative data in the fact table. For instance, a sales star schema might include a central sales fact table with foreign keys linking to dimension tables for products, customers, time, and stores, thereby simplifying complex queries by reducing the number of joins:

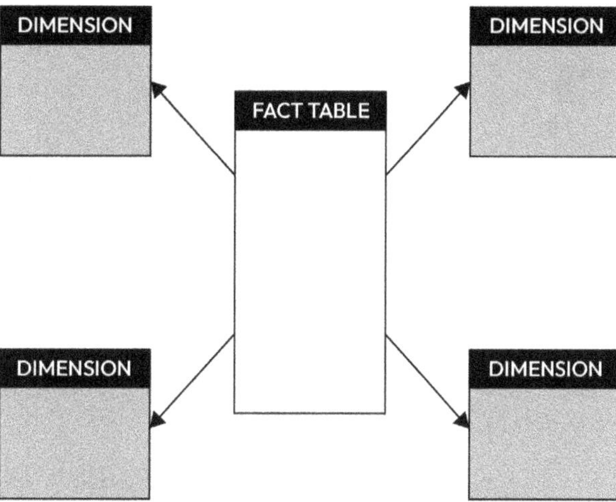

Figure 7.4 – Star schema

On the other hand, the **snowflake schema** is a more normalized version of the star schema and is where dimension tables are further broken down into related sub-tables, resembling a snowflake pattern. This structure reduces data redundancy and can save storage space, although it introduces more complexity in query design due to the additional joins required. For example, a product dimension table in a snowflake schema might be normalized into separate tables for product categories and brands, creating a multi-layered structure that ensures higher data integrity and reduces update anomalies. While the snowflake schema may be slightly more complex to query, it offers benefits in terms of data maintenance and scalability, especially in environments where data consistency and storage optimization are critical:

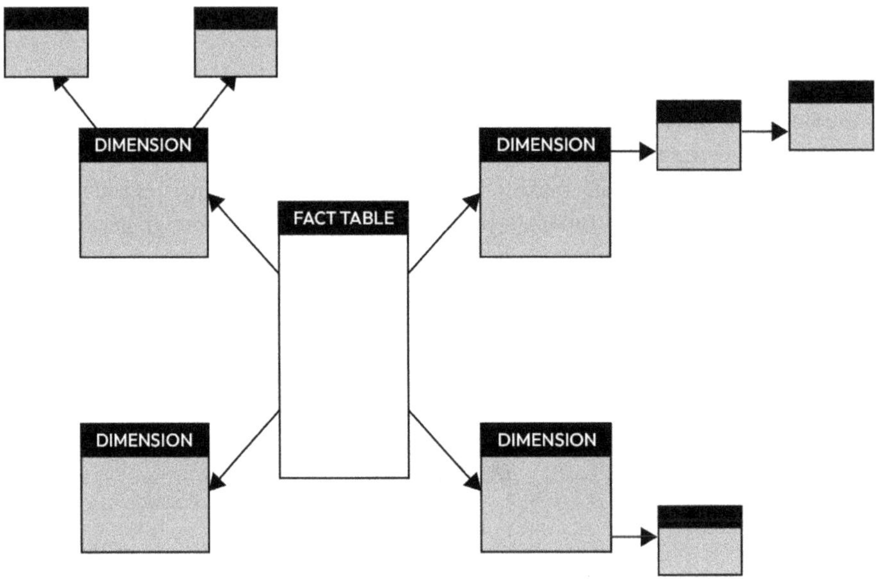

Figure 7.5 – Snowflake schema

Star schemas are often preferred for simpler hierarchies and when query performance is a higher priority, while snowflake schemas may be chosen when more efficient use of storage and achieving a higher level of normalization is essential.

Transitioning from understanding schema types in dimensional modeling, it's essential to explore the diverse options available for implementing and leveraging data warehouses in various organizational contexts.

Data warehouse solutions

There are many different data warehouse options out there. We've summarized the main ones in the following table:

Data Warehouse	Description
Databricks SQL	A cloud-based, serverless data warehouse that brings warehouse capabilities to data lakes. It's also known for its scalability, performance, and parallel processing, as well as its built-in machine learning capabilities.
Amazon Redshift	A fully managed, scalable data warehouse service in the cloud. It's optimized for high-performance analytics.
Snowflake	A cloud-based data warehouse with a multi-cluster, shared architecture that supports diverse workloads.
Google BigQuery	A serverless, highly scalable data warehouse with built-in machine learning capabilities.

Data Warehouse	Description
Teradata	An on-premises or cloud-based data warehouse known for its scalability, performance, and parallel processing.
Microsoft Azure Synapse Analytics	A cloud-based analytics service that offers both on-demand and provisioned resources.

Table 7.2 – Data warehousing solutions

In the next section, we will look at an example of creating a BigQuery table to illustrate the practical application of data warehouses.

An example of a data warehouse

Let's learn how to create a new table in BigQuery. Google Cloud provides a client library for various programming languages, including Python, to interact with BigQuery. To get the ready so that you can run the following example, go to the BigQuery documentation: `https://cloud.google.com/python/docs/reference/bigquery/latest`. Let's deep dive into the example. We will follow the same patterns that have been presented so far in this chapter. Let's get started:

> **Note**
>
> To run this example, you need to have a **Google Cloud Platform** (**GCP**) account and a Google Storage bucket ready.

1. Import the required libraries:

```
from google.cloud import bigquery
from google.cloud.bigquery import SchemaField
```

2. First, we'll set up the project ID. Replace `your_project_id` with your actual values:

```
client = bigquery.Client(project='your_project_id')
```

3. Define the dataset and table name and update the following fields:

```
dataset_name = 'your_dataset'
table_name = 'your_table'
```

4. Check whether the dataset and the table exist:

```
dataset_ref = client.dataset(dataset_name)
table_ref = dataset_ref.table(table_name)
table_exists = client.get_table(
```

```
    table_ref, retry=3, timeout=30, max_results=None
) is not None
```

5. Define the table schema (replace it with your schema if you wish to update the data). In this example, we will create a table with two columns (`column1` and `column2`). The first column will be of the `STRING` type, while the second one will be of the `INTEGER` type. The first column can't contain missing values, whereas the second one can:

```
schema = [
    SchemaField('column1', 'STRING', mode='REQUIRED'),
    SchemaField('column2', 'INTEGER', mode='NULLABLE'),
    # Add more fields as needed
]
```

6. Let's perform the checks for the existence of a table with the same name. If the table doesn't exist, then we create it with the provided name:

```
if not table_exists:
    table = bigquery.Table(table_ref, schema=schema)
    client.create_table(table)
```

7. Let's create some mock data that will be inserted into the table:

```
rows_to_insert = [
    ('value1', 1),
    ('value2', 2),
    ('value3', 3)
]
```

8. Construct the data to be inserted:

```
data_to_insert = [dict(zip([field.name for field in schema],
row)) for row in rows_to_insert]
```

9. Insert the data and check for errors. If everything works as expected, close the connection:

```
errors = client.insert_rows(table, data_to_insert)
```

If any insertion errors occur, print them out:

```
print(f"Errors occurred during data insertion: {errors}")
```

10. Close the BigQuery client:

```
client.close()
```

What's the difference between a container and a table?

A container (called different things in different systems, such as database, dataset, or schema) is a logical grouping mechanism that's used to organize and manage data objects such as tables, views, and related metadata. Containers provide a way to partition and structure data based on specific needs, such as access control, data governance, or logical separation of data domains. On the other hand, a table is a fundamental data structure that stores the actual data records, organized into rows and columns. Tables define the schema (column names and data types) and hold the data values.

At this point, let's transition from understanding the fundamental components of a data warehouse environment to discussing the advantages and disadvantages that data warehouses offer in managing and analyzing large volumes of data efficiently.

Advantages and disadvantages of data warehouses

Let's summarize the advantages and disadvantages of using data warehouses.

The advantages are as follows:

- Optimized for analytical queries and large-scale data processing
- Can handle massive amounts of data
- Integration with other data tools and services

Here are their disadvantages:

- Higher costs for storage and querying compared to traditional databases
- Might have limitations regarding real-time data processing

File types in data warehouses

In terms of file formats, it's accurate to say that many modern data warehouses use **proprietary** internal storage formats for writing data. These proprietary formats are usually columnar storage formats that are optimized for efficient querying and analytics.

Let's have a look at the differences we may find in these proprietary formats:

- Data warehouses often use columnar storage formats such as Parquet, ORC, or Avro. While these formats are open and widely adopted, each data warehouse might have its internal optimizations or extensions.
- The actual implementations of these columnar storage formats may have vendor-specific optimizations or features. For example, how a specific data warehouse handles compression, indexing, and metadata might be specific to the vendor.

- Users interact with data warehouses through standard interfaces, such as SQL queries. The choice of storage format often doesn't affect users, so long as the data warehouse supports common data interchange formats for import and export.

So, while the internal storage mechanisms may have vendor-specific optimizations, the use of well-established, open, and widely adopted columnar storage formats ensures a degree of interoperability and flexibility. Users typically interact with data warehouses using standard SQL queries or data interchange formats such as CSV, JSON, or Avro for data import/export, which adds a layer of standardization to the external-facing aspects of these systems.

Transitioning from traditional data warehouses to data lakes represents a strategic shift that embraces a more flexible and scalable paradigm. In the next section, we will deep dive into data lakes.

Data lakes

The transition from traditional data warehouses to data lakes represents a shift in how organizations handle and analyze their data. Traditional data warehouses are designed to store structured data, which is highly organized and formatted according to a predefined schema, such as tables with rows and columns in relational databases. Structured data is easy to query and analyze using SQL. However, data warehouses struggle with handling unstructured data, which lacks a predefined format or organization. Examples of unstructured data include text documents, emails, images, videos, and social media posts. Data lakes, on the other hand, offer a more flexible and scalable solution by storing both structured and unstructured data in its native format. This allows organizations to ingest and store vast amounts of data without the need for immediate structuring. This transition addresses the limitations of data warehouses, offering a more versatile and future-proof approach to data management.

Overview of data lakes

A data lake is a centralized repository that allows organizations to store vast amounts of raw and unstructured data in any format needed. They're designed to handle diverse data types, such as structured, semi-structured, and unstructured data, and enable data exploration, analytics, and machine learning. Data lakes solve several problems: they enable the consolidation of diverse data sources, break down silos by providing a unified data storage solution, and support advanced analytics by making all types of data easily accessible.

> **Remember**
>
> Think of data lakes as a filesystem, such as for storing data in a file location on your laptop, just on a much larger scale.

Data lake solutions

Here's a summary of the available data lake solutions in the data space:

Data Lake	Description
Amazon S3	A cloud-based object storage service by **Amazon Web Services (AWS)**. It's commonly used as a foundation for data lakes.
Azure Data Lake Storage	A scalable and secure cloud-based storage solution by Microsoft Azure that's designed to support big data analytics and data lakes.
Hadoop Distributed File System (HDFS)	A distributed filesystem that forms the storage layer of Apache Hadoop, an open source big data processing framework.
Google Cloud Storage	Google Cloud's object storage service, often used as part of a data lake architecture in the GCP.

Table 7.3 – Data lake solutions

The shift from traditional data warehouses to data lakes represents a fundamental transformation in how organizations manage and analyze data. This shift was driven by several factors, including the need for greater flexibility, scalability, and the ability to handle diverse and unstructured data types. The following table highlights the key differences between traditional data warehouses and data lakes:

	Data Warehouses	Data Lakes
Data variety and flexibility	Traditional data warehouses are designed to handle structured data and are less adaptable to handle diverse data types or unstructured data.	Data lakes emerged as a response to the increasing volume and variety of data. They provide a storage repository for raw, unstructured, and diverse data types, allowing organizations to store large volumes of data without the need for predefined schemas.
Scalability	Traditional data warehouses often face scalability challenges when dealing with massive amounts of data. Scaling up a data warehouse can be costly and may have limitations.	Data lakes, particularly cloud-based solutions, offer scalable storage and computing resources. They can efficiently scale horizontally to handle growing datasets and processing demands.

	Data Warehouses	Data Lakes
Cost-efficiency	Traditional data warehouses can be expensive to scale, and the cost structure may not be conducive to storing large volumes of raw or less structured data.	Cloud-based data lakes often follow a pay-as-you-go model, allowing organizations to manage costs more efficiently by paying for the resources they use. This is particularly beneficial for storing large amounts of raw data.
Schema-on-dead versus schema-on-write	Follow a schema-on-write approach, where data is structured and transformed before being loaded into the warehouse.	Follow a schema-on-read approach, allowing for the storage of raw, untransformed data. The schema is applied at the time of data analysis, providing more flexibility in data exploration.

Table 7.4 – Data warehouses versus data lakes

The emergence of the Lakehouse architecture further refined the shift away from data warehouses by solving some key challenges associated with data lakes and by bringing features traditionally associated with data warehouses into the data lake environment. Here's a breakdown of the key aspects of this evolution:

- The Lakehouse integrates ACID transactions into the data lake, providing transactional capabilities that were traditionally associated with data warehouses. This ensures data consistency and reliability.
- The Lakehouse supports schema evolution, allowing changes to be made to data schemas over time without requiring a full transformation of existing data. This enhances flexibility and reduces the impact of schema changes on existing processes.
- The Lakehouse introduces features for managing data quality, including schema enforcement and constraints, ensuring that data stored in the lake meets specified standards.
- The Lakehouse aims to provide a unified platform for analytics by combining the strengths of data lakes and data warehouses. It allows organizations to perform analytics on both structured and semi-structured data in a centralized repository.
- The Lakehouse enhances the metadata catalog, providing a comprehensive view of data lineage, quality, and transformations. This facilitates better governance and understanding of the data stored in the lake.

The Lakehouse concept has evolved through discussions in the data and analytics community, with various companies contributing to the development and adoption of Lakehouse principles.

An example of a data lake

Let's explore an example of how to write some Parquet files on S3, AWS's cloud storage. To get everything set up, go to the AWS documentation: `https://docs.aws.amazon.com/code-library/latest/ug/python_3_s3_code_examples.html`. Now, follow these steps:

> **Note**
>
> To run this example, you need to have an AWS account and an S3 bucket ready.

1. We'll start by importing the required libraries:

```
import pandas as pd
import pyarrow.parquet as pq
import boto3
from io import BytesIO
```

2. Now, we're going to create some mock data so that we can write it to S3:

```
data = {'Name': ['Alice', 'Bob', 'Charlie'],
        'Age': [25, 30, 22],
        'City': ['New York', 'San Francisco', 'Los Angeles']}

df = pd.DataFrame(data)
```

3. Next, we must convert the DataFrame into Parquet format:

```
parquet_buffer = BytesIO()
pq.write_table(pq.Table.from_pandas(df), parquet_buffer)
```

4. Update your authentication key and the bucket name in which we're going to write the data:

```
aws_access_key_id = 'YOUR_ACCESS_KEY_ID'
aws_secret_access_key = 'YOUR_SECRET_ACCESS_KEY'
bucket_name = 'your-s3-bucket'
file_key = 'example_data.parquet'  # The key (path) of the file
in S3
```

5. Connect to the S3 bucket using the connection details from the previous step:

```
s3 = boto3.client('s3', aws_access_key_id=aws_access_key_id,
aws_secret_access_key=aws_secret_access_key)
```

6. Upload the Parquet file to S3:

```
s3.put_object(Body=parquet_buffer.getvalue(), Bucket=bucket_
name, Key=file_key)
```

As discussed previously, Lakehouses come with their own set of advantages and disadvantages.

The advantages are as follows:

- Lakehouses provide a unified platform that integrates the strengths of both data lakes and data warehouses. This allows organizations to leverage the flexibility of data lakes and the transactional capabilities of data warehouses in a single environment.
- Lakehouses follow a schema-on-read approach, allowing for the storage of raw, untransformed data.
- Lakehouses support diverse data types, including structured, semi-structured, and unstructured data.
- Lakehouses integrate ACID transactions, providing transactional capabilities that ensure data consistency and reliability. This is particularly important for use cases where data integrity is critical.
- Many Lakehouse solutions offer time travel capabilities, allowing users to query data at specific points in time. Versioning of data provides historical context and supports audit requirements.
- Lakehouses often implement optimized storage formats (e.g., Delta and Iceberg) that contribute to storage efficiency and improved query performance, especially for large-scale analytical workloads.

The disadvantages are as follows:

- Users and administrators may need to adapt to a new way of working with data while considering both schema-on-read and schema-on-write paradigms. This may require training and education.
- Depending on the implementation and cloud provider, costs associated with storing, processing, and managing data in a Lakehouse architecture may vary. Organizations need to carefully manage costs to ensure efficiency.

As we've discussed, Lakehouses have the amazing advantage of allowing any data type to be ingested and stored, from structured to semi-structured to unstructured. This means we can find any file type in the ingestion part of the process, from CSVs to Parquets and Avros. While in the ingestion part, we can see any file type, on the write part, we can take advantage of the flexibility the Lakehouse offers and store the data in optimized open table file formats. Open table format is a file format that's used to store tabular data in a way that's easily accessible and interoperable across various data processing and analytics tools.

File types in data lakes

In Lakehouse architecture, we have three prominent formats: Delta, Apache Iceberg, and Apache Hudi. These formats provide features such as ACID transactions, schema evolution, incremental data processing, and read and write optimizations. Here's a brief overview of these formats:

- Delta Lake is an open source storage layer designed to enhance the reliability and performance of data processing in data lakes. It is well-suited for building data lakes on infrastructure such as S3 or Azure Storage and has strong support for ACID transactions and data versioning.
- Apache Iceberg is another open source table format that's optimized for fast query performance. It's a good choice when query efficiency is required and it has excellent support for schema evolution and versioning.

- Apache Hudi (Hadoop Upserts, Deletes, and Incrementals) is another open source data lake storage format that provides great support for real-time data processing and streaming features. While it may not be as widely known as Delta Lake or Apache Iceberg, Hudi is gaining traction, especially in Apache Spark and Hadoop ecosystems.

In general, all these formats were built to solve the same challenges, which is why they have a lot of common features. Thus, before choosing the best one for your workload, there are a couple of things you should consider to ensure you're going in the right direction:

- Consider the compatibility of each technology with your existing data processing ecosystem and tools
- Evaluate the level of community support, ongoing development, and adoption within the data community for each technology
- Assess the performance characteristics of each technology concerning your specific use case, especially in terms of read and write operations

Ultimately, the choice between Delta Lake, Apache Iceberg, and Apache Hudi should be driven by the specific requirements and priorities of your data lake or lakehouse environment. It's also beneficial to experiment and benchmark each solution with your data and workloads to make an informed decision.

The last sink technology we're going to discuss is streaming data sinks.

Streaming data sinks

The transition from batch and micro-batch processing to streaming technologies marks a significant evolution in data processing and analytics. Batch processing involves collecting and processing data in large, discrete chunks at scheduled intervals, which can lead to delays in data availability and insights. Micro-batch processing improves on this by handling smaller batches at more frequent intervals, reducing latency but still not achieving real-time data processing. Streaming technologies, however, enable the continuous ingestion and processing of data in real-time, allowing organizations to immediately analyze and act on data as it arrives. This shift to streaming technologies addresses the growing need for real-time analytics and decision-making in today's fast-paced business environments, providing a more dynamic and responsive approach to data management.

Overview of streaming data sinks

Streaming data sinks are components or services that consume and store streaming data in real time. They act as the endpoint where streaming data is ingested, processed, and persisted for further analysis or retrieval. Here's an overview of streaming data sinks and their main components:

- **Ingestion component**: This is responsible for receiving and accepting incoming data streams
- **Processing logic**: This is a bespoke logic that may include components for data enrichment, transformation, or aggregation

- **Storage component**: This persists streaming data for future analysis or retrieval
- **Connectors**: Their main role is to interact with various data processing or storage systems

We usually implement streaming sinks in the following areas:

- In real-time analytics systems. This allows organizations to gain insights into their data as events occur.
- In systems monitoring, where streaming data sinks capture and process real-time metrics, logs, or events, enabling immediate alerting and responses to issues or anomalies.
- In financial transactions or e-commerce. Here, streaming data sinks can be used for real-time fraud detection by analyzing patterns and anomalies in transaction data.
- In the **Internet of Things** (**IoT**) scenarios, streaming data sinks handle the continuous flow of data from sensors and devices, supporting real-time monitoring and control.

Now, let's have a look at the available options we have for streaming sinks.

Streaming sinks solutions

Many cloud platforms offer managed streaming data services that act as sinks, such as Amazon Kinesis, Azure Event Hubs, and Google Cloud Dataflow, as shown in the following table:

Streaming Data Sink	Description
Amazon Kinesis	A fully managed service for real-time stream processing in AWS. It supports data streams, analytics, and storage.
Azure Event Hub	A cloud-based real-time analytics service in Azure for processing and analyzing streaming data.
Google Cloud Dataflow	A fully managed stream and batch processing service on GCP.
Apache Kafka	A distributed streaming platform that can act as both a source and a sink for streaming data.
Apache Spark Streaming	A real-time data processing framework that's part of the Apache Spark ecosystem.
Apache Flink	A stream processing framework that supports event time processing and various sink connectors.

Table 7.5 – Different streaming data services

In the next section, we will use one of the most popular streaming sinks, Kafka, to get an idea of what writing in a streaming sink looks like.

An example of a streaming data sink

First things first, let's get an initial understanding of the main components of Apache Kafka:

- **Brokers** are the core servers that make up a Kafka cluster. They handle the storage and management of messages. Each broker is identified by a unique ID. Brokers are responsible for replicating data across the cluster for fault tolerance.
- **Topics** are the primary abstractions in Kafka for organizing and categorizing messages. They are like tables in a database or folders in a filesystem. Messages are published to and read from specific topics. Topics can be partitioned for scalability and parallel processing.
- **Partitions** are the units of parallelism in Kafka. Each topic is divided into one or more partitions, which allow for distributed storage and processing of data. Messages within a partition are ordered and immutable.
- **Producers** are client applications that publish (write) messages to Kafka topics. They can choose which partition to send messages to or use a partitioning strategy. Producers are responsible for serializing, compressing, and load balancing data among partitions.
- **Consumers** are client applications that subscribe to (read) messages from Kafka topics. They can read from one or more partitions of a topic and keep track of which messages they have already consumed.
- **ZooKeeper** is used for managing and coordinating Kafka brokers. It maintains metadata about the Kafka cluster. Newer versions of Kafka are moving toward removing the ZooKeeper dependency.

Now that we have a better understanding of the main Kafka components, let's start with our step-by-step guide. We will need to install a couple of components for this example, so stay with me as we go through the process. To simplify this, we will use Docker as it allows you to define the entire environment in a `docker-compose.yml` file, making it easy to set up Kafka and Zookeeper with minimal configuration. This eliminates the need to manually install and configure each component on your local machine. Follow these steps:

1. Download Docker by following the public documentation: `https://docs.docker.com/desktop/install/mac-install/`.
2. Next, set up Kafka with Docker. For this, let's have a look at the `docker-compose.yml` file at `https://github.com/PacktPublishing/Python-Data-Cleaning-and-Preparation-Best-Practices/blob/main/chapter07/setup/docker-compose.yml`.
3. This Docker Compose configuration sets up a simple Kafka and Zookeeper environment using version 3 of the Docker Compose file format. The configuration defines two services: `zookeeper` and `kafka`.

4. Zookeeper uses the `confluentinc/cp-zookeeper:latest` image. It maps the host machine's port, `2181`, to the container's port, `2181`, for client connections. The `ZOOKEEPER_CLIENT_PORT` environment variable is set to `2181`, which specifies the port Zookeeper will listen on for client requests:

```
version: '3'
services:
    zookeeper:
        image: confluentinc/cp-zookeeper:latest
        ports:
            - "2181:2181"
        environment:
            ZOOKEEPER_CLIENT_PORT: 2181
```

5. Kafka uses the `confluentinc/cp-kafka:latest` image. It maps the host machine's port, `9092`, to the container's port, `9092`, for external client connections:

```
kafka:
    image: confluentinc/cp-kafka:latest
    ports:
        - "9092:9092"
    environment:
        KAFKA_BROKER_ID: 1
        KAFKA_ZOOKEEPER_CONNECT: zookeeper:2181
        KAFKA_ADVERTISED_LISTENERS: PLAINTEXT://
kafka:29092,PLAINTEXT_HOST://localhost:9092
        KAFKA_LISTENER_SECURITY_PROTOCOL_MAP:
PLAINTEXT:PLAINTEXT,PLAINTEXT_HOST:PLAINTEXT
        KAFKA_INTER_BROKER_LISTENER_NAME: PLAINTEXT
        KAFKA_OFFSETS_TOPIC_REPLICATION_FACTOR: 1
```

Here are some key environment variables that configure Kafka:

- `KAFKA_BROKER_ID` is set to `1`, identifying this broker uniquely in a Kafka cluster
- `KAFKA_ZOOKEEPER_CONNECT` points to the Zookeeper service (`zookeeper:2181`), allowing Kafka to connect to Zookeeper for managing cluster metadata
- `KAFKA_ADVERTISED_LISTENERS` advertises two listeners:
    - `PLAINTEXT://kafka:29092` for internal Docker network communication
    - `PLAINTEXT_HOST://localhost:9092` for connections from outside the Docker network (e.g., from the host machine)
- `KAFKA_LISTENER_SECURITY_PROTOCOL_MAP` ensures both advertised listeners use the `PLAINTEXT` protocol, meaning no encryption or authentication

- KAFKA_INTER_BROKER_LISTENER_NAME is set to PLAINTEXT, specifying which listener Kafka brokers will use to communicate with each other
- KAFKA_OFFSETS_TOPIC_REPLICATION_FACTOR is set to 1, meaning the offsets topic (used for storing consumer group offsets) will not be replicated across multiple brokers, which is typical for a single-broker setup

This setup is ideal for local development or testing, where you need a simple, single-node Kafka environment without the complexities of a multi-node, production-grade cluster. Now, we're ready to run the container.

6. Let's run the Docker container to start Kafka and Zookeeper. In your terminal, enter the following command:

```
docker-compose up -d
```

This will take the Kafka and Zookeeper images and install them in your environment. You should see something similar to the following printed on your terminal:

```
[+] Running 3/3
 ✓ Network setup_default           Created          0.0s
 ✓ Container setup-kafka-1         Started          0.7s
 ✓ Container setup-zookeeper-1     Started          0.6s
```

Kafka producer

Now, let's go back to our Python IDE and look at how we can push data to a Kafka producer. For this, we're going to read the data written in MongoDB and produce it in Kafka. You can find the code here: https://github.com/PacktPublishing/Python-Data-Cleaning-and-Preparation-Best-Practices/blob/main/chapter07/4a.kafka_producer.py. Let's get started:

1. First, let's import the necessary libraries:

```
from pymongo import MongoClient
from confluent_kafka import Producer
import json
```

2. Next, define the MongoDB connection:

```
mongo_client = MongoClient('mongodb://localhost:27017')
db = mongo_client['no_sql_db']
collection = db['best_collection_ever']
```

Here, `MongoClient('mongodb://localhost:27017')` connects to a MongoDB instance running on localhost at the default port of `27017`. This creates a client object that allows interaction with the database. Then, `db = mongo_client['no_sql_db']` selects the `no_sql_db` database from the MongoDB instance. Finally, `collection = db['best_collection_ever']` selects the `best_collection_ever` collection from the `no_sql_db` database.

3. Let's perform the Kafka producer configuration that creates a Kafka producer object with the specified configuration. This producer will be used to send messages (in this case, MongoDB documents) to a Kafka topic:

```
kafka_config = {
    'bootstrap.servers': 'localhost:9092'
}
producer = Producer(kafka_config)
```

4. The following function is a callback function that will be called when the Kafka producer finishes sending a message. It checks whether there was an error during the message delivery and prints a message indicating success or failure. This function provides feedback on whether messages were successfully sent to Kafka, which is useful for debugging and monitoring:

```
def delivery_report(err, msg):
    if err is not None:
        print(f'Message delivery failed: {err}')
    else:
        print(f'Message delivered to {msg.topic()} [{msg.
partition()}]')
```

5. Read from MongoDB and produce to Kafka for the document in `collection.find()`:

```
    message = json.dumps(document, default=str)
    producer.produce('mongodb_topic',
                     alue=message.encode('utf-8'),
                     callback=delivery_report)
    producer.poll(0)
```

The preceding code iterates over each document in the `best_collection_ever` collection. The `find()` method retrieves all documents from the collection. Then, `message = json.dumps(document, default=str)` converts each MongoDB document (a Python dictionary) into a JSON string. The `default=str` parameter handles data types that aren't JSON serializable by converting them into strings. Next, `producer.produce('mongodb_topic', value=message.encode('utf-8'), callback=delivery_report)` sends the JSON string as a message to the `mongodb_topic` Kafka topic. The message is encoded in UTF-8, and the `delivery_report` function is set as a callback to handle delivery confirmation. Finally, `producer.poll(0)` ensures that the Kafka producer processes delivery reports and other events. This is necessary to keep the producer active and responsive.

6. This ensures that all messages in the producer's queue are sent to Kafka before the script exits. Without this step, there might be unsent messages remaining in the queue:

```
producer.flush()
```

After running this script, you should see the following print statements:

```
Message delivered to mongodb_topic [0]
Message delivered to mongodb_topic [0]
Message delivered to mongodb_topic [0]
Message delivered to mongodb_topic [0]
Message delivered to mongodb_topic [0]
Message delivered to mongodb_topic [0]
Message delivered to mongodb_topic [0]
Message delivered to mongodb_topic [0]
```

So far, we've connected to a MongoDB database, read the documents from a collection, and sent these documents as messages to the Kafka topic.

Kafka consumer

Next, let's run the consumer so that it can consume the messages from the Kafka producer. The full code can be found at `https://github.com/PacktPublishing/Python-Data-Cleaning-and-Preparation-Best-Practices/blob/main/chapter07/4b.kafka_consumer.py`:

1. Let's start by importing the libraries:

```
from confluent_kafka import Consumer, KafkaError
import json
import time
```

2. Next, we must create the Kafka consumer configuration that specifies the Kafka broker(s) to connect to. Here, it's connecting to a Kafka broker running on localhost at port `9092`. In this case, `group.id` sets the consumer group ID, which allows multiple consumers to coordinate and share the work of processing messages from a topic. Messages will be distributed among consumers in the same group. Next, `auto.offset.reset` defines the behavior when there is no initial offset in Kafka or if the current offset doesn't exist. Setting this to `earliest` means the consumer will start reading from the earliest available message in the topic:

```
consumer_config = {
    'bootstrap.servers': 'localhost:9092',
    'group.id': 'mongodb_consumer_group',
    'auto.offset.reset': 'earliest'
}
```

3. Now, we will instantiate a Kafka consumer with the configuration specified earlier. Here, `consumer.subscribe(['mongodb_topic'])` subscribes the consumer to the `mongodb_topic` Kafka topic. This means the consumer will receive messages from this topic:

   ```
   consumer = Consumer(consumer_config)
   consumer.subscribe(['mongodb_topic'])
   ```

4. Set the duration for which the consumer should run (in seconds):

   ```
   run_duration = 10 # For example, 10 seconds
   start_time = time.time()
   print("Starting consumer...")
   ```

5. The following code begins an infinite loop that will run until it's explicitly broken out of. This loop continuously polls Kafka for new messages. Here, `if time.time() - start_time > run_duration` checks whether the consumer has been running for longer than the specified `run_duration`. If so, it prints a message and breaks out of the loop, stopping the consumer:

   ```
   while True:
       if time.time() - start_time > run_duration:
           print("Time limit reached, shutting down consumer.")
           break
       msg = consumer.poll(1.0)
       if msg is None:
           continue
       if msg.error():
           if msg.error().code() == KafkaError._PARTITION_EOF:
               print('Reached end of partition')
           else:
               print(f'Error: {msg.error()}')
       else:
           document = json.loads(msg.value().decode('utf-8'))
           print(f'Received document: {document}')

   consumer.close()
   print("Consumer closed.")
   ```

After running the preceding code, you should see the following print statements:

```
Starting consumer...
Received document: {'_id': '66d833ec27bc08e40e0537b4', 'name':
'Alice', 'age': 25}
Received document: {'_id': '66d833ec27bc08e40e0537b5', 'name': 'Bob',
'age': 30}
Received document: {'_id': '66d833ec27bc08e40e0537b6', 'name':
'Charlie', 'age': 22}
```

```
Received document: {'_id': '66d835aa1798a2275cecaba8', 'name':
'Alice', 'age': 25, 'email': 'alice@example.com'}
Received document: {'_id': '66d835aa1798a2275cecaba9', 'name': 'Bob',
'age': 30, 'address': '123 Main St'}
Received document: {'_id': '66d835aa1798a2275cecabaa', 'name':
'Charlie', 'age': 22, 'hobbies': ['reading', 'gaming']}
Received document: {'_id': '66d835aa1798a2275cecabab', 'name':
'David', 'age': 40, 'email': 'david@example.com', 'address': '456 Elm
St', 'active': True}
Received document: {'_id': '66d835aa1798a2275cecabac', 'name': 'Eve',
'age': 35, 'email': 'eve@example.com', 'phone': '555-1234'}
```

The goal of this example was to show you how data can be continuously read from a NoSQL database such as MongoDB and then streamed in real-time to other systems using Kafka. Kafka acts as a messaging system that allows data producers (e.g., MongoDB) to be decoupled from data consumers. This example also illustrates how data can be processed in stages, allowing for scalable and flexible data pipelines.

In terms of a real use case scenario, imagine that we are building a ride-sharing app. Handling events such as ride requests, cancellations, and driver statuses in real-time is crucial for efficiently matching riders with drivers. MongoDB stores this event data, such as ride requests and driver locations, while Kafka streams the events to various microservices. These microservices then process the events to make decisions, such as assigning a driver to a rider. By using Kafka, the system becomes highly responsive, scalable, and resilient as it decouples event producers (such as ride requests) from consumers (such as the driver assignment logic).

To summarize what we have seen so far, in contrast to relational sinks, which involve structured data with defined schemas, Kafka can serve as a buffer or intermediary for data ingestion, allowing for decoupled and scalable data pipelines. NoSQL sinks often handle unstructured or semi-structured data, similar to Kafka's flexibility with message formats. Kafka's ability to handle high-throughput data streams complements NoSQL databases' scalability and flexibility.

To clean all the resources that have been used so far, execute the cleaning script: `https://github.com/PacktPublishing/Python-Data-Cleaning-and-Preparation-Best-Practices/blob/main/chapter07/setup/cleanup_script.sh`.

In the next section, we will deep dive into the file format seen in streaming data sinks.

File types in streaming data sinks

Streaming data sinks primarily deal with messages or events rather than traditional file storage. The data that's transmitted through streaming data sinks is often in formats such as JSON, Avro, or binary. These formats are commonly used for serializing and encoding data in streaming scenarios. They are efficient and support schema evolution. In the *NoSQL databases* section of this chapter, we deep-dived in the JSON file format. Here, we'll look at Avro and binary.

Apache Avro is a binary serialization format developed within the Apache Hadoop project. It uses a schema to define data structures, allowing for efficient serialization and deserialization. Avro is known for its compact binary representation, providing fast serialization and efficient storage. In streaming scenarios, minimizing data size is crucial for efficient transmission over the network. Avro's compact binary format reduces data size, improving bandwidth utilization. Avro also supports schema evolution, allowing for changes in data structures over time without requiring all components to be updated simultaneously. Avro's schema-based approach enables interoperability between different systems and languages, making it suitable for diverse ecosystems. Let's see an example of an Avro file:

```
{
  "type": "record",
  "name": "SensorData",
  "fields": [
    {"name": "sensor_id", "type": "int"},
    {"name": "timestamp", "type": "long"},
    {"name": "value", "type": "float"},
    {"name": "status", "type": "string"}
  ]
}
```

Binary formats use a compact binary representation of data, resulting in efficient storage and transmission. Various binary protocols can be employed based on specific requirements, such as Google's **Protocol Buffers** (**protobuf**) or Apache Thrift. Binary formats minimize the size of the transmitted data, reducing bandwidth usage in streaming scenarios. Binary serialization and deserialization are generally faster than text-based formats, which is crucial in high-velocity streaming environments. Let's have a look at a binary file in protobuf:

```
syntax = "proto3";

message SensorData {
  int32 sensor_id = 1;
  int64 timestamp = 2;
  float value = 3;
  string status = 4;
}
```

In streaming sinks, the choice between JSON, Avro, or binary depends on the specific requirements of the streaming use case, including factors such as interoperability, schema evolution, and data size considerations.

So far, we've discussed the most common data sinks used by data engineers and data scientists, as well as the different file types we usually encounter with those sinks. In the following sections, we will provide a summary of all the discussed data sinks and file types, as well as their pros and cons and when to best use them.

Which sink is the best for my use case?

Let's summarize what we've learned regarding the different data sinks and get a deeper understanding of when to use which:

Technology	Pros	Cons	When to Choose	Use Case
Relational database	ACID properties ensure data consistency. Mature query languages (SQL) for complex queries. Support for complex transactions and joins.	Limited scalability for read-heavy workloads. Schema changes may be challenging and downtime-prone. May not scale well horizontally.	Structured data with a well-defined schema. When you're maintaining relationships between data entities.	Transactional applications Enterprise applications with structured data.
NoSQL database	Flexible schema, suitable for semi-structured or unstructured data. Scalability – horizontal scaling is often easier. High write throughput for certain workloads.	Lack of standardized query language may require learning a specific API. May lack ACID compliance in favor of eventual consistency. Limited support for complex transactions.	Dynamic or evolving data schema. Rapid development and iteration. Handling large volumes of data with varying structures.	Document databases for content management. Real-time a pplications with variable schema. JSON data storage for web applications
Data warehouse	Optimized for complex analytics and reporting. Efficient data compression and indexing. Scalability for read-heavy analytical workloads.	May not be cost-effective for high-volume transactional workloads. May have higher latency for real-time queries. May require specialized skills for maintenance and optimization.	Analytical processing on large datasets. Aggregating and analyzing historical data.	Business intelligence and reporting tools. Running complex queries on terabytes of data.

Technology	Pros	Cons	When to Choose	Use Case
Lakehouse	Unified platform combining data lake and data warehouse features. Offers scalable storage and computing resources. It can efficiently scale horizontally to handle growing datasets and processing demands. Pay-as-you-go model, allowing organizations to manage costs more efficiently by paying for the resources they use. This is particularly beneficial for storing large amounts of raw data. Follow a schema-on-read approach, allowing for the storage of raw, untransformed data.	Complexity in managing both schema-on-read and schema-on-write. Depending on the implementation and cloud provider, costs associated with storage, processing, and managing data in a Lakehouse architecture may vary. Organizations need to carefully manage costs to ensure efficiency.	A balance between flexibility and transactional capabilities.	Real-time analytics with long-term storage. Any engineering, machine learning, and analytics use case
Streaming sinks	Enables real-time processing and analysis of streaming data. Scales horizontally to handle high volumes of incoming data. Integral to building event-driven architectures.	Implementing and managing streaming data sinks can be complex. The processing and persistence of streaming data introduces some latency. Depending on the chosen solution, infrastructure costs may be a consideration.	Continuous ingestion and processing of data in real-time	IoT Real-time analytical use cases Systems monitoring

Table 7.6 – Summary table of all the data sinks, as well as their pros, cons, and use cases

In *Table 7.8*, the **Use Case** column provides more context and practical examples of how each data sink technology can be effectively applied in real-world scenarios.

Moving from selecting the right data sink technology to choosing the appropriate file type is a crucial step in designing an effective data processing pipeline. Once you've determined where your data will be stored (data sink), you need to consider how it will be stored (file type). The choice of file type can impact data storage efficiency, query performance, data integrity, and interoperability with other systems.

Decoding file types for optimal usage

Choosing the right file type when selecting a data sink is crucial for optimizing data storage, processing, and retrieval. One of the file types that we haven't discussed so far but is very important since it's used as an underline format for other file formats is the Parquet file.

Parquet is a columnar storage file format designed for efficient data storage and processing in big data and analytics environments. It is an open standard file format that provides benefits such as high compression ratios, columnar storage, and support for complex data structures. Parquet is widely adopted in the Apache Hadoop ecosystem and is supported by various data processing frameworks.

Parquet stores data in a columnar format, which means values from the same column are stored together. This design is advantageous for analytics workloads where queries often involve selecting a subset of columns. Parquet also supports different compression algorithms, allowing users to choose the one that best suits their requirements. This contributes to reduced storage space and improved query performance. Parquet files can handle schema evolution as well, making it possible to add or remove columns without requiring a complete rewrite of the dataset. This feature is essential for scenarios where the data schema evolves. Due to its advantages, Parquet has become a widely adopted and standardized file format in the big data ecosystem, forming the basis for other optimized formats, such as Delta and Iceberg.

Having discussed Parquet files, we can now compare the common file types, along with their pros and cons, and provide guidance on when to choose each type for different data sinks:

File Type	**Pros**	**Cons**	**When to Choose**
JSON	Human-readable	Larger file size compared to binary formats Slower serialization/ deserialization	Semi-structured or human-readable data is required
BSON	Compact binary format Supports richer data types	May not be as human-readable as JSON, with limited adoption outside MongoDB	Efficiency in storage and transmission is crucial
Parquet	Columnar storage, which is efficient for analytics Compression and encoding lead to smaller file sizes	Not as human-readable as JSON You can't update tables – you need to rewrite them	Analytical processing, data warehousing
Avro	Compact binary serialization Schema-based and supports schema evolution Interoperable across different systems	Slightly less human-readable compared to JSON	Bandwidth-efficient streaming and diverse language support
Delta	ACID transactions for data consistency Efficient storage format for data lakes Schema evolution and time-travel queries	Larger size than Parquet	Real-time analytics with long-term storage
Hudi	Efficient incremental data processing ACID transactions for real-time data	Larger size than Parquet	Streaming data applications and change data capture
Iceberg	Schema evolution, ACID transactions Optimized storage formats such as Parquet	Larger size than Parquet	Time-travel queries and evolving schemas

File Type	Pros	Cons	When to Choose
Binary format	Compact and efficient storage Fast serialization and deserialization	Not human-readable Limited support for schema evolution	Efficiency is crucial in bandwidth usage and processing speed

Table 7.7 – A summary table of all the file formats, as well as their pros, cons, and use cases

In the next section, we're going to discuss **partitioning**, an important concept in the context of data storage, especially in distributed storage systems. While the concept of partitioning itself is more closely associated with data lakes, data warehouses, and distributed filesystems, its relevance extends to the broader discussion of data sinks.

Navigating partitioning

Data partitioning is a technique that's used to divide and organize large datasets into smaller, more manageable subsets called partitions. When writing data to sinks, such as databases or distributed storage systems, employing appropriate data partitioning strategies is crucial for optimizing query performance, data retrieval, and storage efficiency. Partitioning in data storage systems, including time-based, geographic, and hybrid partitioning, offers several benefits in terms of read operations, updates, and writes:

- When querying the data, partitioning allows the system to skip irrelevant data quickly. For example, in **time-based partitioning**, if you're interested in data for a specific date, the system can directly access the partition corresponding to that date, leading to faster query times. It ensures that only the necessary partitions are scanned, reducing the amount of data to process.
- Partitioning can simplify updates, especially when the updates are concentrated in specific partitions. For example, if you need to update data for a specific date or region, the system can isolate the affected partition, reducing the scope of the update operation.
- Partitioning can enhance the efficiency of write operations, particularly when appending data. New data can be written to the appropriate partition without affecting the existing data, leading to a more straightforward and faster write process.
- Partitioning supports parallel processing. Different partitions can be read or written concurrently, enabling better utilization of resources and faster overall processing times.
- Partitioning provides a logical organization of data. It simplifies data management tasks such as archiving old data, deleting obsolete records, or migrating specific partitions to different storage tiers based on access patterns.

- With partitioning, you can optimize storage based on usage patterns. For example, frequently accessed partitions can be stored in high-performance storage, while less frequently accessed partitions can be stored in lower-cost storage.
- Partitioning supports pruning, where the system can eliminate entire partitions from consideration during query execution. This pruning mechanism further accelerates query performance.

Let's have a closer look at the different partitioning strategies.

Horizontal versus vertical partitioning

When discussing partitioning strategies in the context of databases or distributed systems, we generally refer to two main types: **horizontal partitioning** and **vertical partitioning**. Each approach organizes data differently to improve performance, scalability, or manageability. Let's start with horizontal partitioning.

Horizontal partitioning, or sharding, involves dividing a table's rows into multiple partitions, each containing a subset of the data. This approach is commonly used to scale out databases by distributing data across multiple servers, where each shard maintains the same schema but holds different rows. For example, a user table in a large application could be sharded by user IDs, with IDs 1 to 10,000 in one partition and IDs 10,001 to 20,000 in another. This strategy enables the system to handle larger datasets than a single machine could manage, enhancing performance in large-scale applications.

Vertical partitioning, on the other hand, involves splitting a table's columns into different partitions, where each partition contains a subset of columns but includes all rows. This strategy is effective when different columns are accessed or updated at varying frequencies as it optimizes performance by minimizing the amount of data that's processed during a query. For example, in a user profiles table, basic information such as name and email could be stored in one partition, while a large binary data column, such as a profile picture, is stored in another. This allows queries targeting specific columns to access a smaller, more efficient dataset, thereby enhancing performance.

Both strategies can be used in combination to meet the specific needs of a database system, depending on the data structure and access patterns. The reality is that in the data field, **horizontal partitioning** is more commonly seen and widely adopted than vertical partitioning. This is particularly true in large-scale, distributed databases and applications that need to handle vast amounts of data, high traffic, or geographically dispersed users. In the next section, we will see some examples of horizontal partitioning.

Time-based partitioning

Time-based partitioning involves organizing data based on timestamps. Each partition represents a specific time interval, such as a day, hour, or minute. It allows for efficient retrieval of historical data and time-based aggregations. It also facilitates data retention and archiving policies.

In this example, you'll learn how to create time-based partitioning on your local laptop using Parquet files. You can find the full code here: `https://github.com/PacktPublishing/Python-Data-Cleaning-and-Preparation-Best-Practices/blob/main/chapter07/5.time_based_partitioning.py`. Follow these steps:

1. Import the required libraries:

    ```
    import os
    import pandas as pd
    import pyarrow as pa
    import pyarrow.parquet as pq
    from datetime import datetime
    ```

2. Define a sample dataset with two columns: `timestamp` and `value`. This dataset represents time series data with timestamps and corresponding values:

    ```
    data = {
        "timestamp": ["2022-01-01", "2022-01-01", "2022-01-02"],
        "value": [10, 15, 12]
    }
    ```

3. Create a pandas DataFrame:

    ```
    df = pd.DataFrame(data)
    ```

4. Convert the `timestamp` column into a `datetime` type. This ensures that the timestamps are treated as datetime objects for accurate time-based operations:

    ```
    df["timestamp"] = pd.to_datetime(df["timestamp"])
    ```

5. Update the path to store the data. Use an existing path:

    ```
    base_path = " path_to_write_data"
    ```

6. Iterate through the DataFrame, grouping rows by the `date` component of the `timestamp` column. Convert each group into a PyArrow table and write it to the corresponding partition path in Parquet format:

    ```
    for timestamp, group in df.groupby(df["timestamp"].dt.date):
    ```

7. Create the directory if it doesn't exist:

    ```
    os.makedirs(base_path, exist_ok=True)
    partition_path = os.path.join(base_path, str(timestamp))
    table = pa.Table.from_pandas(group)
    pq.write_table(table, partition_path)
    ```

After executing this script, you'll see two Parquet files being created in your base directory – one for each day of the week:

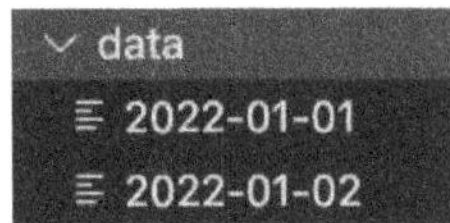

Figure 7.6 – Time-based partitioning output

Let's have a look at another common partitioning strategy, known as geographic partitioning.

Geographic partitioning

Geographic partitioning involves dividing data based on geographical attributes such as regions, countries, or cities. This strategy is valuable when you're dealing with geospatial data or location-based analytics. It enables fast and targeted retrieval of data related to specific geographic areas, thus supporting spatial queries and analysis.

Here's an example of how you can create geographic-based partitioning in your local laptop using Parquet files. You can find the full code here: `https://github.com/PacktPublishing/Python-Data-Cleaning-and-Preparation-Best-Practices/blob/main/chapter07/6.geo_partitioning.py`. Follow these steps:

1. Create a base directory for storing partitioned data:

```
base_directory = "/geo_data"
os.makedirs(base_directory, exist_ok=True)
```

2. Convert each group (region-specific data) into a PyArrow table. Then, write the tables to the corresponding paths:

```
geo_data = {"region": ["North", "South", "East"],
            "value": [10, 15, 12]}
geo_df = pd.DataFrame(geo_data)

for region, group in geo_df.groupby("region"):
```

3. Create a directory for each region within the base directory:

```
region_path = os.path.join(base_directory, region)
```

4. Convert the group into a PyArrow table and write it to the partition path:

```
table = pa.Table.from_pandas(group)
pq.write_table(table, region_path)
```

5. After executing this script, you will see three Parquet files being created in your base directory – one for each geographic location available in the data:

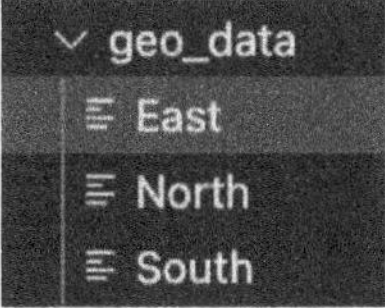

Figure 7.7 – Geographic-based partitioning output

Let's have a look at the last common partitioning strategy, known as hybrid partitioning.

> **Note**
>
> Geographic partitioning is a specialized form of category partitioning that organizes data based on geographical attributes or spatial criteria. **Category partitioning** is a fundamental strategy in data organization that involves grouping data based on specific categories or attributes, such as customer demographics, product types, or transactional characteristics.

Hybrid partitioning

Hybrid partitioning involves combining multiple partitioning strategies to optimize data organization for specific use cases. For instance, you might partition data first by time and then further partition each time interval by a key or geographic location. It offers flexibility for addressing complex querying patterns and diverse data access requirements.

Here's an example of how you can create hybrid partitioning on your local laptop using Parquet files. You can find the full code here: `https://github.com/PacktPublishing/Python-Data-Cleaning-and-Preparation-Best-Practices/blob/main/chapter07/7.hybrid_partitioning.py`. Follow these steps:

1. Create a base directory for storing partitioned data:

```
base_directory = "/hybrid_data"
```

2. Perform hybrid partitioning:

```
hybrid_data = {
    "timestamp": ["2022-01-01", "2022-01-01", "2022-01-02"],
    "region": ["North", "South", "East"],
    "value": [10, 15, 12]}
hybrid_df = pd.DataFrame(hybrid_data)

for (timestamp, region), group in hybrid_df.groupby(
    ["timestamp", "region"]):
```

3. Create a directory for each timestamp and region combination within the base directory:

```
timestamp_path = os.path.join(base_directory, str(timestamp))
os.makedirs(timestamp_path, exist_ok=True)
timestamp_region_path = os.path.join(
    base_directory, str(timestamp), str(region))
```

4. Convert the group into a PyArrow table and write it to the partition path:

```
table = pa.Table.from_pandas(group)
pq.write_table(table, timestamp_region_path)
```

5. After executing this script, you will see three Parquet files being created in your base directory – two locations for January 1, 2022, and one for January 2, 2022:

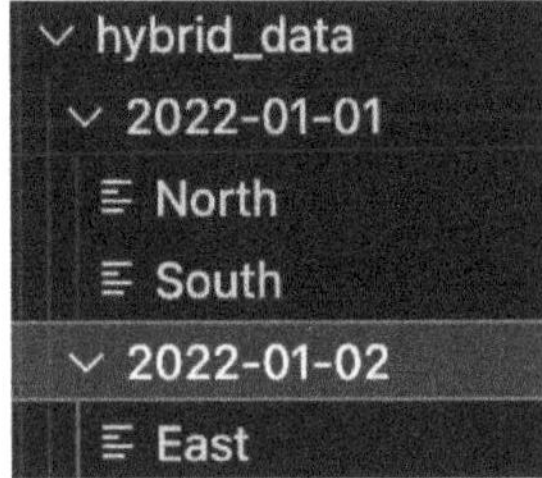

Figure 7.8 – Hybrid partitioning output

> **Remember**
>
> So far, we've explored various types of partitioning, such as time-based and geographic. However, remember you can use any column that makes sense in your data, your use case, and the query patterns for the table as partitioning column(s).

Now that we've discussed different partitioning strategies, it's time to talk about how to choose the column you will partition your data on.

Considerations for choosing partitioning strategies

Choosing the right partitioning strategy for your data involves considering various factors to optimize performance, query efficiency, and data management. Here are some key considerations for choosing partitioning strategies:

- **Query patterns**: Select partitioning strategies based on the types of queries your application or analytics platform will perform most frequently.

- **Data distribution**: Ensure partitions are distributed evenly to prevent data hotspots and resource contention.
- **Data size**: Consider the volume of data that will be stored in each partition. Smaller partitions can improve query performance, but too many small partitions might impact management overhead.
- **Query complexity**: Some queries might benefit from hybrid partitioning, especially if they involve multiple attributes.
- **Scalability**: Partitioning should allow for future scalability and accommodate data growth over time.

Data partitioning is a key architectural decision that can significantly impact the efficiency and performance of your data processing pipeline. By employing appropriate data partitioning strategies, you can ensure that your data is organized in a way that aligns with your querying patterns and maximizes the benefits of your chosen data sink technology.

In the next section, we're going to put everything we've learned in this chapter into practice by describing a real-world case scenario and going through all the logical steps for defining the best strategy associated with data sinks and file types.

Designing an online retail data platform

An online retailer wants to create an analytics platform to collect and analyze all the data generated by their e-commerce website. This platform aims to provide capabilities that allow for real-time data processing and analytics to improve customer experiences, optimize business operations, and drive strategic decision-making for the online retail business.

After long discussions with the team, we identified four main requirements to consider:

- **Handle large volumes of transaction data**: The platform needs to efficiently ingest and transform large volumes of transaction data. This needs to be done by accounting for scalability, high throughput, and cost-effectiveness.
- **Provide real-time insights**: Business analysts require immediate access to real-time insights derived from transaction data. The platform should support real-time data processing and analytics to enable timely decision-making.
- There's a need to combine batch and streaming data ingestion to handle both the real-time website data and the batch customer data, which is updated slowly.
- **Use AWS as the cloud provider**: The choice of the cloud provider (AWS) comes from the fact that the retailer is currently using other AWS services and wants to stick with the same provider.

Let's have a quick look at how we can solve these requirements:

1. Choose the right data sink technology:
   - **Thinking process**: A Lakehouse architecture is an ideal solution for the data platform requirements due to its ability to handle large volumes of data with scalability, high throughput, and cost-effectiveness. It leverages distributed storage and compute resources, allowing for efficient data ingestion and transformation. Additionally, the architecture supports real-time data processing and analytics, enabling business analysts to access immediate insights from transaction data for timely decision-making. By combining batch and streaming data ingestion, the Lakehouse seamlessly integrates real-time website data with batch-updated customer data.
   - **Choice**: A Lakehouse solution on AWS is selected for its scalability, cost-effectiveness, and seamless integration with other AWS services. AWS is compatible with a Lakehouse architecture.
2. Evaluate and choose the data file format:
   - **Data characteristics**: The customer data consists of structured transaction records, including customer IDs, product IDs, purchase amounts, timestamps, and geolocation. The streaming data includes customer IDs and other web metrics, such as what each customer is currently browsing on the website.
   - **Choice**: Delta file format is selected for its transactional capabilities and ACID compliance. It also supports batch and streaming workloads.
3. Implement data ingestion for batch and streaming data:
   - **Data ingestion**: ETL processes are designed to transform incoming transaction data into Delta files. Real-time transaction data is streamed from AWS Kinesis for immediate processing and stored as Delta files while batch data coming from different other systems is integrated.
   - **Partitioning logic**: Batch and streaming data are being processed and stored in Delta files. The streaming data is *partitioned by date when written out*. Next, transformations and data consolidation happen before it's stored as the final analytical tables.
4. Define a partitioning strategy for analytical tables:
   - **Query patterns**: Analysts often query data based on certain periods of time and some tables based on product categories.
   - **Choice**: As we learned in the *Considerations for choosing partitioning strategies* section, we need to take into account the way users are querying the table. To get the best read performance out of the queries, time-based and category-based partitioning must be implemented. Data is partitioned by date and further partitioned by product category in analytics tables that the users query often.

5. Monitor and optimize:

    - **Performance monitoring**: Regularly monitor query performance, streaming throughput, and resource utilization using AWS monitoring and logging services
    - **Optimization**: Continuously optimize both batch and streaming components based on observed performance and changing data patterns
    - **Schema evolution**: Ensure that the Delta schema accommodates streaming data changes and maintains compatibility with existing batch data

With this architecture, the online retail analytics platform gains the capability to process both batch and real-time data in an effective and cost-optimized way.

Summary

Throughout this chapter, we focused on the components of designing and optimizing data write operations. We discussed how to choose the right data sink technology, how file formats significantly impact storage efficiency and query performance, and why it matters to choose the right one for your use case. Finally, we discussed why data partitioning is crucial for optimizing query performance and resource utilization.

In the next chapter, we will start transforming the data that's been written on the data sink to better prepare it for downstream analytics by detecting and handling outliers and missing values.

Part 2: Downstream Data Cleaning – Consuming Structured Data

This part delves into the processes required for cleaning and preparing structured data for analysis, focusing on handling common data challenges that occur in more refined datasets. It provides practical techniques for managing missing values and outliers, ensuring data consistency through normalization and standardization, and effectively processing categorical features. Additionally, it introduces specialized methods for working with time series data, a common yet complex data type. By mastering these downstream cleaning and preparation techniques, readers will be well-equipped to turn structured data into actionable insights for advanced analytics.

This part has the following chapters:

- *Chapter 8, Detecting and Handling Missing Values and Outliers*
- *Chapter 9, Normalization and Standardization*
- *Chapter 10, Handling Categorical Features*
- *Chapter 11, Consuming Time Series Data*

8

Detecting and Handling Missing Values and Outliers

This chapter discusses the techniques of handling missing values and outliers, two critical challenges that can significantly impact the integrity and accuracy of our data products. We will explore a wide range of techniques to identify and manage these data irregularities, ranging from statistical methods to advanced machine learning models. Through practical examples and real-world datasets, we will present strategies to tackle these issues head-on, ensuring that our analyses are robust, reliable, and capable of generating meaningful insights.

The key points for the chapter are as follows:

- Detecting and handling missing data
- Detecting univariate and multivariate outliers
- Handling univariate and multivariate outliers

Technical requirements

You can find all the code for the chapter in the link that follows:

`https://github.com/PacktPublishing/Python-Data-Cleaning-and-Preparation-Best-Practices/tree/main/chapter08`

The different code files follow the names of the different parts of the chapters. Let's install the following library:

```
pip install spacy==3.7.5
```

Detecting missing data

Missing data is a common and inevitable issue in real-world datasets. It occurs when one or more values are absent in a particular observation or record. This data gap can greatly impact the validity and reliability of any analysis or model built with those data. As we say in the data world: *garbage in, garbage out*, meaning that if your data is not correct, then the models or analysis created with that data will not be correct either.

In the following parts, we will use a scenario to demonstrate how to detect missing data and how the different imputation methods work. The scenario is the following:

Imagine you are analyzing a dataset containing information about students, including their ages and test scores. However, due to various reasons, some ages and test scores are missing.

The code for this section can be found at `https://github.com/PacktPublishing/Python-Data-Cleaning-and-Preparation-Best-Practices/blob/main/chapter08/1.detect_missing_data.py`.

In this script, we create the data that we will use across the chapter. Let's start with the import statements:

```
import pandas as pd
```

Let's generate student data with missing ages and test scores. This dictionary data contains two keys, `Age` and `Test_Score`, each with a list of values. Some of these values are `None`, indicating missing data:

```
data = {
    'Age': [18, 20, None, 22, 21, 19, None, 23, 18, 24, 40, 41, 45,
None, 34, None, 25, 30, 32, 24, 35, 38, 76, 90],
    'Test_Score': [85, None, 90, 92, None, 88, 94, 91, None, 87, 75,
78, 80, None, 74, 20, 50, 68, None, 58, 48, 59, 10, 5]}
df = pd.DataFrame(data)
```

The first five rows of the dataset are as follows:

```
    Age  Test_Score
0  18.0        85.0
1  20.0         NaN
2   NaN        90.0
3  22.0        92.0
4  21.0         NaN
```

As we can see, there are NaN values in both columns of the dataset. To understand the extent of the missing values in the dataset, let's count how many we have across the whole DataFrame:

```
missing_values = df.isnull()
```

The `df.isnull()` method creates a `missing_values` DataFrame of the same shape as `df`, where each cell is `True` if the corresponding cell in `df` is `None` (missing value) and `False` otherwise, as shown:

```
     Age  Test_Score
0  False       False
1  False        True
2   True       False
3  False       False
4  False        True
```

In the previous DataFrame, any cell that contained a `NaN` value is now replaced with `True`. Having the data in that format helps us calculate how many `NaN` values we have:

```
null_rows_count = missing_values.any(axis=1).sum()

print("Count of Rows with at least one Missing Value:", null_rows_
count)
print(8/len(df))
```

The `missing_values.any(axis=1)` argument checks each row to see whether it contains any missing values, returning a Series of `True` or `False` for each row. Then the `.sum()` counts the number of `True` values in this Series, giving the number of rows with at least one missing value:

```
Count of Rows with at least one Missing Value: 8
% of rows with at least one missing value: 33%
```

Now we know how much data is missing from our dataset. The next goal of this exercise is to find the best imputation method to fill those.

Handling missing data

Addressing missing data involves making careful decisions to minimize its impact on analyses and models. The most common strategies include the following:

- Removing records with missing values
- Filling in missing values using various techniques such as mean, median, mode imputation, or more advanced methods such as regression-based imputation or k-nearest neighbors imputation
- Introducing binary indicator variables to flag missing data; this can inform models about the presence of missing values
- Leveraging subject matter expertise to understand the reasons for missing data and make informed decisions about how to handle it

Let's deep dive into each of these methods and observe in detail the results on the dataset presented in the previous part.

Deletion of missing data

One approach to handling missing data is to simply remove records (rows) that contain missing values. It is a quick and simple strategy, and is generally more suitable when the percentage of missing data is *low* and the missing data appears in random places.

Before we start deleting data, we need to understand our dataset a bit better. Continuing on the data created in the previous example, let's print the descriptive statistics first before we start deleting data points. The code for this part can be found at `https://github.com/PacktPublishing/Python-Data-Cleaning-and-Preparation-Best-Practices/blob/main/chapter08/2.delete_missing_data.py`.

> **Note**
>
> To keep the chapter to a nice number of pages, we have only presented the key code snippets. To see all the examples, please go to the repository.

To create the descriptive statistics, we can simply call the `.describe()` method in pandas:

```
print(df.describe())
```

The descriptive statistics are presented here:

```
             Age  Test_Score
count  20.000000   19.000000
mean   33.750000   65.894737
std    18.903843   27.989869
min    18.000000    5.000000
25%    21.750000   54.000000
50%    27.500000   75.000000
75%    38.500000   87.500000
max    90.000000   94.000000
```

Let's also create the distribution plots for each column of the dataset.

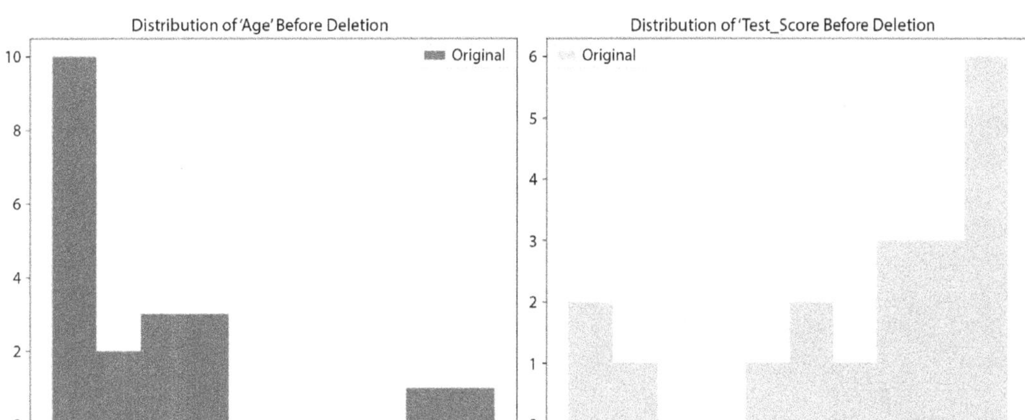

Figure 8.1 – Distribution of features before any alteration

With this analysis done, we can get some key insights into the dataset. For `Age`, with a count of 20, the average age is approximately 33.7 years, with a standard deviation of 18.9 years, showing moderate variability. Ages range from 18 to 90 years, with the middle 50% of ages falling between 21.75 and 38.5 years. For `Test_Score`, based on 19 values, the mean score is around 65.8, with a higher standard deviation of 27.9, indicating more variability in scores. Test scores range from 5 to 94, with the **Interquartile Range** (**IQR**) spanning from 54 to 87.5.

Now, let's have a look at how to delete the missing data. Let's pay attention to how the dataset changes:

```
df_no_missing = df.dropna()
```

Let's explore the distribution of the features after the data deletion:

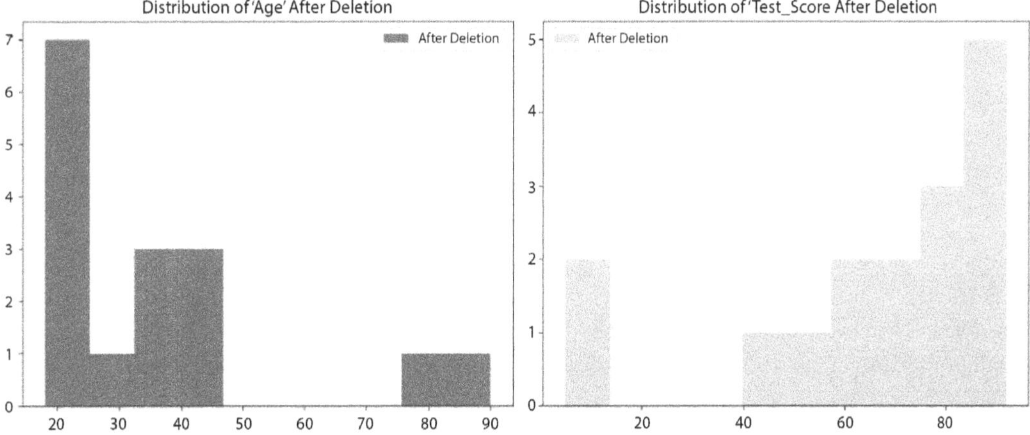

Figure 8.2 – Distribution of features after the data deletion

Let's also have a look at the summary statistics for the altered dataset:

```
print(df_no_missing.describe())
```

```
             Age  Test_Score
count  16.000000   16.000000
mean   36.500000   65.500000
std    20.109699   26.610775
min    18.000000    5.000000
25%    23.750000   56.000000
50%    32.000000   74.500000
75%    40.250000   85.500000
max    90.000000   92.000000
```

Having seen the descriptive statistics for both datasets, the observed changes are presented here:

- **Count change**: The count of observations has decreased from 20 to 16 for both, age and test scores, after deleting rows with missing values.
- **Mean change**: The mean age has increased from 33.75 to 36.50, while the mean test score has slightly decreased from 65.89 to 65.50. This change reflects the values present in the remaining dataset after the deletion.
- **Standard deviation change**: The standard deviation for age has increased from 18.90 to 20.11, indicating a greater spread in age, while the standard deviation for test scores has decreased from 27.99 to 26.61.
- **Minimum and maximum values**: The minimum age remains the same at 18, but the minimum test score remains at 5. The maximum values for both age and test scores have slightly changed, with the maximum test score decreasing from 94 to 92.
- **Percentile changes**: The percentile values (25%, 50%, 75%) have shifted due to the altered dataset:
  - The 25th percentile for age has increased from 21.75 to 23.75, and for test scores, from 54.00 to 56.00.
  - The median (50th percentile) for age has increased from 27.50 to 32.00, while for test scores, it decreased slightly from 75.00 to 74.50.
  - The 75th percentile for age has increased from 38.50 to 40.25, while for test scores, it decreased from 87.50 to 85.50.

The deletion of rows with missing values has led to a smaller dataset, and the remaining data now has *different statistical properties*. This method is suitable when the missing values are deemed to be a small proportion of the dataset and removing them does not significantly impact the data.

What is a small proportion though?

A common rule of thumb is that if less than 5% of the data is missing, it is often considered a small proportion, and deletion might not significantly impact the analysis. The significance of the change caused by deleting data points can be assessed by comparing the results of analyses with and without the missing data. If the results are consistent, the deletion might not be significant.

In these cases of substantial missing data, other imputation methods or advanced techniques may be more appropriate as we will explore in the next part.

Imputation of missing data

Imputation is often used when removing missing records would result in significant information loss. Imputation involves filling in missing values with estimated or calculated values. Common imputation methods include mean, median, and mode imputation, or using more advanced techniques.

Let's have a look at the different imputation methods for our scenario.

Mean imputation

Mean imputation fills missing values with *the mean of the observed values* in the variable. It is a very simple method, and it does not introduce bias when the values missing are completely random. However, this method is sensitive to outliers, and it may distort the distribution of the feature. You can find the code for this part in the repo at `https://github.com/PacktPublishing/Python-Data-Cleaning-and-Preparation-Best-Practices/blob/main/chapter08/3.mean_imputation.py`.

Let's see the code example for mean imputation. For this example, we will use the same dataset as explained before:

```
df_mean_imputed = df.copy()
df_mean_imputed['Age'].fillna(round(df['Age'].mean()), inplace=True)
```

The preceding line fills any missing values in the `Age` column with the mean of the `Age` column from the original `df` DataFrame. The `df['Age'].mean()` argument calculates the mean of the `Age` column, and rounds this mean to the nearest whole number. The `fillna()` method then replaces any `NaN` values in the `Age` column with this rounded mean. The `inplace=True` argument ensures that the changes are made directly in `df_mean_imputed` without creating a new DataFrame:

```
df_mean_imputed['Test_Score'].fillna(df['Test_Score'].mean(),
inplace=True)
```

Similarly, the preceding line fills any missing values in the `Test_Score` column of `df_mean_imputed` with the mean of the `Test_Score` column from the original `df` DataFrame.

Let's have a look at the dataset after the imputation:

```
print(df_mean_imputed)
     Age  Test_Score
0   18.0   85.000000
1   20.0   65.894737
2   34.0   90.000000
3   22.0   92.000000
4   21.0   65.894737
5   19.0   88.000000
6   34.0   94.000000
7   23.0   91.000000
8   18.0   65.894737
9   24.0   87.000000
10  40.0   75.000000
```

As we can see, the rounded mean has replaced all the `NaN` values for the age feature, whereas the absolute mean (abs mean) has replaced the `NaN` values for the `Test_Score` column. We rounded up the mean for the `Age` column to make sure it represents something meaningful.

The updated distributions are presented here:

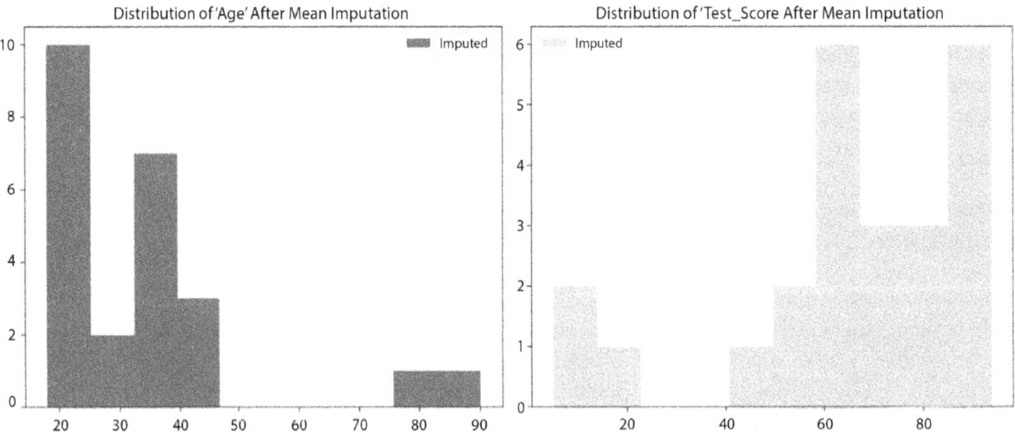

Figure 8.3 – Distribution of features after mean imputation

We can see from the graphs that the distributions have slightly changed for both of the variables. Let's have a look at the descriptive statistics of the imputed dataset:

```
print(df_mean_imputed.describe())

              Age  Test_Score
count   24.000000   24.000000
```

```
mean    33.791667    65.894737
std     17.181839    24.761286
min     18.000000     5.000000
25%     22.750000    58.750000
50%     33.000000    66.947368
75%     35.750000    85.500000
max     90.000000    94.000000
```

Having seen the descriptive statistics for both datasets, the observed changes are as follows:

- **Count increase**: The count for both `Age` and `Test_Score` increased from 20 to 24 after the imputation, indicating that missing values were successfully imputed.
- **Mean and median changes**: The mean age remained stable, increasing slightly from 33.75 to 33.79. The mean test score stayed the same at 65.89. The median age increased from 27.50 to 33.00, reflecting the changes in the distribution of ages. The median test score slightly decreased from 75.00 to 66.95.
- **Standard deviation changes**: The standard deviation for `Age` decreased from 18.90 to 17.18, indicating reduced variability in ages after imputation. The standard deviation for `Test_Score` also decreased from 27.99 to 24.76, reflecting less variability in test scores.
- **Quartile changes**: The **First Quartile** (**Q1**) (25%) for `Age` increased from 21.75 to 22.75, and Q1 for `Test_Score` increased from 54.00 to 58.75. The **Third Quartile** (**Q3**) (75%) for `Age` slightly decreased from 38.50 to 35.75, and Q3 for `Test_Score` remained relatively stable, decreasing slightly from 87.50 to 85.50.

The mean imputation maintained the overall mean values and increased the dataset size by filling in missing values. However, it has reduced the variability (as indicated by the decreased standard deviation for `Age` and `Test_Score`) and altered the distribution of the data (particularly in the quartiles). These changes are typical of mean imputation, as it tends ot underestimate variability and smooth out differences in the data, which can impact certain analyses that are sensitive to data distribution.

Now let's move on to the median imputation to see how this affects the dataset.

Median imputation

Median imputation fills the missing values with the median, the middle value of the dataset when it is ordered. Median imputation is robust in the presence of outliers and can be a good choice when the distribution is skewed. It can preserve the shape of the distribution unless dealing with complex distribution. The code for this part can be found at `https://github.com/PacktPublishing/Python-Data-Cleaning-and-Preparation-Best-Practices/blob/main/chapter08/4.median_imputation.py`.

Let's have a look at the code example for the median imputation:

```
df_median_imputed = df.copy()
```

This following line fills any missing values in the `Age` column of the `df_median_imputed` DataFrame with the median of the `Age` column from the original `df` DataFrame. The `df['Age'].median()` argument calculates the median (the middle value) of the `Age` column). The `fillna()` method then replaces any `NaN` values in the `Age` column with this median. The `inplace=True` argument ensures that the changes are made directly within `df_median_imputed`, without creating a new DataFrame:

```
df_median_imputed['Age'].fillna(df['Age'].median(), inplace=True)
```

Similarly, the following line fills any missing values in `Test_Score`:

```
df_median_imputed['Test_Score'].fillna(df['Test_Score'].median(),
inplace=True)
```

Let's have a look at the dataset after median imputation:

```
print(df_median_imputed)
     Age  Test_Score
0   18.0        85.0
1   20.0        75.0
2   27.5        90.0
3   22.0        92.0
4   21.0        75.0
5   19.0        88.0
6   27.5        94.0
7   23.0        91.0
8   18.0        75.0
9   24.0        87.0
10  40.0        75.0
```

As we can see, the median has replaced all the `NaN` values for the `Age` feature (27.5) and for the `Test_Score` column (75). The updated distributions are as follows.

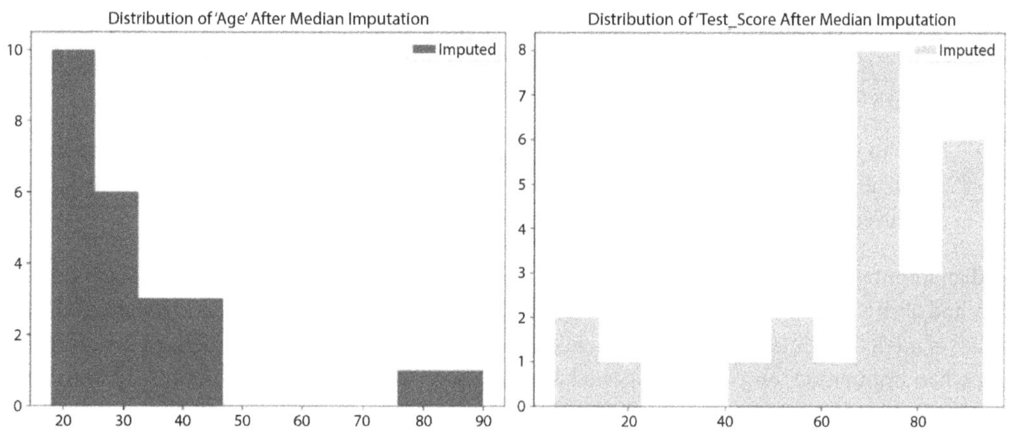

Figure 8.4 – Distribution of features after median imputation

We can see from the graphs that the distributions have slightly changed for both of the variables. Let's have a look at the descriptive statistics of the imputed dataset:

```
print(df_median_imputed.describe())
            Age  Test_Score
count  24.000000   24.000000
mean   32.708333   67.791667
std    17.345540   25.047744
min    18.000000    5.000000
25%    22.750000   58.750000
50%    27.500000   75.000000
75%    35.750000   85.500000
max    90.000000   94.000000
```

Having seen the descriptive statistics for both datasets, the observed changes are presented here:

- **Count increase**: The count for both `Age` and `Test_Score` increased from 20 (for age) and 19 (for test score) to 24 for both variables after median imputation, indicating that missing values were successfully imputed.
- **Mean changes**: The mean age decreased from 33.75 to 32.71 after imputation. The mean test score increased slightly from 65.89 to 67.79. These changes reflect the nature of the data remaining after the imputation.
- **Standard deviation changes**: The standard deviation for `Age` decreased from 18.90 to 17.35, indicating a reduction in variability for age. The standard deviation for `Test_Score` also decreased from 27.99 to 25.05, reflecting less variability in the test scores after imputation.

- **Quartiles changes**: Q1 (25%) for `Age` increased slightly from 21.75 to 22.75, and the Q1 for `Test_Score` increased from 54.00 to 58.75. Q3 (75%) for `Age` decreased from 38.50 to 35.75, and Q3 for `Test_Score` decreased slightly from 87.50 to 85.50.
- **Median (50%) changes**: The median for `Age` remained stable at 27.50, while the median for `Test_Score` also remained stable at 75.00 highlighting the central tendency of the data was preserved after imputation.

Median imputation has successfully filled in the missing values while preserving the median for both `Age` and `Test_Score`. It resulted in a slight change in the mean and reduced variability, which is typical of median imputation. The central tendency (median) was maintained, which is a key advantage of median imputation, especially in skewed distributions. But it also reduces the spread of the data which may be relevant for certain types of analysis.

In the next part, we will use what we learned so far on the imputation. We will also add an extra step, which involves marking where the missing values exist in the dataset for later reference.

Creating indicator variables

Indicator variable imputation, also known as flag or dummy variable imputation, involves creating a binary indicator variable that flags whether an observation has a missing value in a particular variable. This separate dummy variable takes the value of 1 for missing values and 0 for observed values. Indicator variable imputation can be useful when there is a pattern to the missing values, and you want to explicitly model and capture the missingness. Remember here that *we are adding a completely new variable, creating a higher dimensional dataset*. After we build the indicator variables, *whose role is to remind us which values were imputed and which were not*, we go ahead and impute the dataset with any method we want such as median or mean.

Let's see the code example for this imputation method. As always, you can see the full code in the repository:

`https://github.com/PacktPublishing/Python-Data-Cleaning-and-Preparation-Best-Practices/blob/main/chapter08/5.indicator_imputation.py`

Also, remember that we are using exactly the same DataFrame across the chapter, so we have skipped the DataFrame creation here:

```
df['Age_missing'] = df['Age'].isnull().astype(int)
df['Test_Score_missing'] = df['Test_Score'].isnull().astype(int)
```

The code creates new columns in the `df` DataFrame that indicate whether a value is missing (`NaN`) in the `Age` and `Test_Score` columns. `df['Age'].isnull()` checks each value in the `Age` column to see whether it is `NaN` (missing). It returns a Boolean series where `True` indicates a missing value, and `False` indicates a non-missing value. The `.astype(int)` method converts the Boolean series into an integer series where `True` becomes `1` (indicating a missing value) and `False` becomes `0` (indicating no missing value). The `df['Age_missing']` DataFrame stores this integer series in a new column named `Age_missing`.

Similarly, `df['Test_Score_missing']` is created to indicate missing values in the `Test_Score` column:

```
df_imputed['Age'].fillna(df_imputed['Age'].mean(), inplace=True)
df_imputed['Test_Score'].fillna(df_imputed['Test_Score'].mean(),
inplace=True)
```

This code fills in the missing values in the `Age` and `Test_Score` columns of the `df_imputed` DataFrame with the mean of the respective columns, as we learned in the previous part. Let's have a look at the dataset after the indicator variable imputation:

```
print(df_imputed)
```

```
      Age  Test_Score  Age_missing  Test_Score_missing
0   18.00   85.000000            0                   0
1   20.00   65.894737            0                   1
2   33.75   90.000000            1                   0
3   22.00   92.000000            0                   0
4   21.00   65.894737            0                   1
5   19.00   88.000000            0                   0
6   33.75   94.000000            1                   0
7   23.00   91.000000            0                   0
8   18.00   65.894737            0                   1
9   24.00   87.000000            0                   0
10  40.00   75.000000            0                   0
```

As you can see from the imputed dataset, we added two indicator variables (`Age_missing` and `Test_Score_missing`) that take the value of 1 if the corresponding variable is missing and 0 otherwise. So, we mainly flag *which values from the original rows were imputed.*

Let's see how the distribution of the indicator variables looks:

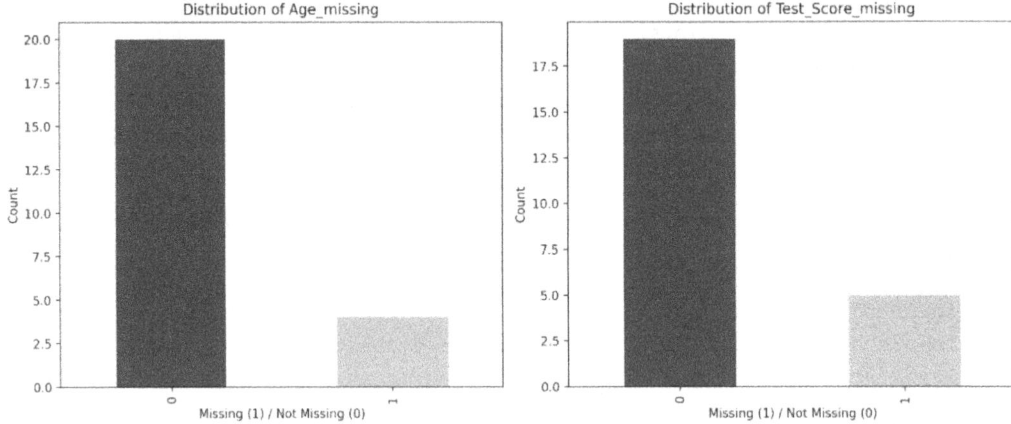

Figure 8.5 – Distribution of indicator variables

Now, let's explore the relationship between the indicator variables and other features in your dataset by building some box plots:

```
import seaborn as sns
import matplotlib.pyplot as plt

plt.figure(figsize=(12, 5))
plt.subplot(1, 2, 1)
sns.boxplot(x='Age_missing', y='Test_Score', data=df_imputed)
plt.title("Boxplot of Test_Score by Age_missing")
plt.subplot(1, 2, 2)
sns.boxplot(x='Test_Score_missing', y='Age', data=df_imputed)
plt.title("Boxplot of Age by Test_Score_missing")
plt.tight_layout()
plt.show()
```

The created box plots can be seen in *Figure 8.6*:

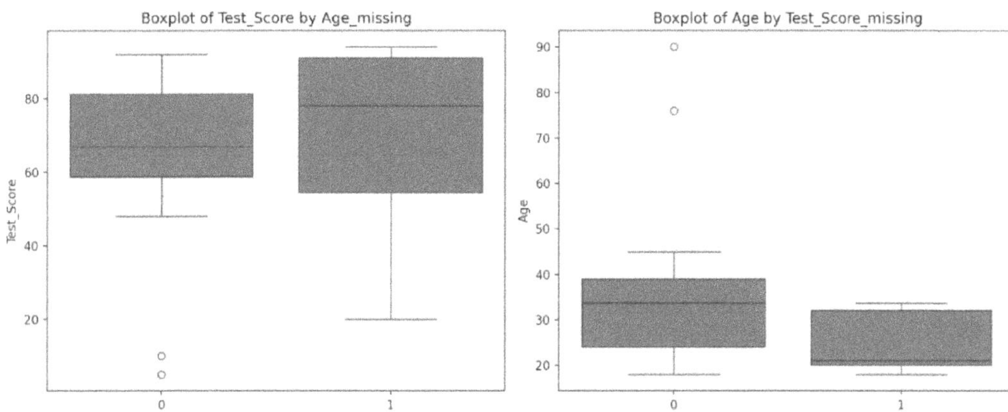

Figure 8.6 – Box plots comparing the relationship between indicator variables and the rest of the features

> **Reminder – how to read the box plots**
>
> **Box extent**: The box in a box plot represents the IQR, which contains the central 50% of the data. Values within the box are considered typical or within the normal range.
>
> **Whiskers**: Whiskers extend from the box and show the range of typical values. Outliers are often defined as values outside a certain multiple (e.g., 1.5 times) of the IQR.
>
> **Outliers**: Individual data points beyond the whiskers are considered potential outliers. Outliers are plotted as individual points or asterisks.
>
> **Suspected outliers**: Sometimes, points just beyond the whiskers may be plotted as suspected outliers, marked separately to indicate they are potential outliers but not extreme.

Back to our example, the box plot of `Test_Score` by `Age Missing` shows that when the age is missing in the data, the mean of `Test_Score` is around 80 and the distribution values are between 55 and 85. When `Age` is not missing, the mean is around 65, with most of the values being around 60 and 80, with some outliers around 20. Now, when the score is missing, the mean age of the students is around 20, whereas for the students with scores, the mean age is around 35.

> **Note**
>
> When building predictive models, include the indicator variables as additional features to capture the impact of missing values on the target variable. Evaluate the performance of models with and without the indicator variables to assess their contribution.

Comparison between imputation methods

The following table provides a guide for selecting the appropriate imputation method based on the data's characteristics and objectives of the task at hand.

Remember that there is no one-size-fits-all solution!

Imputation method	Use cases	Pros	Cons
Mean imputation	Normally distributed data Missing values are MCAR or MAR	Simple and easy to implement Preserves the mean of the distribution	Sensitive to outliers May distort the distribution if missingness is not random
Median imputation	Skewed or non-normally distributed data Presence of outliers	Robust to outliers Preserves the median of the distribution	Ignores potential relationships between variables May be less precise for non-skewed data
Indicator variable imputation	Systematic pattern in missing data	Captures missingness pattern	Increases dimensionality Assumes meaningful missingness pattern, which may not always be the case
Deletion of rows	MCAR or MAR missingness mechanism Presence of outliers	Preserves the existing data structure Can be effective when missingness is random	Reduces the sample size May lead to biased results if missingness is not completely random

Table 8.1 – Comparison between the various imputation methods

In the examples provided, we consistently applied the same imputation method to each column of the dataset. However, as demonstrated, our analysis and considerations were tailored to each column individually. This implies that we have the flexibility to tailor our imputation strategy to the specific characteristics and requirements of *each column*. As a practical exercise, take some time to experiment with different imputation methods for various columns in your dataset and observe how these choices impact your results.

To build on the foundation we've established with our imputation strategies, it's essential to recognize that data cleaning doesn't stop with handling missing values. Another critical aspect of data preprocessing is identifying and managing outliers. In the next part, we will dive deeper into detecting and handling outliers, ensuring our dataset is as accurate and reliable as possible.

Detecting and handling outliers

Outliers are data points that significantly deviate from the general pattern or trend shown by most of the data points in a dataset. They lie at an unusually distant location from the center of the data distribution and can have a significant impact on statistical analyses, visualizations, and model performance. Defining outliers involves recognizing data points that do not conform to the expected behavior of the data and understanding the context in which they occur.

Impact of outliers

Outliers, while often a small fraction of a dataset, wield a disproportionate influence that can disrupt the integrity of a dataset. Their presence has the potential to distort statistical summaries, mislead visualizations, and negatively impact the performance of models.

Let's go deeper into the various ways in which outliers distort the truth:

- **Distorted summary statistics**: Outliers can significantly skew summary statistics, giving a misleading impression of the central tendencies of the data:
    - **Mean and median**: The mean, a common measure of central tendency, can be greatly affected by outliers. An outlier with a value much higher or lower than the rest can pull the mean in its direction. On the other hand, the median is determined by the middle value of a sorted dataset. It effectively serves as the central point that divides the data into two equal halves, making it less susceptible to the influence of extreme values.
    - **Variance and standard deviation**: Outliers can inflate the variance and standard deviation, making the data appear more spread out than it actually is. This can misrepresent the variability of the majority of the data.
- **Misleading visualizations**: Outliers can distort the scale and shape of visualizations, leading to misinterpretation:
    - **Box plots**: Outliers can cause box plots to extend excessively, making the bulk of the data appear compressed. This can make the distribution seem less spread out than it actually is.
    - **Histograms**: Outliers might lead to the creation of bins that capture only a few extreme values, causing other bins to seem disproportionately small and the distribution shape to be distorted.

- **Influence on model performance**: Outliers can negatively affect the performance of predictive models:
    - **Regression**: Outliers can heavily influence the slope and intercept of the regression line, leading to models that are overly influenced by extreme values.
    - **Clustering**: Outliers can affect the centroids and boundaries of clusters, potentially leading to the creation of clusters that do not accurately represent the data distribution.

Outliers can be categorized based on dimensions as univariate versus multivariate. In the next section, we will use the example presented in the first part to see how we can handle the univariate outliers.

Identifying univariate outliers

Univariate outliers occur when an extreme value is observed in a single variable, regardless of the values of other variables. They are detected based on the distribution of a single variable and are often identified using visualizations or statistical methods such as Z-score or IQR.

In the next part, we will build one of the most common visualizations to identify outliers.

Classic visualizations for identifying outliers

Before going deeper into the statistical methods to identify outliers, there are a couple of easy visualizations we could build to spot them. The data example we have been using so far will still be used for this part; you can find the full code at `https://github.com/PacktPublishing/Python-Data-Cleaning-and-Preparation-Best-Practices/blob/main/chapter08/6.outliers_visualisation.py`.

Let's start with the first visualization, the box plot, where outliers are depicted as dots on the left or right of the whisker. The following code snippet creates the box plots for each variable:

```
plt.figure(figsize=(12, 5))
plt.subplot(1, 2, 1)
plt.title("Box Plot for 'Age'")
plt.boxplot(df['Age'].dropna(), vert=False)

plt.subplot(1, 2, 2)
plt.title("Box Plot for 'Test_Score'")
plt.boxplot(df['Test_Score'].dropna(), vert=False)
plt.tight_layout()
plt.show()
```

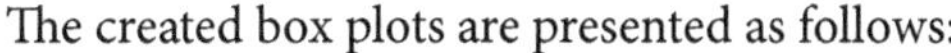

The created box plots are presented as follows:

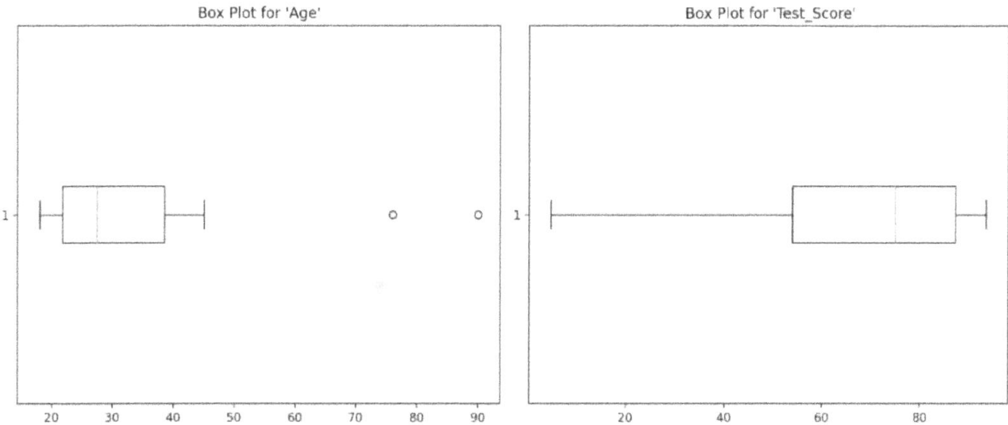

Figure 8.7 – Box plots to spot outliers

In our example, we can see that the `Age` feature has some clear outliers.

Another classic plot is the violin chart, as shown in *Figure 8.8*. Violin plots are a powerful visualization tool that combines aspects of box plots and kernel density plots. To create the violin plots, run the following code snippet:

```
plt.figure(figsize=(12, 5))
plt.subplot(1, 2, 1)
plt.title("Violin Plot for 'Age'")
plt.violinplot(df['Age'].dropna(), vert=False)

plt.subplot(1, 2, 2)
plt.title("Violin Plot for 'Test_Score'")
plt.violinplot(df['Test_Score'].dropna(), vert=False)
plt.tight_layout()
plt.show()
```

The created violin plots are as follows:

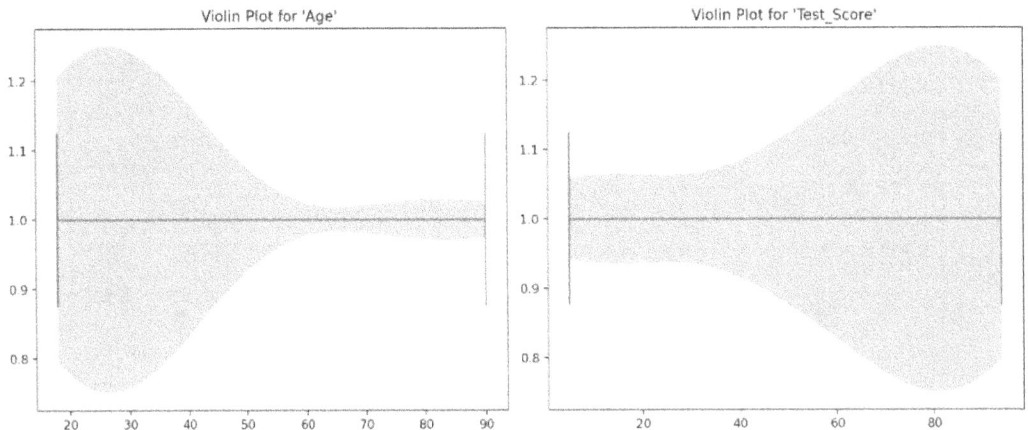

Figure 8.8 – Violin plots to spot outliers

> **Reminder – how to read the violin plots:**
>
> **Width of the violin**: The width of the violin represents the density of the data at different values. A wider section indicates a higher density of data points at that specific value, meaning a higher probability that members of the population will have the given value; the skinnier sections represent a lower probability.
>
> **Box-and-whisker elements**: Inside the violin, you may see a box-and-whisker plot, similar to what you would see in a traditional box plot. The box represents the IQR, and the median is usually displayed as a horizontal line inside the box. Whiskers extend from the box to indicate the range of the data.
>
> **Kernel Density Estimation (KDE)**: The entire shape of the violin is a mirrored representation of the KDE. The KDE provides a smooth representation of the data distribution, allowing you to see peaks and valleys in the data.
>
> **Outliers**: Outliers may be visible as points beyond the ends of the whiskers or outside the overall shape of the violin.

Now having seen these charts, we are starting to form some hypotheses about the existence of outliers specifically in the `Age` column. The next step is to use some statistical methods to validate these hypotheses starting with the Z-score method.

Z-score method

The Z-score method is a statistical technique used to identify univariate outliers in a dataset by measuring how far individual data points deviate from the mean in terms of standard deviations. The Z-score for a data point is calculated using the following formula:

$$Z = (X - Mean)/Standard\ Deviation$$

Here, X is the data point, *Mean* is the average of the dataset, and *Standard Deviation* quantifies the dispersion of the data.

Typically, a threshold Z-score is chosen to determine outliers. Commonly used thresholds are $Z > 3$ or $Z < -3$, indicating that data points deviating more than three standard deviations from the mean are considered outliers.

Let's go back to our code example to calculate the Z-score for the `Age` and `Test_Score` columns. We will continue with the example we started before. You can find the full code at `https://github.com/PacktPublishing/Python-Data-Cleaning-and-Preparation-Best-Practices/blob/main/chapter08/7.identify_univariate_outliers.py`.

Let's calculate the Z-score:

```
z_scores_age = np.abs(stats.zscore(df['Age'].dropna()))
```

The `stats.zscore(df['Age'].dropna())` function calculates the Z-scores for the `Age` column. A Z-score represents how many standard deviations a data point is from the mean. The `dropna()` function is used to exclude `NaN` values before calculating the Z-scores:

```
z_scores_test_score = np.abs(stats.zscore(df['Test_Score'].dropna()))
```

The `np.abs()` function takes the absolute value of the Z-scores. This is done because Z-scores can be negative (indicating a value below the mean) or positive (indicating a value above the mean). By using the absolute value, we're only concerned with the magnitude of deviation from the mean, regardless of direction.

```
z_threshold = 3
outliers_age = np.where(z_scores_age > z_threshold)[0]
outliers_test_score = np.where(z_scores_test_score > z_threshold)[0]
```

`np.where(z_scores_age > z_threshold)[0]` identifies the indices of the data points in the `Age` column that have Z-scores greater than the threshold of `3`. The `[0]` at the end is used to extract the indices as an array. The `outliers_age` and `outliers_test_score` variables store the indices of the outlier data points in the `Age` and `Test_Score` columns, respectively.

If we plot the Z-scores for each observation and feature of the data, we can start spotting some outliers already.

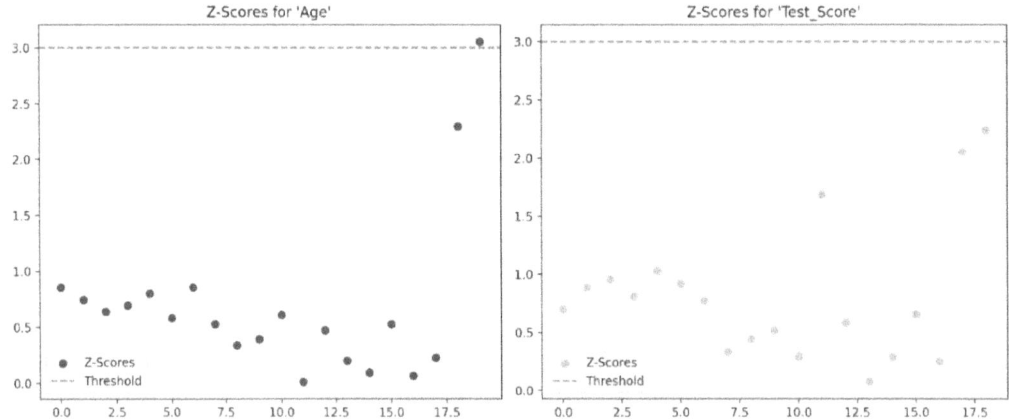

Figure 8.9 – Outlier detection with Z-score

In these scatter plots of Z-scores, each point represents the Z-score of an individual data point. The red dashed line indicates the chosen Z-score threshold (in this case, `3`). Outliers are identified as points above this threshold. As we can see, in `Age`, there is an outlier clearly captured.

How to choose the right threshold for the z score?

A Z-score tells you how many standard deviations a data point is from the mean. In a normal distribution, the following is true:

- Approximately 68% of data falls within *one standard deviation* of the mean.
- Approximately 95% of data falls within *two standard deviations*
- Approximately 99.7% of data falls within *three standard deviations*

This means that a Z-score threshold of `3` is often used because it captures values that are *extremely far from the mean*, identifying the most extreme outliers. In a perfectly normal distribution, only 0.3% of data points will have a Z-score greater than 3 or less than -3. This makes it a reasonable threshold for detecting outliers that are unlikely to be part of the normal data distribution.

Now, apart from the Z-score, another common method is the IQR, which we will discuss in the following part.

IQR method

The IQR is a measure of statistical dispersion, representing the range between Q1 and Q3 in a dataset. The IQR is a robust measure of *spread* because it is less sensitive to outliers. At this point, it is clear that the IQR is based on quartiles. Quartiles divide the dataset into segments, and since Q1 and Q3 are less sensitive to extreme values, the IQR is not heavily influenced by outliers. On the other hand, standard deviation is influenced by each data point's deviation from the mean. Outliers with large deviations can disproportionately impact the standard deviation.

> **Reminder – how to calculate the IQR**
>
> **Calculate Q1 (25th percentile)**: Identify the value below which 25% of the data falls.
>
> **Calculate Q3 (75th percentile)**: Identify the value below which 75% of the data falls.
>
> **Calculate IQR**: IQR = Q3 - Q1.

To identify potential outliers using IQR, do the following:

1. Calculate the lower and upper bounds as follows: Lower Bound = Q1 - 1.5 * IQR, Upper Bound = Q3 + 1.5 * IQR.
2. Any data point outside the lower and upper bounds is considered a potential outlier.

It's important to note that the choice of the multiplier (in this case, `1.5`) is somewhat arbitrary but has been widely adopted in practice. Adjusting this multiplier can make the method more or less sensitive to potential outliers. For example, using a larger multiplier would result in broader boundaries, potentially identifying more data points as potential outliers, while a smaller multiplier would make the method less sensitive.

We'll be using the same script as before, which can be found at `https://github.com/PacktPublishing/Python-Data-Cleaning-and-Preparation-Best-Practices/blob/main/chapter08/7.identify_univariate_outliers.py`. Let's have a look at how to calculate the IQR and identify the outliers:

```
def identify_outliers(column):
    Q1 = df[column].quantile(0.25)
    Q3 = df[column].quantile(0.75)
    IQR = Q3 - Q1
    lower_bound = Q1 - 1.5 * IQR
    upper_bound = Q3 + 1.5 * IQR
    outliers = df[(df[column] < lower_bound) | (df[column] > upper_
bound)]
    return outliers
```

This code defines a function to identify outliers in any column of a DataFrame using the IQR method. It calculates the IQR, sets upper and lower bounds for normal data, and then filters out the rows where the values in the column fall outside these bounds.

Then, let's identify and print outliers for `Age`:

```
age_outliers = identify_outliers('Age')
print("Outliers in 'Age':")
print(age_outliers)
```

Identify and print outliers for `Test_Score`:

```
test_score_outliers = identify_outliers('Test_Score')
print("\nOutliers in 'Test_Score':")
print(test_score_outliers)
```

After running this code, we can see in the print statement the identified outliers/rows based on the `Age` column:

```
  Age  Test_Score
 76.0        10.0
 90.0         5.0
```

As discussed, the simplicity of the IQR, as well as its robustness to outliers, contributes to its popularity in various analytical scenarios. However, it comes with certain drawbacks. One limitation is the loss of information, as IQR only considers the central 50% of the dataset, disregarding the entire range. Additionally, IQR's sensitivity to sample size, especially in smaller datasets, can affect its accuracy in reflecting the true spread of the data.

Finally, we will quickly discuss leveraging domain knowledge to identify outliers.

Domain knowledge

To better understand the use of domain knowledge in outlier detection, let's use the test scores example. Suppose the dataset represents student test scores, and based on educational standards, test scores are expected to fall within a range of 0 to 100. Any score outside this range could be considered an outlier. By leveraging domain knowledge in education, we can set these boundaries to identify potential outliers. For instance, if a test score is recorded as 120, it would likely be flagged as an outlier because it exceeds the maximum possible score of 100. Similarly, negative scores or scores below 0 would be considered outliers. Integrating domain knowledge in this manner allows us to establish meaningful thresholds for outlier detection, ensuring that the analysis aligns with the expected norms within the educational context.

Handling univariate outliers

Handling univariate outliers refers to the process of identifying, assessing, and managing data points in individual variables that deviate significantly from the typical patterns or distribution of the dataset. The goal is to mitigate the impact of these extreme values on data products.

There are several approaches to handling univariate outliers. We will start with deletions, always working on the example presented at the beginning of the chapter.

Deletion of outliers

Deleting outliers refers to the process of removing data points in a dataset that are considered unusually extreme or deviant from the overall pattern of the data. The deletion of outliers comes with trade-offs. On the one hand, it is the simplest way to deal with extreme values. On the other hand, it leads to a reduction in the sample size and potential loss of valuable information. Additionally, if outliers are not genuine errors but rather reflect legitimate variability in the data, their removal can introduce bias.

Back to our example, after having imputed the missing data with the mean and calculated the IQRs, we dropped the outliers that passed the outlier threshold. Let's see the code that performs these steps; you can also find it in the repository at `https://github.com/PacktPublishing/Python-Data-Cleaning-and-Preparation-Best-Practices/blob/main/chapter08/8.handle_univariate_outliers_deletions.py`.

Let's calculate the IQR and use it to set lower and upper bounds for what is considered a normal range of data:

```
(IQR) Q1 = df['Test_Score'].quantile(0.25)
Q3 = df['Test_Score'].quantile(0.75)
IQR = Q3 - Q1
outlier_threshold = 1.5
```

Let's define the lower and upper outlier bounds. Any value outside this range is flagged as an outlier:

```
lower_bound = Q1 - outlier_threshold * IQR
upper_bound = Q3 + outlier_threshold * IQR
```

The last line filters the DataFrame (`df`) to include only rows where the `Test_Score` values fall within the calculated lower and upper bounds:

```
df_no_outliers = df[(df['Test_Score'] >= lower_bound) & (df['Test_
Score'] <= upper_bound)].copy()
```

In the following charts, we can see the updated distribution charts, after the removal of the outliers.

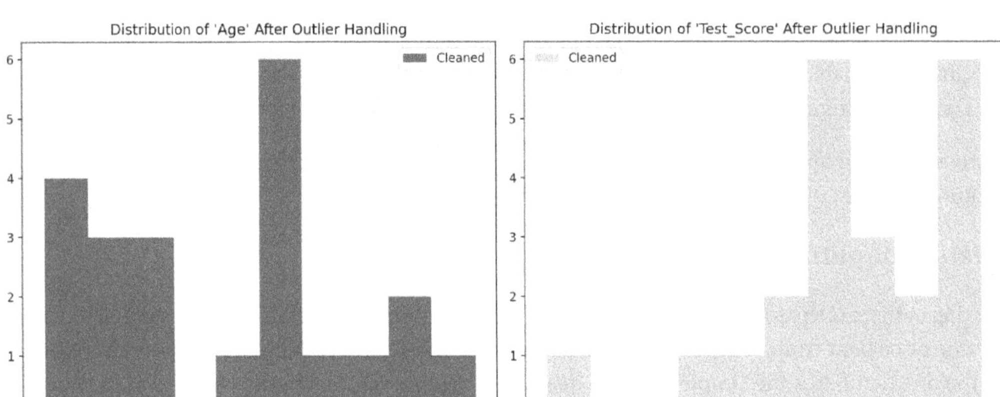

Figure 8.10 – Distribution charts after deletion of outliers

Let's have a look at the descriptive statistics after the outlier deletion:

```
             Age  Test_Score
count  22.000000   22.000000
mean   29.272727   71.203349
std     8.163839   17.794339
min    18.000000   20.000000
25%    22.250000   65.894737
50%    31.000000   71.000000
75%    33.937500   86.500000
max    45.000000   94.000000
```

The changes observed after the deletion of outliers are described as follows:

- **Mean age change**: The mean age, after deleting outliers, decreased slightly from 33.75 to approximately 29.27. This reduction suggests that the removed outliers were older individuals.
- **Standard deviation change for age**: The standard deviation for age decreased from 17.18 to 8.16, indicating that the spread of ages became slightly narrower after outlier removal, which were likely contributing to greater variability in the original dataset.
- **Minimum and maximum age values**: The minimum age remained the same, at 18, while the maximum age decreased from 90 to 45, indicating that older individuals (potential outliers) were removed during outlier handling.
- **Mean test score change**: The mean test score increased slightly from 65.89 to 71.20 after removing outliers, suggesting that the deleted outliers were lower test scores that were pulling down the original mean.

- **Standard deviation change for test scores**: The standard deviation decreased from 24.76 to 17.79, indicating a narrower spread of test scores.
- **Minimum and maximum test scores**: The minimum test score increased from 5.00 to 20.00, while the maximum test scores remained the same at 94.00. This indicates that extremely low scores were removed as part of the outlier handling.

The removal of outliers led to a decrease in both – the mean age and standard deviation, as well as a slight increase in the mean test score. While removing outliers can improve data quality, especially when the outliers are due to data entry errors or measurement inaccuracies, it also reduces the dataset's variability. If the outliers represent true variability in the population, removing them could distort the overall picture of the data. Therefore, careful consideration should be given to whether the outliers represent genuine data points or errors.

> **Note**
>
> Some statistical models assume normality and can be sensitive to outliers. Removing outliers may help meet the assumptions of certain models. So before deleting, you need to better understand the problem you are solving and the techniques to be used.

There are other ways to deal with outliers in case you don't want to drop them completely from the data. In the next part, we will discuss the trimming and winsorizing of outliers.

Trimming

Trimming involves removing a certain percentage of data from both ends of a distribution and then calculating the mean. For the trimming, we need to define a trimming fraction, which represents the proportion of data to be trimmed from *both tails of the distribution* when calculating the trimmed mean. It is used to exclude a certain percentage of extreme values (outliers) from the calculation of the mean. The trimming fraction is a value between 0 and 0.5, where the following is true:

- 0 means no trimming (include all data points)
- 0.1 means trim 10% of data from each tail
- 0.2 means trim 20% of data from each tail
- 0.5 means trim 50% of data from each tail (exclude the most extreme values)

In our given scenario, our analysis indicates that the `Age` column exhibits the most significant outliers. In response, we have decided to trim our dataset by excluding the top and bottom percentiles specific to the `Age` column. The following example code demonstrates this trimming process. We are still working on the same dataset so we will skip the creation of the DataFrame here. However, you can see the whole code at the following link: `https://github.com/PacktPublishing/Python-Data-Cleaning-and-Preparation-Best-Practices/blob/main/chapter08/9.trimming.py`.

Let's have a look at the following code snippet that creates a new DataFrame (`df_trimmed`), which includes only the rows where the `Age` value is between the 10th and 90th percentiles. This effectively drops the lowest 10% and highest 10% of values in the `Age` column:

```
df_trimmed = df[(df['Age'] >= df['Age'].quantile(0.1)) & (df['Age'] <=
df['Age'].quantile(0.9))]
```

Let's now calculate the trimmed mean for each column:

```
df_trimmed_mean = df_trimmed.mean()
```

After trimming the data, the last line calculates the mean for each column in the `df_trimmed` DataFrame. The mean calculated after trimming the data is known as the *trimmed mean*. It represents the average value of the central 80% of the data, excluding the most extreme 20% (10% from each side).

> **Note**
>
> Keep in mind that the trimming fraction is a way to balance the robustness of the trimmed mean against the amount of data excluded. You may need to experiment with different fractions to find a suitable balance for your data.

Let's have a look at the updated distribution after the trimming:

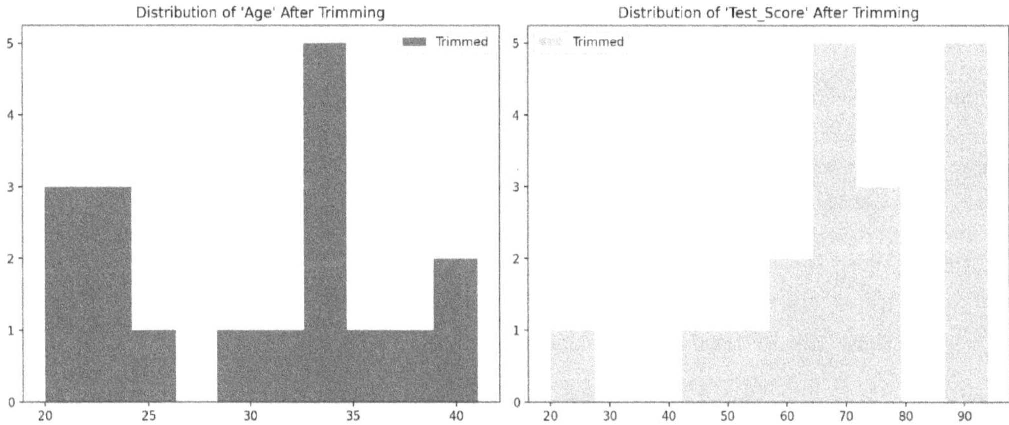

Figure 8.11 – Distribution charts after trimming of outliers at 10% threshold

Let's also have a look at the updated statistics of the data:

```
             Age  Test_Score
count  18.000000   18.000000
mean   30.222222   69.309942
std     6.757833   18.797436
```

```
min     20.000000   20.000000
25%     24.000000   60.723684
50%     32.875000   66.947368
75%     33.937500   84.750000
max     41.000000   94.000000
```

In the original dataset, the `Age` column displayed a mean of 33.75 with a standard deviation of 17.18, while the trimmed data exhibited a higher mean of 30.22 and a much lower standard deviation of 6.76. The minimum age value increased from 18 to 20 in the trimmed data, indicating the removal of lower outliers. The maximum age value decreased from 90 to 41, suggesting the exclusion of higher outliers.

For the `Test_Score` column, the mean in the original dataset was 65.89, and the standard deviation was 24.76. In the trimmed data, the mean increased to 69.31, and the standard deviation decreased to 18.80 indicating a narrower spread of test scores. The minimum test score increased from 5 to 20, indicating the removal of lower outliers, while the maximum test score stayed at 94.

Overall, the deletion of outliers led to changes in the central tendency (mean) and the spread (standard deviation) of the data for both `Age` and `Test_Score`. This indicates that the trimmed dataset has become more concentrated around the middle values, with extreme values removed.

Remember!

While trimming can help in reducing the influence of extreme values, it also involves discarding a portion of the data. This may result in information loss, and the trimmed variable may not fully represent the original dataset.

In the next section, we will present a slightly different way to deal with the outlier called **winsorizing**.

Winsorizing

Instead of removing extreme values outright as with trimming, winsorizing involves *replacing them with less extreme values*. The extreme values are replaced with values closer to the center of the distribution, often at a specified percentile. Winsorizing can be useful when you want to *retain the size of your dataset* and helps preserve the overall shape of the data distribution.

Let's go back to our example use case and have a look at the code. You can find the full code at `https://github.com/PacktPublishing/Python-Data-Cleaning-and-Preparation-Best-Practices/blob/main/chapter08/10.winsorizing.py`:

```
winsorizing_fraction = 0.1
```

`winsorizing_fraction` is set to `0.1`, representing the proportion of data to be adjusted at each end of the distribution. It is specified as a percentage, and its value typically ranges between 0 and 50%. The process of coming up with the winsorizing fraction involves considering the desired amount of influence you want to reduce from extreme values. A common choice is to winsorize a certain percentage from both tails, such as 5% or 10%.

Another thing to know here is that the winsorizing process is performed at each column *separately and independently of the others*. Remember: we are dealing with outliers in a *univariate* way here:

```
df_winsorized = df.apply(lambda x: mstats.winsorize(x,
limits=[winsorizing_fraction, winsorizing_fraction]))
```

The `limits=[winsorizing_fraction, winsorizing_fraction]` argument specifies the proportion of data to be winsorized from each end of the distribution. Here, 10% from the lower end and 10% from the upper end are adjusted. Extreme values (the lowest 10% and highest 10%) are replaced with the nearest values within the specified limits, thereby reducing their influence on statistical measures.

Here, the updated distributions after the winsorizing are presented:

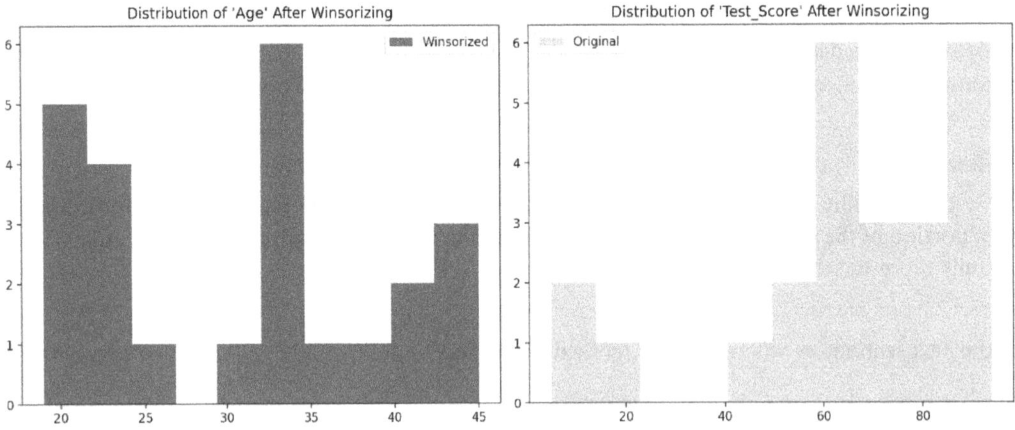

Figure 8.12 – Distribution charts after winsorizing of outliers

Let's also have a look at the updated statistics of the data:

```
            Age  Test_Score  Age_Winsorized
count  24.000000   24.000000       24.000000
mean   33.750000   65.894737       30.666667
std    17.181575   24.761286        8.857773
min    18.000000    5.000000       19.000000
25%    22.750000   58.750000       22.750000
50%    32.875000   66.947368       32.875000
75%    35.750000   85.500000       35.750000
max    90.000000   94.000000       45.000000
```

The mean for the `Age` column decreased from 33.75 to 30.67 after winsorizing, indicating a shift toward lower values as extreme high values were adjusted. The standard deviation also decreased significantly from 17.18 to 8.86, suggesting reduced variability in the dataset. The minimum value increased slightly from 18 to 19, and the maximum value decreased from 90 to 45, reflecting the capping of extreme values.

As for `Test_Score`, the mean remained the same at 65.89 after winsorizing. The standard deviation stayed constant at 24.76, indicating that variability in test scores was not affected by the winsorizing process. The maximum value stayed the same at 94, showing no changes to the upper extreme values.

Overall, winsorizing the `Age` column resulted in a more concentrated distribution of values, as evidenced by the decreased standard deviation. Winsorizing successfully reduced the impact of extreme values in the `Age` column, making the data more focused around the middle range. For `Test_Score`, winsorizing did not affect the distribution, likely because the extreme values were already within the accepted range.

Next, we will explore how we can apply mathematical transformations to the data to minimize the effect of the outliers.

Data transformation

Applying mathematical transformations such as logarithm or square root is a common technique to handle skewed data or stabilize variance.

> **Reminder**
>
> Skewness is a measure of the asymmetry in a distribution. A positive skewness indicates a distribution with a tail on the right side, while a negative skewness indicates a tail on the left side.

When data is right-skewed (positive skewness), meaning that most of the data points are concentrated on the left side with a few larger values on the right side, applying a logarithmic transformation compresses larger values, making the distribution more symmetric and closer to a normal distribution.

Similar to logarithmic transformation, square root transformation is used to mitigate the impact of larger values and make the distribution more symmetric. It is particularly effective when the right tail of the distribution contains extreme values.

Another thing to note is that when the variance of the data increases with the mean (heteroscedasticity), logarithmic and square root transformation can compress the larger values, reducing the impact of extreme values and stabilizing the variance.

Let's go back to our example and perform a log transformation on both columns of our dataset. As always, you can find the full code at `https://github.com/PacktPublishing/Python-Data-Cleaning-and-Preparation-Best-Practices/blob/main/chapter08/11.data_transformation.py`.

Let's apply a logarithmic transformation to `Age` and `Test_Score`:

```
df_log_transformed = df.copy()
df_log_transformed['Age'] = np.log1p(df_log_transformed['Age'])
df_log_transformed['Test_Score'] = np.log1p(df_log_transformed['Test_
Score'])
```

`np.log1p` is a NumPy function that computes the natural logarithm of $1 + x$ for each value in the `Age` and `Test_Score` columns. The `log1p` function is used instead of the simple logarithm (`np.log`) to handle zero and negative values in a dataset without errors. It's particularly useful when dealing with data that includes zero values or very small numbers. The transformation reduces skewness and can make the distribution more normal, which is useful for various statistical techniques that assume normally distributed data.

> **More transformations implemented**
>
> In the code, you'll find both logarithmic and root transformations applied to the data. Take some time to explore and understand the differences between these two methods. Evaluate which transformation better suits your data by considering how each affects the distribution and variance of your dataset.

The updated distributions are presented in the following plot after log transforming the `Age` column and applying a square root transformation to the `Test_Score` column:

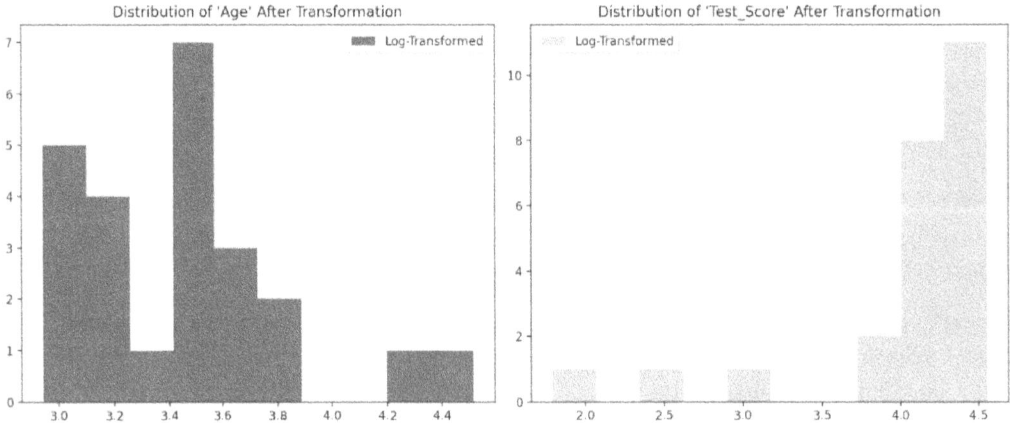

Figure 8.13 – Distribution charts after log and square route transformation

Let's also have a look at the updated statistics of the data:

```
             Age  Test_Score
count  24.000000   24.000000
```

```
mean     3.462073    4.059624
std      0.398871    0.687214
min      2.944439    1.791759
25%      3.167414    4.090143
50%      3.522344    4.218613
75%      3.603530    4.460095
max      4.510860    4.553877
```

The descriptive statistics illustrate the impact of applying logarithmic transformation to the `Age` variable and square root transformation to the `Test_Score` variable. Before the transformations, the original dataset displayed a right-skewed distribution for `Age` with a mean of 33.75 and a wide standard deviation of 17.18. `Test_Score` had a mean of 65.89, ranging from 5 to 94, with a high standard deviation of 24.76, indicating a large spread in the test scores.

After applying the transformations, the distributions of both variables were visibly altered:

- The logarithmic transformation on `Age` reduced the spread of values, bringing the standard deviation down to 0.40 as compared to the original 17.18. The transformed values now range from 2.94 to 4.51, showing a compression of extreme values.
- For `Test_Score`, the logarithmic transformation resulted in a much more evenly distributed set of values, with the standard deviation decreasing from 24.76 to 0.69. The values became more compact and symmetric, ranging from 1.79 to 4.55.

The transformations had a clear leveling effect on both variables, reducing skewness and variability. This is evident in the reduction of standard deviations and narrower ranges, making the data more symmetric and closer to a normal distribution.

However, it's important to note that transformations, especially logarithmic ones, compress the scale of values and may affect interpretability. While they can help meet the assumptions of statistical methods by reducing skewness and heteroscedasticity, the transformed data may be less intuitive to understand compared to the original scale. Despite this, such transformations are often useful when preparing data for regression models or other analyses that assume normally distributed data.

> **Note**
>
> Keep in mind that log transformation is not suitable for data that contains zero or negative values, as the logarithm is undefined for such values.

To conclude this section of the chapter, we have compiled a summary table with the various methods discussed for handling outliers. This table highlights the optimal scenarios to employ each technique and provides an overview of their respective pros and cons.

Technique	When to use it	Pros	Cons
Trimming	Mild outliers, preserving overall data structure	Retains majority of the dataset, maintains data integrity	Reduces sample size, may impact representativeness, arbitrary choice of trimming percentage
Winsorizing	Moderate outliers, preserving overall data	Preserves data distribution, mitigates the impact of extreme values	Alters data values; may distort distribution; requires specifying trimming limits
Deleting Data	Severe outliers	Removes the influence of extreme values, simplifies analysis	Reduces sample size, potential loss of information; may bias results toward a central tendency
Transformation	Skewed or non-normal distributions	Stabilizes variance, makes the data more symmetric and amenable to traditional statistical techniques	Interpretation challenges, results may be less intuitive, choice of transformation method is subjective

Table 8.2 – Summary of the univariate methods to deal with outliers

After exploring various techniques for addressing univariate outliers, ranging from simpler to more complex methods, the upcoming section will deep dive into the different statistical measures that are generally preferable when working with data containing outliers.

Robust statistics

Using robust statistical measures such as median and **Median Absolute Deviation** (**MAD**) instead of mean and standard deviation can reduce the influence of outliers.

When dealing with datasets that contain outliers or skewed distributions, choosing robust statistical measures becomes crucial for obtaining accurate and representative summaries of the data. The use of robust measures, such as the median and MAD, proves advantageous in scenarios where the presence of extreme values could impact traditional measures such as the mean and standard deviation. The median, being the middle value when data is ordered, is less sensitive to outliers, providing a more reliable measure of central tendency. Additionally, MAD, which assesses the spread of data while being robust to outliers, further ensures a more accurate representation of the dataset's variability.

> **MAD**
>
> MAD is a measure of statistical dispersion that quantifies the dispersion or spread of a dataset. It is calculated as the median of the absolute differences between each data point and the median of the dataset.

This table summarizes the key considerations, pros, and cons of using median and MAD versus mean and standard deviation:

Criteria	**Median and MAD**	**Mean and standard deviation**
When to use	Presence of outliers	Normal or symmetric distributions
	Skewed distributions	Precision in measurement
Pros	Robustness against outliers	Efficiency for normal distributions
	Applicability to skewed data	Ease of interpretation
Cons	Loss of sensitivity without outliers	Sensitivity to outliers
		Not robust in the presence of outliers
Considerations	Useful when the central tendency needs stability	Suitable for datasets with minimal or no outliers
		Provides precise measures in a normal distribution

Table 8.3 – Which statistical methods work better with outliers

In the next section of this chapter, we will discuss how to identify multivariate outliers.

Identifying multivariate outliers

Multivariate outliers occur when an observation is extreme in the context of multiple variables simultaneously. These outliers cannot be detected by analyzing individual variables alone; rather, they require consideration of interactions between variables. Detecting multivariate outliers involves assessing data points in higher dimensional space. In the following parts, we will outline different methods to identify multivariate outliers, along with code examples for each.

Mahalanobis distance

Mahalanobis distance is a statistical measure used to identify outliers in multivariate data. It accounts for the correlation between variables and calculates the distance of each data point from the mean of the dataset in a scaled space. This distance is then compared to a threshold to identify observations that deviate significantly from the multivariate mean.

For this example, we have created a new dataset with some multivariate student data so that we can showcase the technique in the best way possible. The code can be fully seen in the repository at `https://github.com/PacktPublishing/Python-Data-Cleaning-and-Preparation-Best-Practices/blob/main/chapter08/12.mahalanobis_distance.py`. The key steps of the process are as follows:

1. Let's import the required libraries first:

```
import pandas as pd
import numpy as np
import matplotlib.pyplot as plt
from scipy.stats import chi2
from mpl_toolkits.mplot3d import Axes3D
```

2. Let's generate multivariate student data:

```
np.random.seed(42)
data = np.random.multivariate_normal(mean=[0, 0], cov=[[1, 0.5],
[0.5, 1]], size=100)
```

We generate a dataset of 100 samples from a multivariate normal distribution with a specified mean vector of `[0, 0]` and a covariance matrix of `[[1, 0.5], [0.5, 1]]`.

3. Let's introduce outliers and create the DataFrame:

```
outliers = np.array([[8, 8], [9, 9]])
data = np.concatenate([data, outliers])
df = pd.DataFrame(data, columns=['X1', 'X2'])
```

4. The following function calculates the Mahalanobis distance for each data point based on the mean and the inverse of the covariance matrix:

```
def mahalanobis_distance(x, mean, inv_cov_matrix):
    centered_data = x - mean
    mahalanobis_dist = np.sqrt(np.dot(centered_data,
                                np.dot(inv_cov_matrix,
                                centered_data)))
    return mahalanobis_dist
```

5. The mean, covariance matrix, and inverse covariance matrix are calculated for the dataset:

```
mean = np.mean(df[['X1', 'X2']], axis=0)
cov_matrix = np.cov(df[['X1', 'X2']], rowvar=False)
inv_cov_matrix = np.linalg.inv(cov_matrix)
```

6. Mahalanobis distance is calculated for each data point and added as a new column in the DataFrame:

```
df['Mahalanobis_Distance'] = df.apply(lambda row: mahalanobis_
distance(row[['X1', 'X2']], mean, inv_cov_matrix), axis=1)
```

7. Set a significance level for outlier detection:

```
alpha = 0.01
```

The significance level (`alpha`) represents the probability of rejecting the null hypothesis when it is true, which in this context is the probability of incorrectly identifying a data point as an outlier. A common choice for `alpha` is `0.01`, meaning there is a 1% chance of mistakenly classifying a normal data point as an outlier. A lower `alpha` value makes the outlier detection more conservative, reducing false positives (normal points labeled as outliers). Conversely, a higher `alpha` value makes it more permissive, potentially identifying more points as outliers but increasing the chance of false positives.

8. Next, we set the chi-squared threshold:

```
chi2_threshold = chi2.ppf(1 - alpha, df=2) # df is the degrees
of freedom, which is the number of features
```

The chi-squared threshold is a critical value from the chi-squared distribution used to define the cutoff for outlier detection. The `chi2.ppf` function computes the percentile point function (inverse of the cumulative distribution function) for the chi-squared distribution. The degrees of freedom is equal to the number of features or variables used in the Mahalanobis distance calculation. In this case, it's 2 (for X1 and X2). The chi-squared threshold is used to determine the cutoff value beyond which Mahalanobis distances are considered excessively high, indicating that the corresponding data points are outliers. For example, with `alpha = 0.01`, you are finding the threshold above which only 1% of the data points are expected to fall, assuming the data is normally distributed.

9. This step involves comparing each data point's Mahalanobis distance against the chi-squared threshold to determine whether it is an outlier:

```
outliers = df[df['Mahalanobis_Distance'] > chi2_threshold]
df_no_outliers = df[df['Mahalanobis_Distance'] <= chi2_
threshold]
```

Data points with distances greater than the threshold are flagged as outliers and separated from the rest of the data.

10. Let's now visualize the outliers:

```
fig = plt.figure(figsize=(10, 8))
ax = fig.add_subplot(111, projection='3d')
```

```
ax.scatter(df_no_outliers['X1'], df_no_outliers['X2'], df_no_
outliers['Mahalanobis_Distance'], color='blue', label='Data
Points')
ax.scatter(outliers['X1'], outliers['X2'],
outliers['Mahalanobis_Distance'], color='red', label='Outliers')
ax.set_xlabel('X1')
ax.set_ylabel('X2')
ax.set_zlabel('Mahalanobis Distance')
ax.set_title('Outlier Detection using Mahalanobis Distance')
plt.legend()
plt.show()
```

In the following chart, we can see all the data points projected in a 3D space and we can see the outliers marked with *x*:

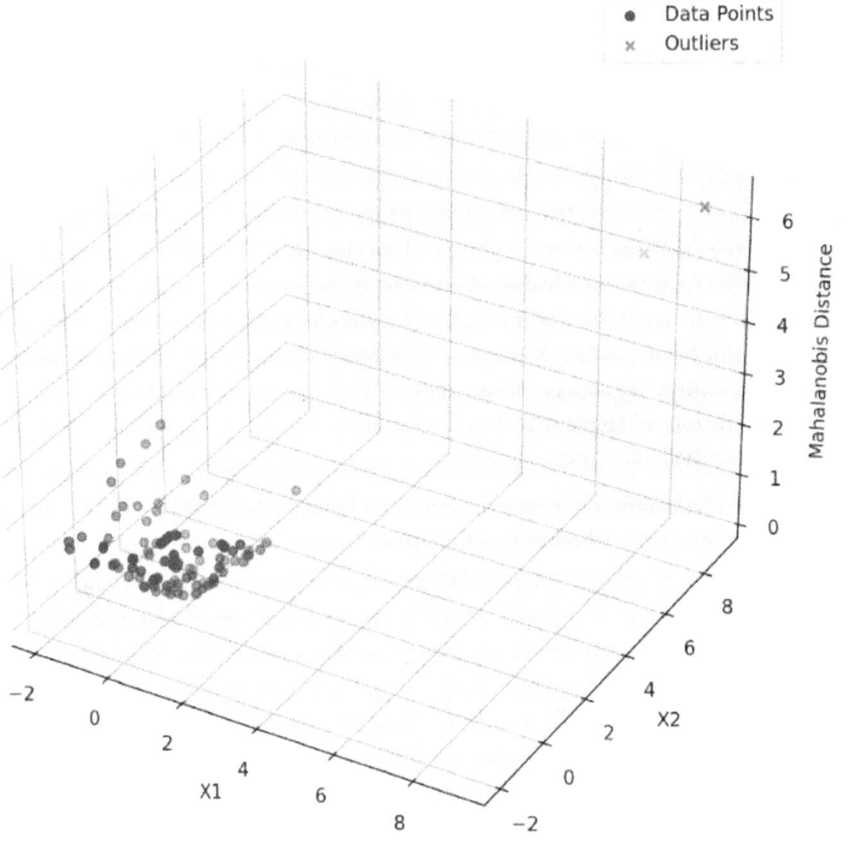

Figure 8.14 – Data plotted with Mahalanobis distance

> **Note**
>
> Run the visualization on your laptop to be able to see this space and move around it in a 3D view; it's cool!

As you can see from the 3D plot, it is very clear to spot the outliers in our data. Mahalanobis distance is most effective when dealing with datasets that involve multiple variables as it takes both the means and covariances among variables into account and allows identifying outliers that may not be apparent when looking at individual variables. In situations where variables have different units or scales, Mahalanobis distance can normalize the distances across variables, providing a more meaningful measure of outliers. Unlike univariate methods, Mahalanobis distance is sensitive to relationships among variables. It captures how far each data point is from the center of the data distribution, considering correlations between variables.

In the next section of the multivariate part, we will discuss how clustering methods can help us detect outliers.

Clustering techniques

Clustering methods such as k-means or hierarchical clustering can be used to group similar data points. Points that do not belong to any cluster or form small clusters might be considered multivariate outliers.

One popular method for outlier detection using clustering is the **Density-Based Spatial Clustering of Applications with Noise** (**DBSACN**) algorithm. DBSCAN can identify clusters of dense data points and classify outliers as noise. DBSCAN is advantageous because it *doesn't require specifying the number of clusters beforehand* and can effectively identify outliers based on density. It's a relatively simple yet powerful method for outlier detection, especially in cases where clusters may not be well-separated or when outliers form isolated points.

Let's deep dive into the code for the DBSCAN. As always, you can find the full code in the repository at `https://github.com/PacktPublishing/Python-Data-Cleaning-and-Preparation-Best-Practices/blob/main/chapter08/13.clustering.py`:

1. Let's import the libraries:

```
import pandas as pd
import numpy as np
import matplotlib.pyplot as plt
from sklearn.cluster import DBSCAN
from sklearn.preprocessing import StandardScaler
```

2. Let's generate the example data for this method. The dataset consists of 100 samples from a multivariate normal distribution with a mean vector of `[0, 0]` and a covariance matrix of `[[1, 0.5], [0.5, 1]]`. This creates a cluster of points that are normally distributed around the origin with some correlation between the features:

```
np.random.seed(42)
data = np.random.multivariate_normal(mean=[0, 0], cov=[[1, 0.5],
[0.5, 1]], size=100)
outliers = np.random.multivariate_normal(mean=[8, 8], cov=[[1,
0], [0, 1]], size=10)
data_with_outliers = np.vstack([data, outliers])
```

3. Let's turn the data into a DataFrame:

```
df = pd.DataFrame(data_with_outliers, columns=['Feature1',
'Feature2'])
```

4. Standardize the data by removing the mean and scaling to unit variance. `StandardScaler` from `sklearn.preprocessing` is used to fit and transform the data. Standardizing ensures that all features contribute equally to distance calculations by scaling them to have a mean of 0 and a standard deviation of 1. This is especially important for distance-based algorithms such as DBSCAN:

```
scaler = StandardScaler()
data_scaled = scaler.fit_transform(df)
```

5. Apply DBSCAN for outlier detection. `eps=0.4` sets the maximum distance between points to be considered in the same neighborhood, and `min_samples=5` specifies the minimum number of points required to form a dense region. DBSCAN is a clustering algorithm that can identify outliers as points that do not belong to any cluster. Points labeled `-1` by DBSCAN are considered outliers. The choice of `eps` and `min_samples` parameters can significantly impact the detection of outliers, and these values might need tuning based on the specific dataset:

```
dbscan = DBSCAN(eps=0.4, min_samples=5)
df['Outlier'] = dbscan.fit_predict(data_scaled)
```

In the following chart, we have plotted all data points in a 2D space and can see the outliers on the right side of the graph:

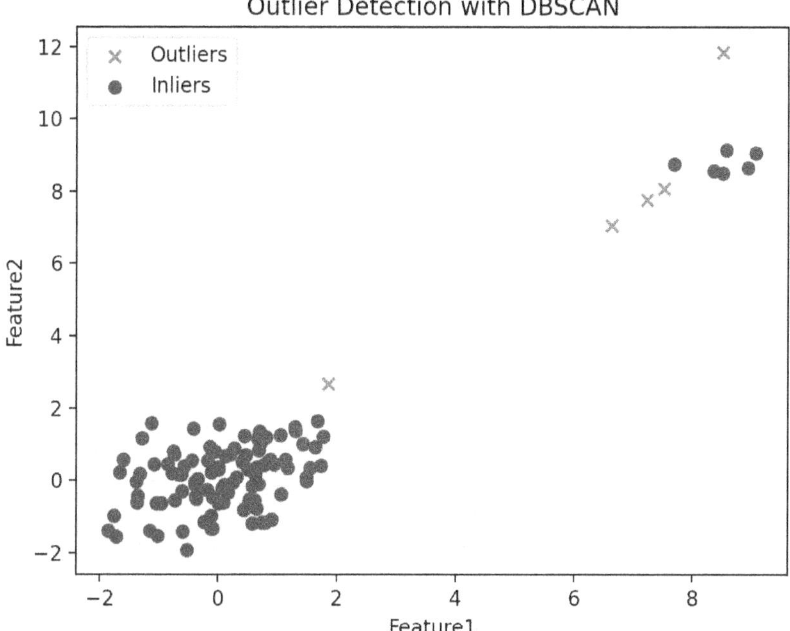

Figure 8.15 – DBSCAN clustering for outlier detection

There is a key parameter that needs to be adjusted in DBSCAN: `eps`. The `eps` (epsilon) parameter essentially defines the radius around a data point, and all other data points within this radius are considered its neighbors.

When performing DBSCAN clustering, the algorithm starts by selecting a data point and identifying all the data points that lie within a distance of `eps` from it. If the number of data points within this distance exceeds a specified threshold (`min_samples`), the selected data point is considered a core point, and all the points within its epsilon-neighborhood become part of the same cluster. The algorithm then recursively expands the cluster by finding the neighbors of the neighbors until no more points can be added.

The choice of `eps` depends on the specific characteristics of the dataset and the desired granularity of clusters. It may require some experimentation and domain knowledge to find the appropriate value for `eps`.

Employing k-means instead of DBSCAN offers a different approach. K-means is a centroid-based clustering algorithm that requires *pre-specifying the number of clusters*, making it essential to have prior knowledge or conduct exploratory analysis to determine an appropriate value for *k*. While it is sensitive to outliers, the simplicity and computational efficiency of k-means make it an attractive choice for certain scenarios. K-means may be well-suited when clusters are well-separated and have a relatively uniform structure. However, it is essential to be aware that k-means may struggle with irregularly shaped or overlapping clusters and can be influenced by outliers in its attempt to minimize the sum of squared distances.

After having spotted the multivariate outliers, we need to decide how we are going to deal with those. This is the focus of the next part.

Handling multivariate outliers

Handling multivariate outliers involves addressing data points that deviate significantly in the context of multiple variables. In this part of the chapter, we will provide explanations and code examples for different methods to handle multivariate outliers.

Multivariate trimming

This method involves limiting extreme values based on a combined assessment of the values across multiple variables. For example, the limits for trimming can be determined by considering the Mahalanobis distance, which accounts for correlations between variables. This technique is particularly useful when dealing with datasets containing outliers present across different variables. The idea is to preserve the overall structure of the data while mitigating the influence of extreme values across variables.

For this example, we are going to continue working on the data from the Mahalanobis distance example, and after we have calculated the Mahalanobis distance, we are going to drop the outliers passing the threshold. You can find the full code in the repository at `https://github.com/PacktPublishing/Python-Data-Cleaning-and-Preparation-Best-Practices/blob/main/chapter08/14.multivariate_trimming.py`:

1. Let's start by importing the libraries:

```
import pandas as pd
import numpy as np
import matplotlib.pyplot as plt
import seaborn as sns
from scipy.stats import chi2
from mpl_toolkits.mplot3d import Axes3D
```

2. Let's generate multivariate student data

```
np.random.seed(42)
data = np.random.multivariate_normal(mean=[0, 0], cov=[[1, 0.5],
[0.5, 1]], size=100)
outliers = np.array([[8, 8], [9, 9]])
```

```
data = np.concatenate([data, outliers])
df = pd.DataFrame(data, columns=['X1', 'X2'])
```

3. Define the function to calculate the Mahalanobis distance, which measures how far a data point is from the mean of the distribution, taking into account the correlation between features:

```
def mahalanobis_distance(x, mean, inv_cov_matrix):
    centered_data = x - mean
    mahalanobis_dist = np.sqrt(np.dot(centered_data, np.dot(inv_
cov_matrix, centered_data)))
    return mahalanobis_dist
```

4. Prepare the data for outlier detection:

```
df[['X1', 'X2']] = df[['X1', 'X2']].astype(float)
mean = np.mean(df[['X1', 'X2']], axis=0)
cov_matrix = np.cov(df[['X1', 'X2']], rowvar=False)
inv_cov_matrix = np.linalg.inv(cov_matrix)
```

5. Calculate the Mahalanobis distance for each data point:

```
df['Mahalanobis_Distance'] = df.apply(lambda row: mahalanobis_
distance(row[['X1', 'X2']], mean, inv_cov_matrix), axis=1)
```

6. Set the threshold for outlier detection:

```
alpha = 0.1
chi2_threshold = chi2.ppf(1 - alpha, df=2)
```

7. Filter the DataFrame to separate outliers from the rest of the data.

```
outliers = df[df['Mahalanobis_Distance'] > chi2_threshold]
df_no_outliers = df[df['Mahalanobis_Distance'] <= chi2_
threshold]
```

Let's present the distribution plots before the outlier handling.

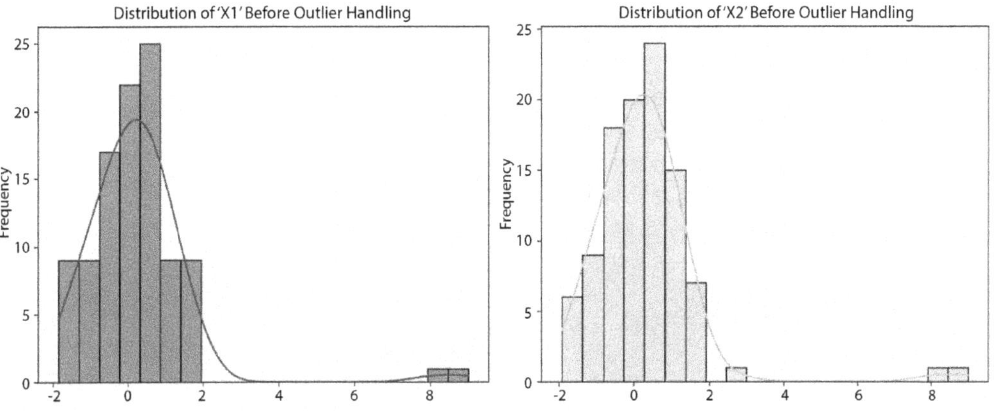

Figure 8.16 – Distribution charts with multivariate outliers

The descriptive statistics of the original data are as follows:

```
              X1          X2
count  102.000000  102.000000
mean     0.248108    0.281463
std      1.478963    1.459212
min     -1.852725   -1.915781
25%     -0.554778   -0.512700
50%      0.108116    0.218681
75%      0.715866    0.715485
max      9.000000    9.000000
```

After having dropped the data that are considered multivariate outliers, we can observe the changes in the following distributions:

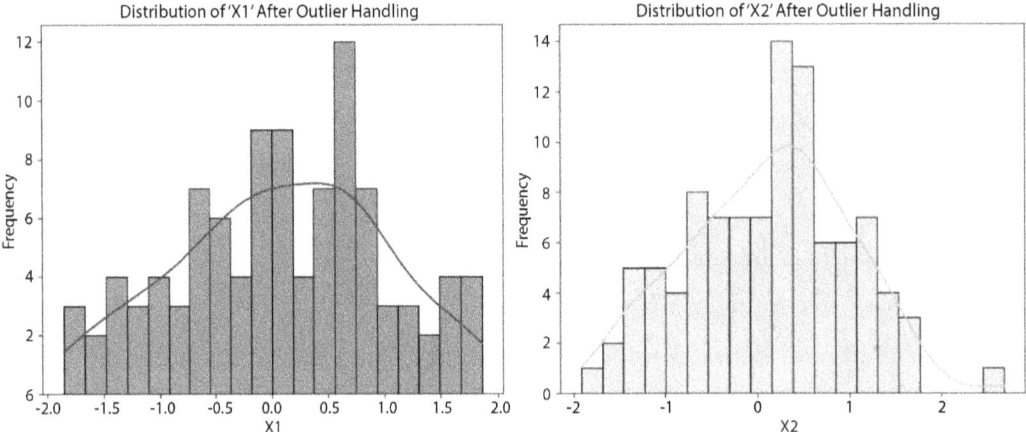

Figure 8.17 – Distribution charts after removing multivariate outliers

Finally, let's have a look at the updated descriptive statistics:

```
              X1          X2  Mahalanobis_Distance
count  100.000000  100.000000            100.000000
mean     0.083070    0.117093              1.005581
std      0.907373    0.880592              0.547995
min     -1.852725   -1.915781              0.170231
25%     -0.574554   -0.526337              0.534075
50%      0.088743    0.200745              0.874940
75%      0.699309    0.707639              1.391190
max      1.857815    2.679717              2.717075
```

Having trimmed the outliers, let's discuss the changes observed in the data:

- The count of observations reduced from 102 to 100 after removing outliers, so we dropped two records
- In the `X1` column, the mean decreased from 0.248 to 0.083, and the standard deviation reduced from 1.479 to 0.907
- In the `X2` column, the mean decreased from 0.281 to 0.117, and the standard deviation reduced from 1.459 to 0.881
- The maximum values for `X1` and `X2` were capped at 1.857815 and 2.679717, respectively, indicating the removal of extreme outliers

Overall, removing outliers has resulted in a dataset with reduced variability, particularly in terms of mean and standard deviation. Extreme values that could potentially skew the analysis have been mitigated.

Let's summarize the key takeaways from this chapter.

Summary

In this chapter, we deep-dived into handling missing values and outliers. We understood that missing values can distort our analyses and learned a range of imputation techniques, from simple mean imputation to advanced machine learning-based strategies. Similarly, we recognized that outliers could skew our results and deep-dived into methods to detect and manage them, both in univariate and multivariate contexts. By combining theory and practical examples, we gained a deeper understanding of the considerations, challenges, and strategies that go into ensuring the quality and reliability of our data.

Armed with these insights, we can now move on to the next chapter, where we will discuss scaling, normalization, and standardization of features.

9

Normalization and Standardization

Feature scaling, normalization, and standardization are essential preprocessing steps that help ensure that machine learning models can effectively learn from data. These techniques address issues related to numerical stability, algorithm convergence, model performance, and more, ultimately contributing to better, more reliable results in data analysis and machine learning tasks.

In this chapter, we will dive deep into the following topics:

- Scaling features to a range
- Z-score scaling
- Robust scaling

Technical requirements

You can find all the code for this chapter at `https://github.com/PacktPublishing/Python-Data-Cleaning-and-Preparation-Best-Practices/tree/main/chapter09`.

The different code files follow the names of the different sections of the chapters.

Scaling features to a range

Feature scaling is a preprocessing technique in machine learning that rescales the *range* of independent variables or features of a dataset. It's used to ensure that all features contribute equally to the model training process by bringing them to a common scale. Feature scaling is particularly important for algorithms that are sensitive to the scale of input features, such as k-nearest neighbors and gradient descent-based optimization algorithms.

> **Note**
>
> When scaling features, we're changing the range of the distribution of the data.

Let's present an example to make the concept of feature scaling easier to grasp. Let's suppose you're working on a machine learning project to predict housing prices based on various features of the houses, such as the following:

- Square footage (in square feet)
- Distance to the nearest school (in miles)
- Distance to the nearest public transportation stop (miles)

Now, let's discuss why feature scaling is important in this context. The **Square footage** feature could range from hundreds to thousands of square feet. The **Distance to the nearest school** and **Distance to the nearest public transportation stop** features might range from a fraction of a mile to several miles. If you don't scale these features, the algorithm may give excessive importance to the larger values, making square footage the dominant factor in predicting house prices. Features such as Distance to the nearest school might be unfairly downplayed.

For all the sections in this chapter, we will use the above example to showcase the different scaling methods. Let's go through the data creation for this example; the code can be found at `https://github.com/PacktPublishing/Python-Data-Cleaning-and-Preparation-Best-Practices/blob/main/chapter09/min_max_scaling.py`:

1. Let's start by importing the required libraries:

```
import pandas as pd
import numpy as np
import matplotlib.pyplot as plt
from sklearn.preprocessing import MinMaxScaler
```

2. Next, we will create a dataset with features related to housing prices:

```
np.random.seed(42)
num_samples = 100
```

We'll create the following features that affect the price of a house:

- Square footage in square feet:

```
square_footage = np.random.uniform(500, 5000, num_samples)
```

- Distance to the nearest school in miles:

```
distance_to_school = np.random.uniform(0.1, 5, num_samples)
```

- Commute distance to work in miles:

```
commute_distance = np.random.exponential(5, num_samples)
```

- Traffic density (skewed feature):

```
traffic_density = np.random.exponential(2, num_samples)
```

3. Then, we create a DataFrame that holds all the features:

```
data = pd.DataFrame({
        'Square_Footage': square_footage,
        'Distance_to_School': distance_to_school,
        'Commute_Distance': commute_distance,
        'Traffic_Density': traffic_density
})
```

4. Finally, we plot the original distributions:

```
plt.figure(figsize=(12, 8))
```

You can see the original distributions of the data here:

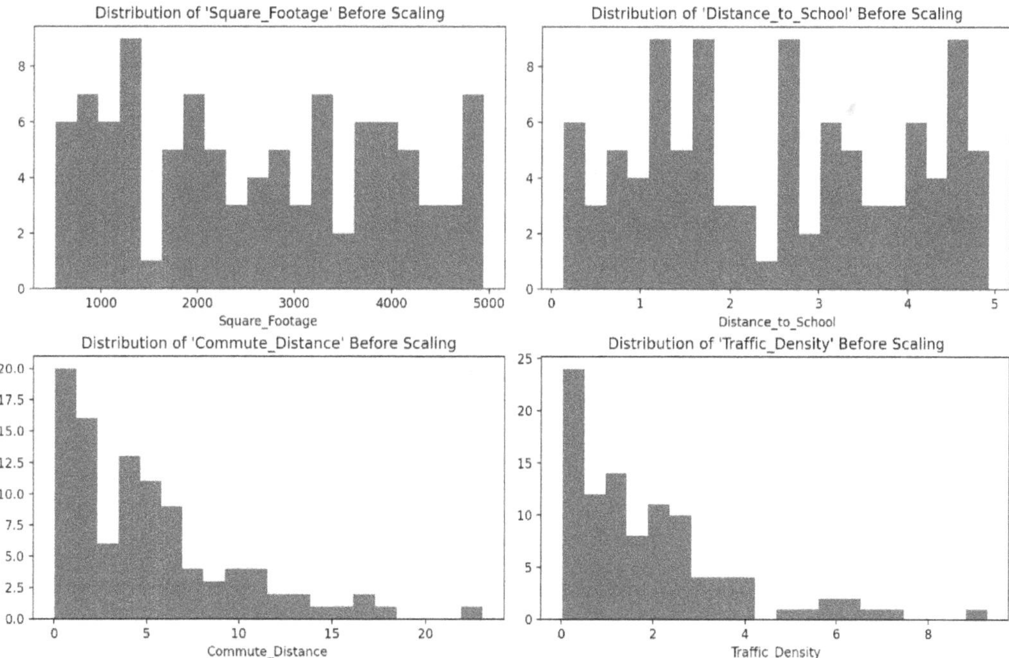

Figure 9.1 – Distribution of house pricing prediction use case

5. Let's display the statistics of our dataset:

```
print("Original Dataset Statistics:")
print(data.describe())
```

This will print the following output:

	Square_Footage	Distance_to_School	Commute_Distance	Traffic_Density
count	100.000000	100.000000	100.000000	100.000000
mean	2615.813345	2.539375	5.137017	1.904488
std	1338.702350	1.436245	4.660039	1.823873
min	524.849527	0.134065	0.025372	0.028996
25%	1369.403423	1.285822	1.621115	0.574340
50%	2588.641046	2.577562	4.134453	1.425731
75%	3785.914036	3.854300	6.983401	2.661949
max	4940.991215	4.929687	23.052849	9.314010

Figure 9.2 – Original dataset statistics

Now let's discuss one of the most common methods for scaling – **min-max scaling**.

Min-max scaling

Min-max scaling, also known as **normalization**, scales the values of a variable to a specific range, typically between 0 and 1. Min-max scaling is useful when you want to ensure that all values in a variable fall within a standardized range, making them directly comparable. It is commonly employed when the distribution of the variable is not assumed to be normal.

Let's have a look at the formula for calculating min-max scaling:

$$X\_scaled = (X - X\_min)/(X\_max - X\_min)$$

As you can see from the formula, min-max scaling preserves the relative ordering of values but *compresses them into a specific range*. One thing to note here is that it is *not a way to deal with outliers* and if outliers exist in the data, these extreme values can disproportionately influence the scaling. So, it is a good practice to deal with outliers first and then proceed to the scaling of features.

Scaling to a specific range is a suitable approach when the following conditions are satisfied:

- You have prior knowledge of the approximate upper and lower bounds of your data
- Your data follows a relatively uniform or bell-shaped distribution across that range
- Your chosen machine learning algorithm or model benefits from having features within a specific range, typically `[0, 1]` or any other desired range

A classic example of this scenario is age. Age values typically span from 0 to 90 and the entire range includes a significant number of individuals. However, scaling something such as income is not really recommended as there are a limited number of individuals with exceptionally high incomes. If you were to apply linear scaling to income, the upper limit of the scale would become exceedingly high, and most data points would be concentrated in a small segment of the scale, leading to loss of information and skewed representations.

Let's have a look at the code for *the house pricing prediction use case* we discussed previously to understand how a min-max scaler can transform the data:

1. First, we'll scale the data using `MinMaxScaler()`:

   ```
   scaler = MinMaxScaler()
   data_scaled = pd.DataFrame(scaler.fit_transform(data),
   columns=data.columns)
   ```

2. We can display the dataset statistics after scaling using the following code:

   ```
   print("\nDataset Statistics After Scaling:")
   print(data_scaled.describe())
   ```

3. Let's plot and observe the distributions after scaling:

   ```
   plt.figure(figsize=(12, 8))
   ```

Let's have a look at the modified data distributions after the scaling:

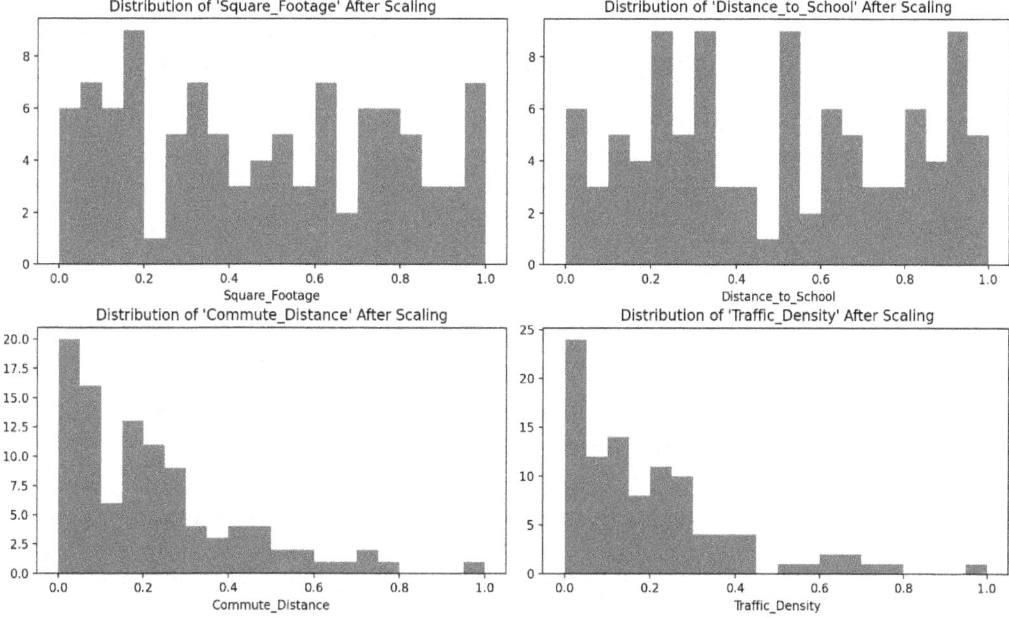

Figure 9.3 – Distribution of house pricing prediction use case after min-max scaling

We applied min-max scaling to transform each column into a standardized range between 0 and 1. The shape of the original feature distributions remained the same after scaling because min-max scaling maintains the relative distances between data points. The scaling had a normalization effect on the data, bringing all features to a common scale. This is important when features have different units or ranges, preventing one feature from dominating others.

> **Note**
>
> If your dataset is sparse (contains many zero values), min-max scaling may not be appropriate, as it can result in loss of information. Alternative methods such as **MaxAbsScaler** or robust scalers may be considered.

In the following section, we will discuss z-score scaling.

Z-score scaling

Z-score scaling, also known as standardization, is applied when you want to transform your data to have a mean of 0 and a standard deviation of 1. Z-score scaling is widely used in statistical analysis and machine learning, especially when algorithms such as k-means clustering or **Principal Component Analysis** (**PCA**) are employed.

Here is the formula for z-score:

$$X\_scaled = (X - mean(X)) / std(X)$$

Let's continue with the house pricing prediction use case to showcase the z-score scaling. The code can be found at `https://github.com/PacktPublishing/Python-Data-Cleaning-and-Preparation-Best-Practices/blob/main/chapter09/zscaler.py`:

1. We first perform z-score scaling:

```
data_zscore = (data - data.mean()) / data.std()
```

2. Then, we print the dataset statistics:

```
print("\nDataset Statistics after Z-score Scaling:")
print(data_zscore.describe())
```

3. Finally, we visualize the distributions:

```
data_zscore.hist(figsize=(12, 10), bins=20, color='green',
alpha=0.7)
plt.suptitle('Data Distributions after Z-score Scaling')
plt.show()
```

Let's have a look at the modified data distributions after the scaling:

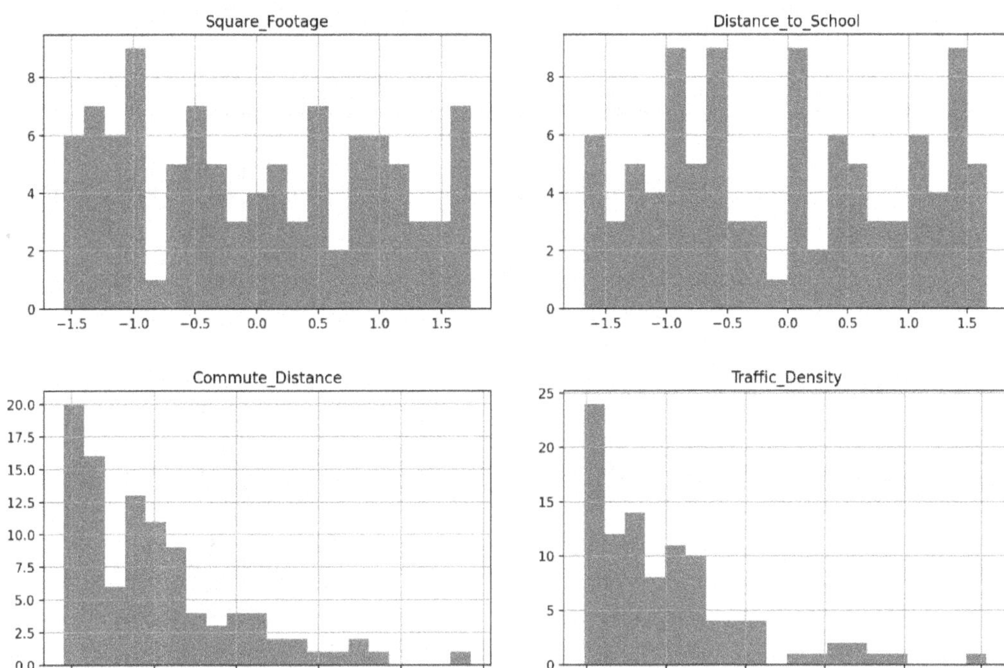

Figure 9.4 – Distribution of house pricing prediction use case after Z-scaling

The mean of each feature after scaling is very close to 0, which is expected in z-score scaling. The data is now centered around 0. The standard deviation for each feature is now approximately 1, making the scales comparable. The minimum and maximum values have been transformed, maintaining the relative distribution of data.

When to use Z-score scaling

Let's also discuss some considerations regarding when to use z-score scaling:

- Z-score scaling assumes that the data is approximately normally distributed, or at least symmetrically distributed, around a central mean value. If your data is highly skewed or has a non-standard distribution, standardization may not be as effective in making the data more Gaussian. As you can see, the `Commute_Distance` and `Traffic_Density` features are skewed and the z-score scaling was not that successful as the data is not centered around the mean compared to the rest of the features.
- This is most applicable to numerical features rather than categorical or ordinal ones. Ensure that the data you are standardizing is quantitative in nature.

- Extreme outliers can have a significant impact on the mean and standard deviation, which are used in z-score scaling. It's important to address outliers before standardization, as they can distort the scaling effect.
- Z-score scaling assumes a linear relationship between variables. If the underlying relationship is non-linear, other scaling methods or transformations may be more appropriate.
- Z-score scaling assumes that the variables are independent or at least not highly correlated. If variables are highly correlated, standardization may not provide additional benefits, and correlation structure should be considered separately.
- Z-score scaling alters the original units of the data, which can affect the interpretability of results. Consider whether maintaining the original units is essential for your analysis.

For small datasets, the impact of z-score scaling can be more pronounced. Be cautious when applying standardization to very small datasets, as it can potentially overemphasize the effects of outliers.

Robust scaling

Robust scaling, also known as **robust standardization**, is a method of feature scaling that is particularly useful when dealing with datasets containing outliers. Unlike min-max scaling and z-score scaling, which can be sensitive to outliers, robust scaling is designed to be robust in the presence of extreme values. It is especially beneficial when you want to normalize or standardize features while minimizing the impact of extreme values. Robust scaling is also suitable for datasets where the features do not follow a normal distribution and may have skewness or heavy tails.

Here is the formula for robust scaling:

$$X\_scaled = (X - median)/IQR$$

Subtracting the median and dividing by the **Interquartile Range** (**IQR**)in the scaling process normalizes the data by centering it around the median and scaling it based on the spread represented by the IQR. This normalization helps to mitigate the impact of extreme values, making the scaled values more representative of the overall distribution.

As we have already discussed in the previous chapter, the median being the middle value when the data is ordered makes robust scaling less sensitive to extreme values or outliers than other scaling methods that rely on the mean. In addition, the IQR, representing the range between the first quartile (Q1) and the third quartile (Q3), is also robust against outliers. Unlike the full range or standard deviation, the IQR focuses on the middle 50% of the data, making it less affected by extreme values.

Here's an example of how to apply robust scaling using Python code:

```
robust_scaler = RobustScaler()
data_scaled = robust_scaler.fit_transform(data)
```

Let's have a look at the modified data distributions after the scaling:

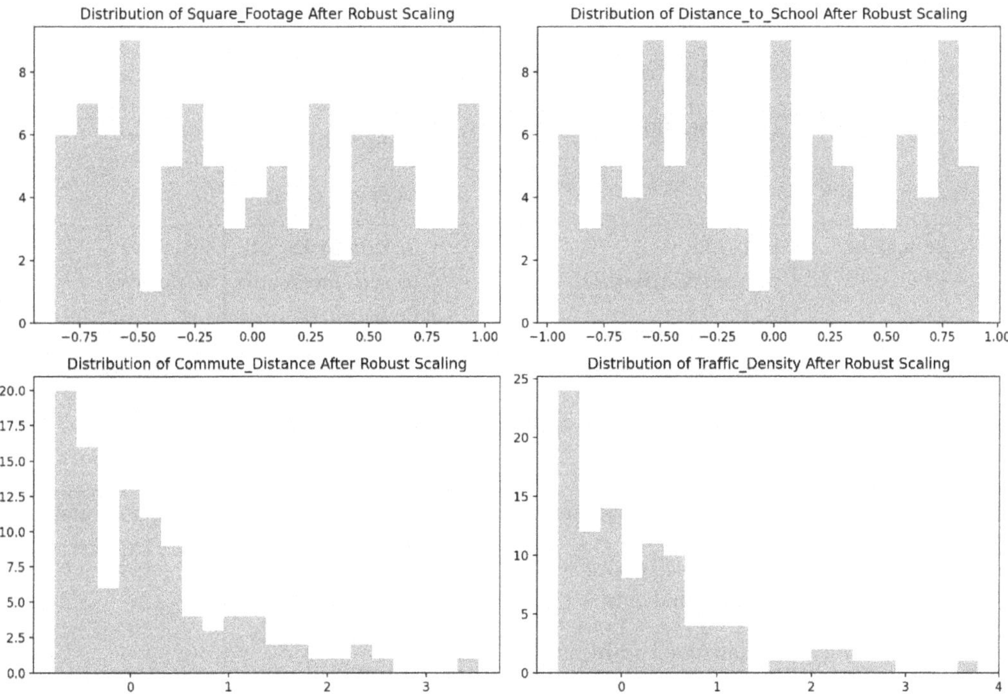

Figure 9.5 – Distribution of house pricing prediction use case after robust scaling

After the robust scaling exercise, we can observe that the central tendency of the data has shifted, and that the mean of each feature is now closer to zero. This is because the robust scaling process subtracts the median. As for the spread of the data, it changed as a result of dividing by the IQR. The variability in each feature is now represented in a more consistent manner, robust to outliers. The range of values for each feature is now compressed, particularly for features with larger initial ranges, preventing dominance by features with extreme values.

To close the chapter, we have created a summary table presenting all the techniques discussed so far, including information on when to use them, as well as the advantages and disadvantages of each.

Comparison between methods

This chart provides guidance on which scaling technique is suitable for various data scenarios.

Scaling method	When to use	Pros	Cons
Min-max scaling	Features have a clear, known range Normal distribution is assumed Data does not contain outliers	Simple and easy to understand Preserves relative relationships Memory efficient	Sensitive to outliers
Z-score scaling	Data follows a normal distribution No strong assumptions about the range Handling outliers is not a priority	Standardizes data to zero mean and deviation of 1	Sensitive to outliers May not be suitable for skewed data
Robust scaling	Data contains outliers Skewed distributions or non-normal data Equalizing feature contributions Resilience to varying feature variances	Less sensitive to outliers Preserves central tendency and spread	Computationally more expensive

Table 9.1 – Comparing different scaling techniques

Computational complexity

Min-max scaling tends to be more memory-efficient, especially when dealing with large datasets. This is because min-max scaling only involves scaling the values based on the minimum and maximum values of each feature, and the computation is relatively straightforward.

On the other hand, z-score scaling and robust scaling both require additional calculations such as mean, standard deviation (for z-score scaling), median, and interquartile range (for robust scaling), which may involve more memory usage. The computational complexity and memory requirements of z-score scaling and robust scaling can become more pronounced, particularly when dealing with large datasets.

Finally, let's summarize our learning for this chapter and get inspired for the next one.

Summary

In this chapter, we explored three common methods for scaling numerical features: min-max scaling, z-score scaling, and robust scaling. Min-max scaling transforms data to a specific range, making it suitable for algorithms sensitive to feature magnitudes. Z-score scaling standardizes data to zero mean and unit variance, providing a standardized distribution. Robust scaling, robust to outliers, employs the median and interquartile range, making it suitable for datasets with skewed distributions or outliers. We also went through different considerations to keep in mind while deciding on the best approach for your use case.

Moving forward, we'll shift our focus to handling categorical features in the next chapter.

10

Handling Categorical Features

Handling categorical features involves representing and processing information that isn't inherently numerical. **Categorical features** are attributes that can take on a limited, fixed number of values or categories, and they often define distinct categories or groups within a dataset, such as types of products, genres of books, or customer segments. Effectively managing categorical data is crucial because most **machine learning** (**ML**) algorithms require numerical inputs.

In this chapter, we will cover the following topics:

- Label encoding
- One-hot encoding
- Target encoding (mean encoding)
- Frequency encoding
- Binary encoding

Technical requirements

The complete code for this chapter can be found in the following GitHub repository:

`https://github.com/PacktPublishing/Python-Data-Cleaning-and-Preparation-Best-Practices/tree/main/chapter10`

Let's install the necessary libraries we will use in this chapter:

```
pip install scikit-learn==1.5.0
pip install matplotlib==3.9.0
pip install seaborn==0.13.2
pip install category_encoders==2.6.3
```

Label encoding

Label encoding is a technique for handling categorical data by converting each category into a unique integer. It's suitable for categorical features with ordinal relationships, where there is a clear ranking or order among the categories.

For example, when dealing with educational levels such as "high school," "bachelor's," "master's," and "Ph.D.," label encoding can be used because there's a clear order from the least to most advanced level of education.

Use case – employee performance analysis

A **Human Resources** (**HR**) department wants to analyze employee performance data to understand the relationship between employee ratings and other factors such as salary, years of experience, and department. They plan to use ML to predict employee ratings based on these factors.

The data

Let's have a quick look at the data we have available for the performance analysis:

- `Employee Rating`: Categorical feature with `Poor`, `Satisfactory`, `Good`, and `Excellent` values
- `Salary`: Numeric feature representing employee salaries
- `Years of Experience`: Numeric feature indicating the number of years an employee has worked
- `Department`: Categorical feature specifying the department in which the employee works

Let's have a look at the original DataFrame before the encoding:

```
Employee    Rating  Salary  Years of Experience  Department
0             Poor  35000                     2          HR
1             Good  50000                     5          IT
2     Satisfactory  42000                     3     Finance
3        Excellent  60000                     8          IT
4             Good  52000                     6   Marketing
```

Having understood the data, we can move to the objective of the use case.

Objective of the use case

The use case's objective is to encode the `Employee Rating` feature using label encoding to prepare the data for ML analysis. Let's see how we can do this using scikit-learn the complete code can be found at `https://github.com/PacktPublishing/Python-Data-Cleaning-and-Preparation-Best-Practices/blob/main/chapter10/1a.label_encoding.py`:

1. Let's import the required libraries:

   ```
   import pandas as pd
   from sklearn.preprocessing import LabelEncoder
   ```

2. Let's create a sample dataset and turn it into a DataFrame:

   ```
   data = {
       'Employee Rating': ['Poor', 'Good', 'Satisfactory',
   'Excellent', 'Good'],
       'Salary': [35000, 50000, 42000, 60000, 52000],
       'Years of Experience': [2, 5, 3, 8, 6],
       'Department': ['HR', 'IT', 'Finance', 'IT', 'Marketing']
   }
   df = pd.DataFrame(data)
   ```

3. Initialize a `LabelEncoder` class:

   ```
   label_encoder = LabelEncoder()
   ```

4. Apply label encoding to the `Employee Rating` column:

   ```
   df['Employee Rating (Encoded)'] = label_encoder.fit_
   transform(df['Employee Rating'])
   ```

Let's have a look at the encoded output we created.

Encoded output

In this use case, label encoding is applied to the `Employee Rating` feature to convert it into numeric values while preserving the ordinal relationship. The following table shows the output of the encoding operation.

	Employee Rating	Salary	Years of Experience	Department	Employee Rating (Encoded)
1	Poor	35000	2	HR	2
2	Good	50000	5	IT	1
3	Satisfactory	42000	3	FINANCE	3
4	Excellent	60000	8	IT	0
5	Good	52000	6	MARKETING	1

Table 10.1 – Output dataset after label encoding

As you can see, an `Employee Rating (Encoded)` feature has been added, and all the items are now numeric. Let's have a look at the distribution graphs for the encoded column:

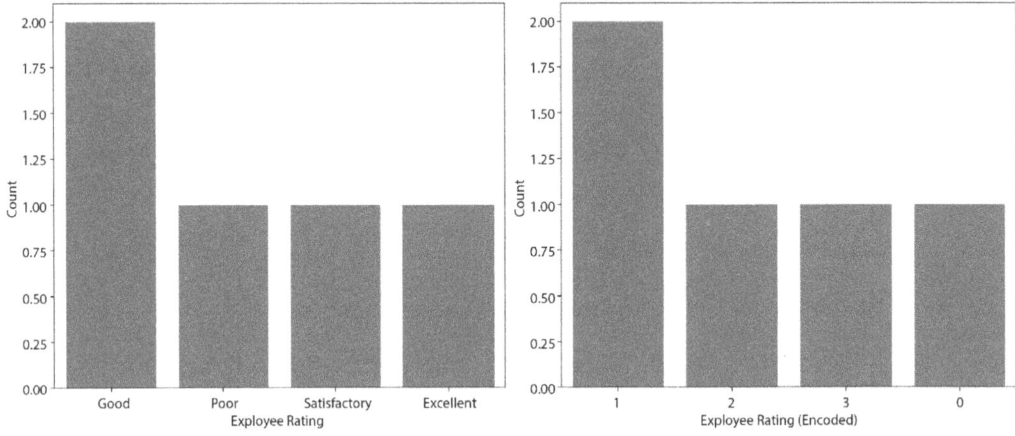

Figure 10.1 – Distribution before and after the encoding

As we can see, there are no changes in the distribution before and after encoding. Label encoding converts categorical labels into numerical values *while preserving the original data distribution*. It simply assigns unique integer values to each category *without altering their frequency*. However, visually, the labels on the *x* axis will change from categorical values to numerical values, but the counts (or frequencies) of each label will remain the same.

> **Note**
>
> If the data is shuffled or the order of the categories changes between different runs of the encoder, the encoded values might differ. This is because the assignment of integers to categories can depend on the order in which they appear. Also, if you initialize a new instance of the label encoder each time, the mapping of categories to integers might vary as well. For consistent results, you should fit the encoder once and then use it for transforming data.

The encoded values can then be used as input features for an ML model to predict employee ratings based on salary, years of experience, and department. Let's now discuss some things to keep in mind when encoding features with a label encoder.

Considerations for label encoding

When performing label encoding, especially on large datasets, there are several important considerations to keep in mind. Ensure that categorical features have a meaningful order. If there's no natural order among the categories, label encoding might not be appropriate. Label encoding assigns *integer values to categories based on alphabetical order*. This introduces a potential issue if the categories do not have an inherent order, but the model might interpret the numerical values as ordered. For example, `Poor`, `Good`, and `Excellent` might be encoded as `2`, `1`, and `0`, respectively, but `Poor` is not inherently greater than `Good`. This is exactly what happened in the use case presented previously. What we could have done to ensure that the label encoding reflects the correct ordinal order (that is, `Poor` < `Satisfactory` < `Good` < `Excellent`) was to manually set the order by specifying the desired mapping, as shown next the complete code can be found at `https://github.com/PacktPublishing/Python-Data-Cleaning-and-Preparation-Best-Practices/blob/main/chapter10/1b.label_encoding_forced.py`:

1. Define the correct order of categories with prefixes:

```
ordered_categories = {
    'Poor': '1.Poor',
    'Satisfactory': '2.Satisfactory',
    'Good': '3.Good',
    'Excellent': '4.Excellent'
}
```

2. Map the `Employee Rating` column to the prefixed categories:

```
df['Employee Rating Ordered'] = df['Employee Rating'].
map(ordered_categories)
```

The resulting DataFrame is presented as follows:

```
  Employee Rating Ordered  Employee Rating (Encoded)
0                    Poor                          0
1                    Good                          2
2            Satisfactory                          1
3               Excellent                          3
4                    Good                          2
```

Always keep consistency in mind when encoding *across training and test datasets*. The encoder should be fitted *on the training data* and used to transform both training and test datasets. This prevents issues where unseen categories in the test set lead to errors or incorrect encoding. Follow the next steps as best practice:

3. Apply label encoding to the `Employee Rating` column:

```
df['Employee Rating (Encoded)'] = label_encoder.fit_
transform(df['Employee Rating'])
```

4. Save the encoder:

```
joblib.dump(label_encoder, 'label_encoder.pkl')
```

5. Load the encoder (in another script or session):

```
loaded_encoder = joblib.load('label_encoder.pkl')
```

6. Transform the new data:

```
df['Employee Rating (Encoded)'] = loaded_encoder.
transform(df['Employee Rating'])
```

The last point to mention, important when dealing with large datasets, is that label encoding is generally more memory-efficient than one-hot encoding, which can create many binary columns.

While label encoding is a straightforward approach for converting categorical data into numerical form, it can inadvertently introduce ordinal relationships between categories that don't inherently exist. To avoid this issue and ensure that each category is treated independently, one-hot encoding is often a more suitable method.

One-hot encoding

One-hot encoding is a technique used to convert categorical data into a binary matrix (1s and 0s). Each category is transformed into a new column, and a 1 is placed in the column corresponding to the category present for each observation, while all other columns get a 0. This method is particularly useful when dealing with categorical data where there is **no ordinal relationship among categories**.

When to use one-hot encoding

One-hot encoding is suitable for categorical data that lacks a natural order or ranking among categories. Here are some scenarios where it is appropriate:

- **Nominal categorical data**: When dealing with nominal data, where categories are distinct and have no inherent order.
- **Algorithms that don't handle ordinal data**: Some ML algorithms (for example, decision trees and random forests) are not designed to handle ordinal data correctly. One-hot encoding ensures that each category is treated as a separate entity.
- **Preventing misinterpretation**: To prevent a model from assuming any ordinal relationship that doesn't exist, one-hot encoding is used to represent categorical data as binary values.

Next, let's look at a use case where we can use one-hot encoding.

Use case – customer churn prediction

A telecommunications company is experiencing high customer churn and wants to build an ML model to predict which customers are likely to leave their service. They have collected data on customer demographics, contract details, and services used.

The data

Let's have a quick look at the data we have available for the analysis:

- `Contract Type`: Categorical feature with `Month-to-Month`, `One Year`, and `Two Year` values
- `Internet Service`: Categorical feature with `DSL`, `Fiber Optic`, and `No Internet Service` values
- `Payment Method`: Categorical feature with `Electronic Check`, `Mailed Check`, `Bank Transfer`, and `Credit Card` values

Let's have a look at the sample data for the use case:

```
   Customer ID  Contract Type Internet Service Payment Method
 0           1 Month-to-Month              DSL  Electronic Check
 1           2       One Year      Fiber Optic  Mailed Check
 2           3 Month-to-Month              DSL Bank Transfer
 3           4       Two Year      Fiber Optic   Credit Card
```

Having understood the data, we can move to the objective of the use case.

Objective of the use case

The objective of the use case is to encode the categorical features using one-hot encoding to prepare the data for ML analysis. The code for this example can be found here: `https://github.com/PacktPublishing/Python-Data-Cleaning-and-Preparation-Best-Practices/blob/main/chapter10/2.one_hot_encoding.py`.

Follow the next steps:

1. Initialize a `OneHotEncoder` class:

```
one_hot_encoder = OneHotEncoder(sparse_output=False,
drop='first')
```

2. Fit and transform the categorical columns:

```
encoded_columns = one_hot_encoder.fit_transform(df[['Contract
Type', 'Internet Service', 'Payment Method']])
```

3. Create a new DataFrame with the one-hot encoded columns:

```
encoded_df = pd.DataFrame(encoded_columns, columns=one_hot_
encoder.get_feature_names_out(['Contract Type', 'Internet
Service', 'Payment Method']))
```

4. Concatenate the one-hot encoded DataFrame with the original DataFrame:

```
df_encoded = pd.concat([df, encoded_df], axis=1)
```

5. Drop the original categorical columns as they are now encoded:

```
df_encoded = df_encoded.drop(['Contract Type', 'Internet
Service', 'Payment Method'], axis=1)
```

Let's have a look at the encoded output we created.

Encoded output

In this use case, we are preparing customer data for a churn prediction model. Categorical features such as `Contract Type`, `Internet Service`, and `Payment Method` are one-hot encoded to convert them into binary representations suitable for ML. These encoded features can be used to train a predictive model that helps the telecommunications company identify customers at risk of churning and take proactive measures to retain them.

Let's see with some plots how the distribution of the features changes when we are applying the encoding. Let's have a look at the original distribution before the encoding first:

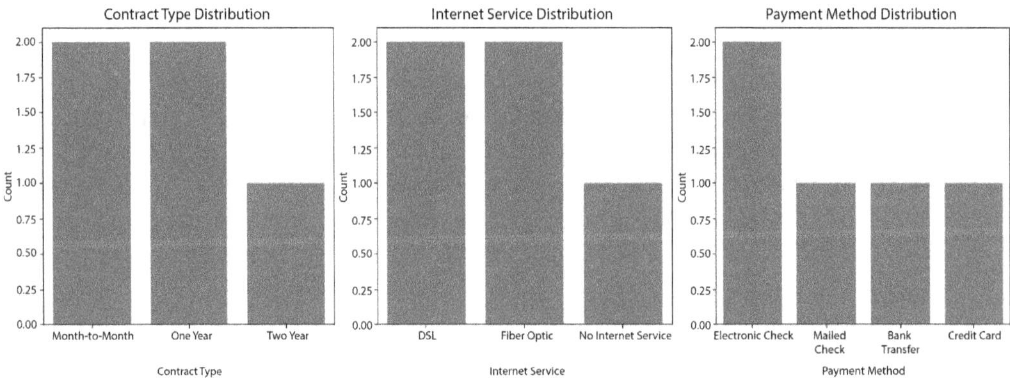

Figure 10.2 – Distribution before the one-hot encoding

After encoding, each value of the categorical variables is turned into a unique column, showcasing binary values (0 or 1) reflecting the presence of that category in each row of the dataset. Let's see the distribution graphs for the `Contract Type` column:

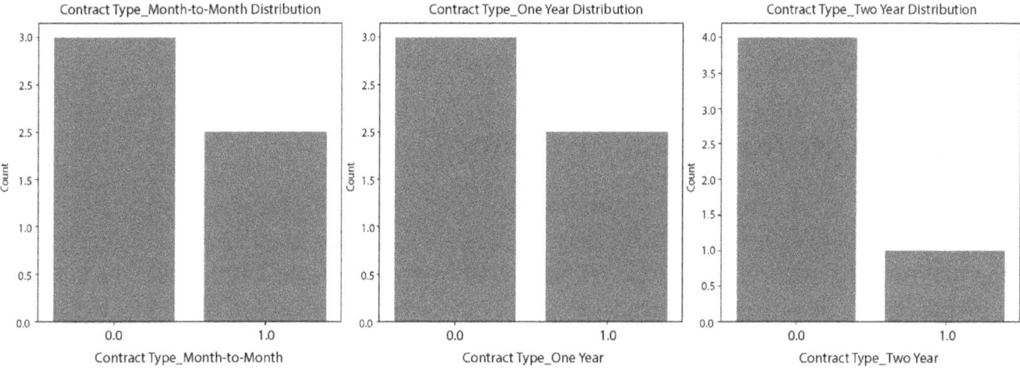

Figure 10.3 – Distribution after the one-hot encoding for the Contract Type feature

> **Note**
>
> Visualizing the original categorical data helps understand the data distribution and identify any imbalances. Visualizing the encoded columns ensures that the transformation has been applied correctly. Each binary column should only have values of 0 or 1.

Let's now discuss some things to keep in mind when encoding features with a one-hot encoder.

Considerations for one-hot encoding

When performing one-hot encoding, especially on large datasets, there are several important considerations to keep in mind:

- One-hot encoding can significantly increase the dimensionality of your dataset, especially when you have many categories. This can lead to the "curse of dimensionality," which can be problematic for some algorithms.
- **Collinearity**: Since each category is represented as a separate binary column, there can be collinearity between these columns. This means some columns might be highly correlated, which can affect the performance of linear models.
- **Handling missing values**: Decide how to handle missing values in categorical features before applying one-hot encoding. You may choose to create a separate column for missing values or use imputation techniques.

- Handling one-hot encoding on large datasets can be challenging due to the increase in the number of features and the potential for high memory usage. Process the data in smaller batches if the dataset is too large to fit into memory.

Moving from one-hot encoding to target encoding can be particularly beneficial when dealing with high-cardinality categorical features. Let's explore target encoding in more detail.

Target encoding (mean encoding)

Target encoding, also known as **mean encoding**, is a technique used for encoding categorical features by replacing each category with the **mean** of the target variable (or another relevant aggregation function) for that category. This method is particularly useful for classification tasks when dealing with **high-cardinality categorical features**, where one-hot encoding would result in a significant increase in dimensionality.

In more detail, target encoding replaces categorical values with the mean (or other aggregation metric) of the target variable for each category. It leverages the relationship between the categorical feature and the target variable to encode the information.

When to use target encoding

When you have categorical features with many unique categories, using one-hot encoding might lead to a high-dimensional dataset. Target encoding can be an effective alternative in such cases.

If there's a strong relationship between the categorical feature and the target variable, target encoding can capture this relationship and potentially improve predictive power.

You can also use target encoding when you have memory constraints and need to reduce the dimensionality of your dataset, as target encoding doesn't create additional columns.

Use case – sales prediction for retail stores

A retail chain with multiple stores wants to build an ML model to predict daily sales for each store. They have collected data on various features, including the `Store Type` feature, which has a high cardinality. Instead of using one-hot encoding, which would result in a large number of features, the retail chain decides to use target encoding to encode the `Store Type` feature.

The data

Let's have a quick look at the data we have available for the analysis:

- `Store Type`: The type of store (categorical variable with `Type A`, `Type B`, `Type C`, and `Type D` values)
- `Number of Employees`: The number of employees working at the store (integer variable)

- `Advertising Budget`: The budget allocated for advertising by the store (continuous variable in dollars)
- `Daily Sales`: The sales made by the store in a day (target variable in dollars)

Let's have a look at a sample of the data for the use case:

```
   Store Type  Number of Employees  Advertising Budget   Daily Sales
0      Type C                   21        23117.964192  16195.682148
1      Type D                   13         9017.567238    851.127834
2      Type A                   37        39945.667889  19274.801963
3      Type C                   24        34990.429063  14670.084345
4      Type C                   17        11817.711027   6442.646360
```

Having understood the data, we can move to the objective of the use case.

Objective of the use case

The use case's objective is to encode the categorical features using target encoding to prepare the data for ML modeling. Let's see how we can do this using scikit-learn. The code for this example can be found here: `https://github.com/PacktPublishing/Python-Data-Cleaning-and-Preparation-Best-Practices/blob/main/chapter10/3.target_encoding.py`.

Make sure you have installed and imported the libraries mentioned in the *Technical requirements* section at the beginning of the chapter. Once that is done, let's begin:

1. Let's create a synthetic dataset of sample size 1000:

   ```
   np.random.seed(42)
   n_samples = 1000
   ```

2. Generate some random data:

   ```
   data = {
     'Store Type': np.random.choice(['Type A', 'Type B', 'Type C',
   'Type D'], size=n_samples),
     'Number of Employees': np.random.randint(5, 50, size=n_
   samples),
     'Advertising Budget': np.random.uniform(1000, 50000, size=n_
   samples),
     'Daily Sales': np.random.uniform(500, 20000, size=n_samples)
   }
   ```

3. Put the data into a DataFrame:

   ```
   df = pd.DataFrame(data)
   ```

4. Define a target variable and features:

```
X = df.drop(columns=['Daily Sales']) # Features
y = df['Daily Sales'] # Target variable
```

5. Split the data into training and testing sets:

```
X_train, X_test, y_train, y_test = train_test_split(X, y, test_
size=0.2, random_state=42)
```

6. Initialize a `TargetEncoder` class:

```
target_encoder = TargetEncoder(cols=['Store Type'])
```

7. Fit and transform on the training data:

```
X_train_encoded = target_encoder.fit_transform(X_train, y_train)
```

> **Note**
>
> In the code provided on GitHub for this section, we are using the data and the encoded feature to train a random forest regressor model and calculate validation metrics. If you are interested, explore the code file here: `https://github.com/PacktPublishing/Python-Data-Cleaning-and-Preparation-Best-Practices/blob/main/chapter10/3.target_encoding.py`.

This encoding technique helps capture the relationship between different store types and daily sales, so let's have a look at the encoded output.

Encoded output

Let's have a look at the encoded data:

```
        Store Type  Number of Employees  Advertising Budget
29    10025.134200                   37        43562.535230
535   10190.055174                   12         1940.421564
695   10025.134200                   14        47945.600526
557   10190.055174                   23        19418.525972
836   10560.489044                   27        35683.919764
```

Let's focus on the `Store Type` column, which is now encoded into numerical values. We can see the difference in more detail before and after the encoding in the following graphs:

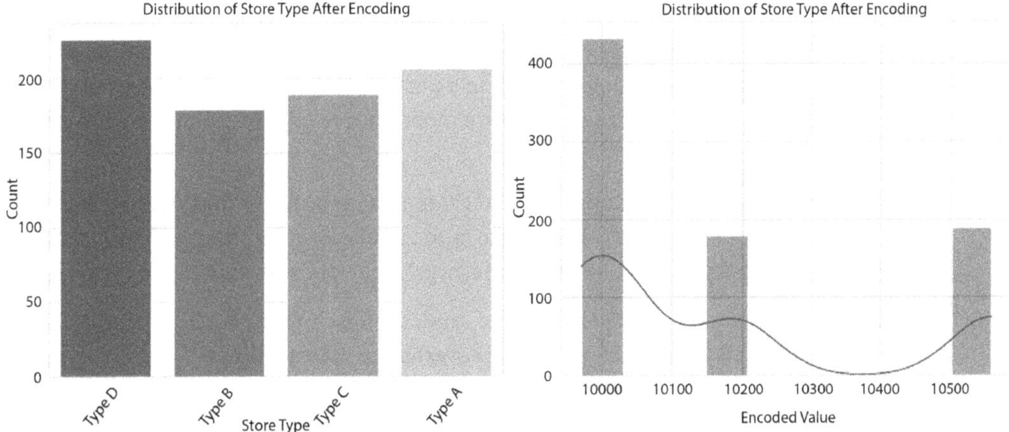

Figure 10.4 – Distribution of store type before and after the encoding

Target encoding can be advantageous in this scenario as it efficiently encodes the categorical feature, making it suitable for regression tasks such as sales prediction while avoiding the dimensionality issues associated with one-hot encoding.

Let's now discuss some things to keep in mind when encoding features with a target encoder.

Considerations for target encoding

When performing target encoding, especially on large datasets, there are several important considerations to keep in mind:

- **Overfitting**: Target encoding can lead to overfitting if not applied carefully or if some categories have only a few samples. To mitigate this, techniques such as smoothing or adding regularization terms are often used.
- **Smoothing (regularization)**: Smoothing involves blending the mean of the target variable for each category with a global mean. This reduces the impact of extreme values or noise in the training data. The formula for smoothed target encoding is often the following:

  $$smoothed\_mean = (n * category\_mean + m * global\_mean)/(n+m)$$

  Here, we have the following:

  - n is the number of observations in the category.
  - m is a hyperparameter that controls the strength of smoothing.

Adjusting the value of m allows you to control the level of regularization. Smaller values of m give more weight to the category's actual mean, while larger values give more weight to the global mean.

- **Cross-validation**: Perform target encoding within each fold of a cross-validation scheme. This helps ensure that the encoding is based on a portion of the data independent of the one being predicted. Cross-validation can provide a more reliable estimate of the target variable's distribution for each category.
- **Leave-one-out encoding**: In this approach, you compute the mean of the target variable for a category excluding the current observation. It can be more robust to overfitting because it considers the effect of the category without including the target value of the instance being encoded.
- **Adding noise**: Introducing a small amount of random noise to the encoded values can help reduce overfitting. This is often referred to as **Bayesian target encoding**.
- Be cautious of data leakage. It's crucial to calculate the mean on the training dataset only and apply the same encoding to the validation and test datasets.
- **Compute encoding statistics on training data only**: Calculate the encoding statistics (for example, mean) based solely on the training dataset. This ensures that the model is trained on unbiased information.
- **Apply the same encoding to all datasets**: Once you've calculated the encoding statistics on the training data, use the same encoding when preprocessing the validation and test datasets. Do not recalculate the statistics separately for these datasets.
- While target encoding can improve model performance, it may reduce the interpretability of your model, as the original categorical values are lost.

After exploring target encoding, another effective technique for handling high-cardinality categorical features is frequency encoding. Frequency encoding replaces each category with its frequency or count in the dataset, which can help capture the inherent importance of each category and maintain the overall distribution of the data. Let's deep dive into frequency encoding and its advantages in processing categorical variables.

Frequency encoding

Frequency encoding, also known as **count encoding**, is a technique used for encoding categorical features by replacing each category with its corresponding frequency or count in the dataset. In this encoding method, the more frequent a category is, the higher its encoded value. Frequency encoding can be a valuable tool in certain situations where the frequency of occurrence of a category carries valuable information.

When to use frequency encoding

Frequency encoding can be considered in the following scenarios:

- **Informative frequency**: The frequency or count of categories is informative and has a direct or indirect relationship with the target variable. For example, in a customer churn prediction problem, the frequency of product purchases by a customer may correlate with their likelihood to churn.
- **Efficiency**: You need an efficient encoding method that requires minimal computational resources and memory compared to one-hot encoding.

This encoding method often works well with tree-based models such as decision trees, random forests, and gradient boosting, as these models can effectively capture the relationship between the encoded frequency and the target variable.

Use case – customer product preference analysis

A retail company wants to analyze customer product preferences based on their purchase history. They have a dataset with information about customer purchases, including the product category they buy the most.

The data

In this example, we will use frequency encoding on the `Product Category` feature to determine the most frequently purchased product categories by customers. This encoding method allows the retail company to analyze customer preferences and understand how to optimize product recommendations or marketing strategies based on popular product categories.

Let's have a look at the sample dataset:

```
   Customer ID Product Category  Total Purchases
0            1      Electronics                5
1            2         Clothing                2
2            3      Electronics                3
3            4            Books                8
4            5            Books                7
5            6         Clothing                4
```

Having understood the data, we can move to the objective of the use case.

Objective of the use case

The use case's objective is to encode the categorical features using frequency encoding to prepare the data for ML modeling. Let's see how we can do that using scikit-learn:

1. Let's create a sample dataset:

```
data = {
    'Customer ID': [1, 2, 3, 4, 5, 6, 7, 8, 9, 10],
    'Product Category': ['Electronics', 'Clothing',
'Electronics', 'Books', 'Books', 'Clothing', 'Electronics',
'Books', 'Clothing', 'Books'],
    'Total Purchases': [5, 2, 3, 8, 7, 4, 2, 5, 1, 6]
}
df = pd.DataFrame(data)
```

2. Define features:

```
X = df[['Customer ID', 'Product Category', 'Total Purchases']]
```

3. Split the data into training and testing sets:

```
X_train, X_test = train_test_split(X, test_size=0.2, random_
state=42)
```

4. Initialize a `CountEncoder` class for `Product Category`:

```
count_encoder = CountEncoder(cols=['Product Category'])
```

5. Fit and transform the training data:

```
X_train_encoded = count_encoder.fit_transform(X_train)
```

The company wants to encode this categorical feature using frequency encoding to understand which product categories are most frequently purchased. Let's have a look at the encoded data.

Encoded output

Let's have a look at the encoded data:

```
   Customer ID  Product Category  Total Purchases
5            6                 1                4
0            1                 3                5
7            8                 4                5
2            3                 3                3
9           10                 4                6
```

Let's focus on the `Product Category` feature, which is now encoded into numerical values based on frequency. We can see the difference in more detail before and after the encoding on the following graphs:

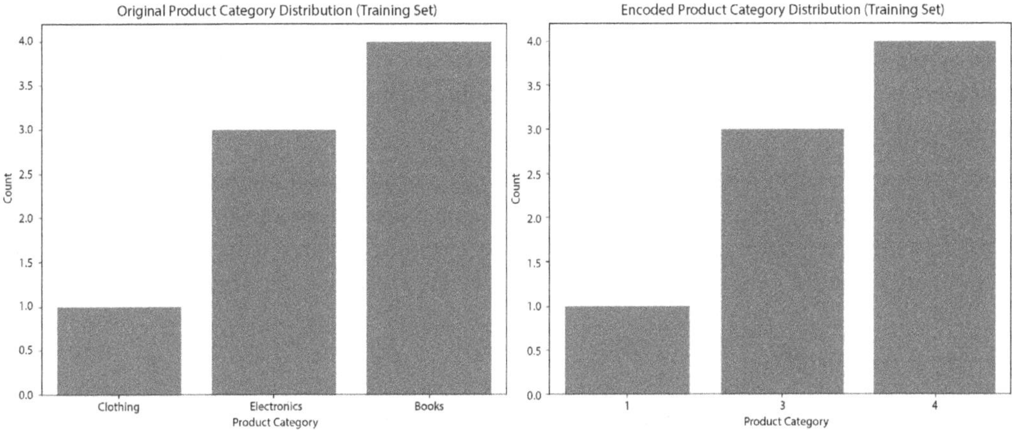

Figure 10.5 – Distribution of Product Category before and after the encoding

The first subplot shows the distribution of `Product Category` in the training set before encoding. The second subplot shows the distribution of the encoded `Product Category` feature in the training set after encoding. As we can see, each category in the `Product Category` column is replaced by the **frequency count** of that category within the training set.

> **Note**
>
> Frequency encoding preserves information about the prevalence of each category in the dataset.

Let's now discuss some things to keep in mind when encoding features with a frequency encoder.

Considerations for frequency encoding

When performing frequency encoding, there are several important considerations to keep in mind:

- Frequency encoding can lead to overfitting, especially if the dataset is small or if there are categories with very few observations. This is because the model might learn to rely too heavily on the frequency counts, which may not generalize well to new data.
- When two or more categories have the same frequency, they will end up with the same encoded value. This can be a limitation if those categories have different effects on the target variable.

- Frequency encoding is generally not suitable for linear models, as it does not create a linear relationship between the encoded values and the target variable. It may be necessary to normalize the encoded values to a similar scale, especially if you're using linear models that are sensitive to feature scaling.

Overall, frequency encoding is straightforward to implement and does not expand the feature space, unlike one-hot encoding, making it efficient for handling high-cardinality features without creating many new columns.

While frequency encoding offers simplicity and efficiency for handling high-cardinality features, another effective technique is binary encoding. Binary encoding represents categories as binary numbers, providing a more compact representation than one-hot encoding and preserving ordinal relationships. Let's explore how binary encoding can further enhance the processing of categorical variables.

Binary encoding

Binary encoding is a technique used for encoding categorical features by converting each category into binary code. Each unique category is represented by a unique binary pattern, where each digit (0 or 1) in the pattern corresponds to the presence or absence of that category. Binary encoding is particularly useful for handling high-cardinality categorical features while reducing dimensionality.

When to use binary encoding

Binary encoding can be considered in the following scenarios:

- **Dimensionality reduction**: You want to reduce the dimensionality of the dataset while still capturing information contained within the categorical feature. Binary encoding is particularly useful in this scenario.
- **Efficiency**: You need an efficient encoding method that results in a compact representation of categorical data and can be easily processed by ML algorithms.

Let's look at a use case.

Use case – customer subscription prediction

A subscription-based service provider wants to predict whether customers will subscribe to a premium plan based on various features, including the `Country` feature, which has a high cardinality. Binary encoding will be used to efficiently encode the `Country` feature.

The data

Let's have a look at the sample dataset:

- `Country`: This categorical feature represents the country of the customers. It helps to understand if the location influences the subscription status.
- `Age`: This numerical feature represents the age of the customers. Age can be a significant factor in determining the likelihood of a customer subscribing to a service.
- `Income`: This numerical feature represents the annual income of the customers. Income can indicate the financial capability of the customers to subscribe to a service.
- `Subscription`: This binary target variable indicates whether a customer has subscribed to a service. This is the target variable that we want to predict using the other features.

Let's have a look at a sample of the data for the use case:

```
  Country  Age  Income  Subscription
0     USA   25   50000             1
1  Canada   30   60000             0
2     USA   35   70000             1
3  Canada   40   80000             0
4  Mexico   45   90000             1
```

The distribution of `Country` can be seen in the following plot:

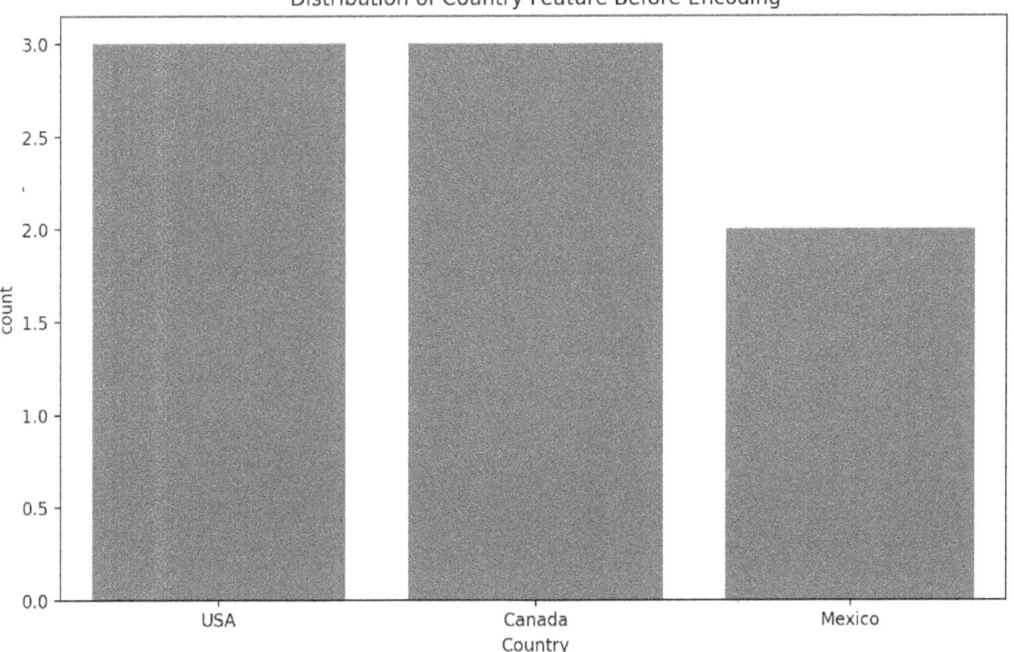

Figure 10.6 – Distribution of Country before and after the encoding

Objective of the use case

The goal of this analysis is to predict the Subscription status of customers based on their country, age, and income. We use binary encoding for the `Country` feature to convert it from a categorical variable to a numerical format that can be used in the ML algorithm. The code for this use case can be found here: `https://github.com/PacktPublishing/Python-Data-Cleaning-and-Preparation-Best-Practices/blob/main/chapter10/5.binary_encoding.py`.

Follow the next steps:

1. Let's create a sample dataset:

```
data = {
    'Country': ['USA', 'Canada', 'USA', 'Canada', 'Mexico',
'USA', 'Mexico', 'Canada'],
    'Age': [25, 30, 35, 40, 45, 50, 55, 60],
    'Income': [50000, 60000, 70000, 80000, 90000, 100000,
110000, 120000],
    'Subscription': [1, 0, 1, 0, 1, 0, 1, 0]
}

df = pd.DataFrame(data)
```

2. Apply binary encoding to the `Country` feature:

```
encoder = BinaryEncoder(cols=['Country'])
df_encoded = encoder.fit_transform(df)
```

3. Display the encoded DataFrame:

```
print(df_encoded)
```

Let's have a look at the encoded data.

Encoded output

In this example, binary encoding is applied to the `Country` feature, as we can see in the following output:

```
   Country_0  Country_1  Age  Income  Subscription
0          0          1   25   50000             1
1          1          0   30   60000             0
2          0          1   35   70000             1
3          1          0   40   80000             0
4          1          1   45   90000             1
5          0          1   50  100000             0
6          1          1   55  110000             1
7          1          0   60  120000             0
```

As we can see from the encoded output, the binary digits are split into separate columns. Let's also have a look at the changes in the distribution after the encoding:

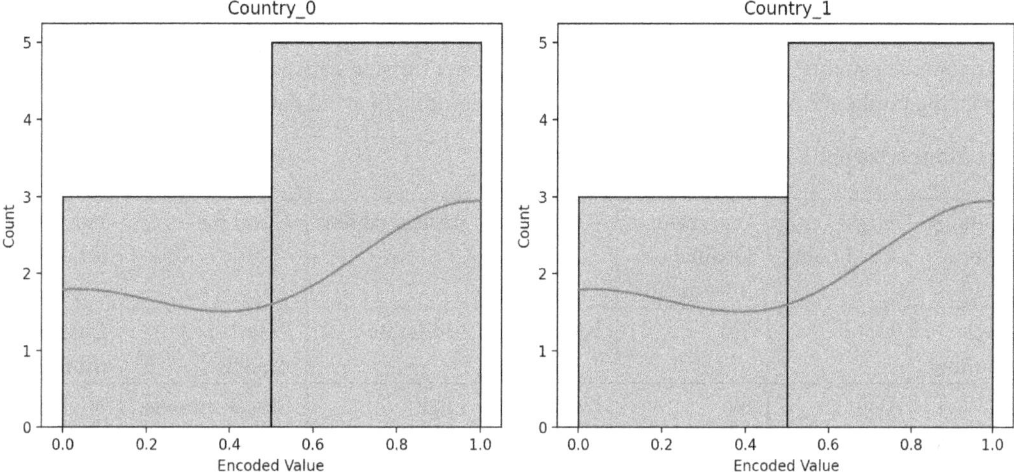

Figure 10.7 – Distribution of Country encoded feature

Let's now discuss some things to keep in mind when encoding features with a binary encoder.

Considerations for binary encoding

When performing binary encoding, there are several important considerations to keep in mind:

- Binary encoding doesn't provide direct interpretability for encoded features. The encoded binary patterns may not have a clear meaning, unlike one-hot encoding, where each binary feature corresponds to a specific category.
- The binary representation can become complex for categories with very high cardinality as the number of binary digits increases logarithmically with the number of categories.
- Some ML algorithms, particularly linear models, may not work well with binary-encoded features. Careful evaluation of algorithm compatibility is necessary.

Now that we've explored the nuances of the different encoding methods, let's transition to summarizing their key differences and considerations for practical application in ML workflows.

Summary

Throughout this chapter, we explored various techniques for encoding categorical variables essential for ML tasks. Label encoding, which assigns unique integers to each category, is straightforward but may inadvertently impose ordinality where none exists. One-hot encoding transforms each category into a binary feature, maintaining categorical independence but potentially leading to high-dimensional datasets. Binary encoding condenses categorical values into binary representations, balancing interpretability, and efficiency particularly well for high-cardinality datasets. Frequency encoding replaces categories with their occurrence frequencies, capturing valuable information about distributional patterns. Target encoding incorporates target variable statistics into categorical encoding, enhancing predictive power while requiring careful handling to avoid data leakage.

Let's summarize our learning in the following table:

Encoding Method	High Cardinality	Preserves Ordinal Information	Collisions	Interpretability	Good for	Not Good for
Label encoding	Good	Yes	No	Moderate	Tree-based models	Linear models
One-hot encoding	Poor	No	No	High	Linear models, **neural networks** (**NNs**)	High-cardinality features
Target encoding	Good	No	Possible	Low	Most algorithms	Small datasets (risk of overfitting)
Frequency encoding	Good	No	Possible	Moderate	Tree-based models	Linear models
Binary encoding	Good	Partially	Possible	Low	Tree-based models	Linear models

Table 10.2 – Comparison of all the encoding techniques

Each method offers distinct advantages depending on the dataset's characteristics and the specific requirements of the modeling task. In the next chapter, we will shift focus to the considerations and methodologies involved in analyzing time series data. Time series data introduces temporal dependencies that require specialized techniques for feature engineering, as we will expand upon in the next chapter.

11

Consuming Time Series Data

In this chapter about time series analysis, we will explore the fundamental concepts, methodologies, and practical applications of time series across various industries. Time series analysis involves studying data points collected over time to identify patterns and trends and make predictions.

In this chapter, we will deep dive into the following topics:

- Understanding the components of time series data
- Types of time series data
- Identifying missing values in time series data
- Handling missing values in time series data
- Analyzing time series data
- Dealing with outliers
- Feature engineering with time series data
- Applying time series techniques in different industries

Technical requirements

The complete code for this chapter can be found in this book's GitHub repository at `https://github.com/PacktPublishing/Python-Data-Cleaning-and-Preparation-Best-Practices/tree/main/chapter11`.

Run the following code to install all the necessary libraries we will use in this chapter:

```
pip install pandas
pip install numpy
pip install matplotlib
pip install statsmodels
```

Understanding the components of time series data

Time series data refers to a sequence of observations or measurements that are collected and recorded *over time*. Unlike non-sequential data, where observations are taken at a single point in time, time series data captures information at multiple points in a sequential order. Each data point in a time series is associated with a specific timestamp, creating a temporal structure that allows trends, patterns, and dependencies to be analyzed over time. Let's discuss the different components of the time series data, starting with the trend.

Trend

The **trend** component represents the long-term movement or direction in the data. It reflects the overall pattern that persists over an extended period, indicating whether the values are generally increasing, decreasing, or remaining relatively constant.

Trends have the following characteristics:

- **Upward trend**: Values systematically increase over time
- **Downward trend**: Values systematically decrease over time
- **Flat trend**: Values remain relatively constant over time

Identifying the trend is essential for making informed decisions about the long-term behavior of the phenomenon being observed. It provides insights into the overall direction and can be valuable for forecasting future trends. In the following section, we will present a use case inspired by the data world that focuses on the trend component.

Analyzing long-term sales trends

In this use case, we aim to analyze a decade-long sales trend to understand the growth pattern of a business from 2010 to 2020. You can find the code for this example in this book's GitHub repository at `https://github.com/PacktPublishing/Python-Data-Cleaning-and-Preparation-Best-Practices/blob/main/chapter11/1.decomposing_time_series/trend.py`. Let's get started:

1. We will start by generating a date range and corresponding sales data for each month over 10 years:

```
date_rng = pd.date_range(start='2010-01-01', end='2020-12-31', freq='M')
sales_data = pd.Series(range(1, len(date_rng) + 1), index=date_rng)
```

2. Then, we must plot the data to visualize the upward trend:

```
plt.figure(figsize=(10, 5))
plt.plot(sales_data, label='Sales Data')
plt.title('Time Series Data with Trend')
plt.xlabel('Time')
plt.ylabel('Sales')
plt.legend()
plt.show()
```

This will result in the following graph:

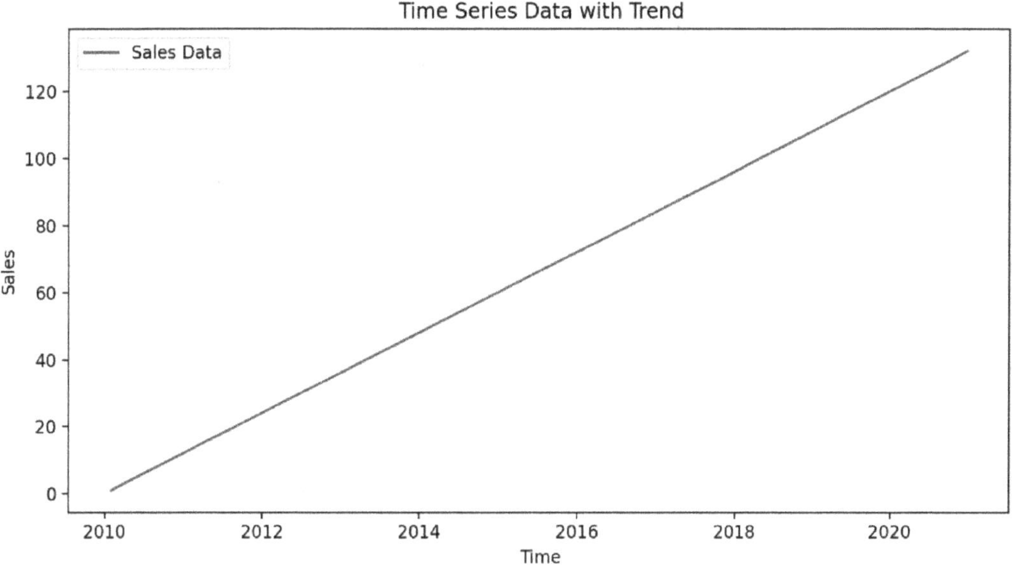

Figure 11.1 – Monthly sales data with upward trend

Figure 11.1 shows a consistent upward trend in sales over the decade. This indicates that the business has been growing steadily.

In our initial analysis, we focused on understanding the overall upward trend in sales data over a decade. This provided us with valuable insights into the long-term growth of the business.

Often, businesses experience fluctuations that recur regularly within specific periods, such as months or quarters. This is known as seasonality. Recognizing these seasonal patterns can be just as crucial as understanding the overall trend as it helps businesses anticipate and prepare for periods of high or low demand. To illustrate this, let's extend our analysis so that it includes seasonality in the sales data.

Seasonality

Seasonality refers to the repetitive and predictable patterns that occur at regular intervals within the time series. These patterns often correspond to specific time frames, such as days, months, or seasons, and can be influenced by external factors such as weather, holidays, or cultural events.

Unlike long-term trends, seasonality spans shorter time frames, exerting a short-term influence on the data. This recurring nature of seasonality allows businesses to anticipate and plan for fluctuations in demand, thereby optimizing their operations and strategies.

> **Important note**
>
> Understanding seasonality helps in identifying recurring patterns and predicting when certain behaviors or events are likely to occur. This information is crucial for accurate forecasting and planning.

In the following section, we will extend the sales use case presented previously while focusing on the seasonal component.

Analyzing long-term sales trends with seasonality

In this part of the use case, we aim to analyze a decade-long sales trend that includes seasonal variations. You can find the full code example here: `https://github.com/PacktPublishing/Python-Data-Cleaning-and-Preparation-Best-Practices/blob/main/chapter11/1.decomposing_time_series/seasonality.py`. Let's get started:

1. We will start by generating a date range and corresponding sales data for each month over 10 years:

```
date_rng = pd.date_range(start='2010-01-01', end='2020-12-31',
freq='M')
seasonal_data = pd.Series([10, 12, 15, 22, 30, 35, 40, 38, 30,
22, 15, 12] * 11, index=date_rng)
```

2. Then, we must plot the data to visualize the seasonal component:

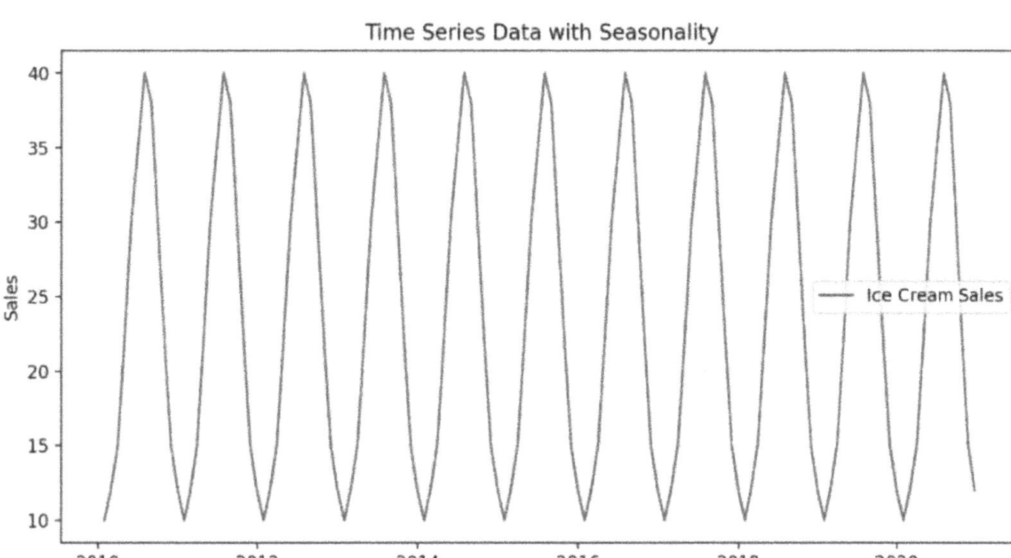

Figure 11.2 – Monthly sales data with seasonality

Figure 11.2 reveals a repeating pattern every 12 months, indicating a clear seasonality in sales. Sales peak around mid-year and drop toward the end and beginning of the year, suggesting higher sales in summer and lower sales in winter. This pattern's consistency over the years can aid in predicting future sales cycles. Understanding these seasonal trends is valuable for inventory management, marketing campaigns, and resource allocation during peak sales periods, allowing businesses to optimize their strategies accordingly.

While identifying trends and seasonality provides valuable insights into sales patterns, real-world data often contains another critical component: noise. In the following section, we will deep dive into noise and extend the sales use case so that it explores how noise affects sales.

Noise

Noise, also known as residuals or errors, represents the random fluctuations or irregularities in the time series data that cannot be attributed to the trend or seasonality. It reflects the variability in the data that is not explained by the underlying patterns.

> **Important note**
>
> While noise is often considered unwanted, it is a natural part of any real-world data. Recognizing and isolating noise is essential for building accurate models and understanding the inherent uncertainty in the time series.

In the following section, we will extend the sales use case presented previously while focusing on noise.

Analyzing sales data with noise

In this use case, we aim to analyze sales data that includes noise, in addition to trends and seasonality. This will help us understand how random fluctuations impact our ability to identify underlying patterns. To follow along with this example, take a look at the following code: `https://github.com/PacktPublishing/Python-Data-Cleaning-and-Preparation-Best-Practices/blob/main/chapter11/1.decomposing_time_series/noise.py`. Let's get started:

1. Let's import the required libraries:

```
import pandas as pd
import matplotlib.pyplot as plt
```

2. We will start by generating a date range for each month over 10 years:

```
date_rng = pd.date_range(start='2010-01-01', end='2020-12-31',
freq='M')
```

3. Then, we must create sales data with noise:

```
np.random.seed(42)
noise_data = pd.Series(np.random.normal(0, 2, len(date_rng)),
index=date_rng)
```

4. Now, we must plot the data to visualize the noise:

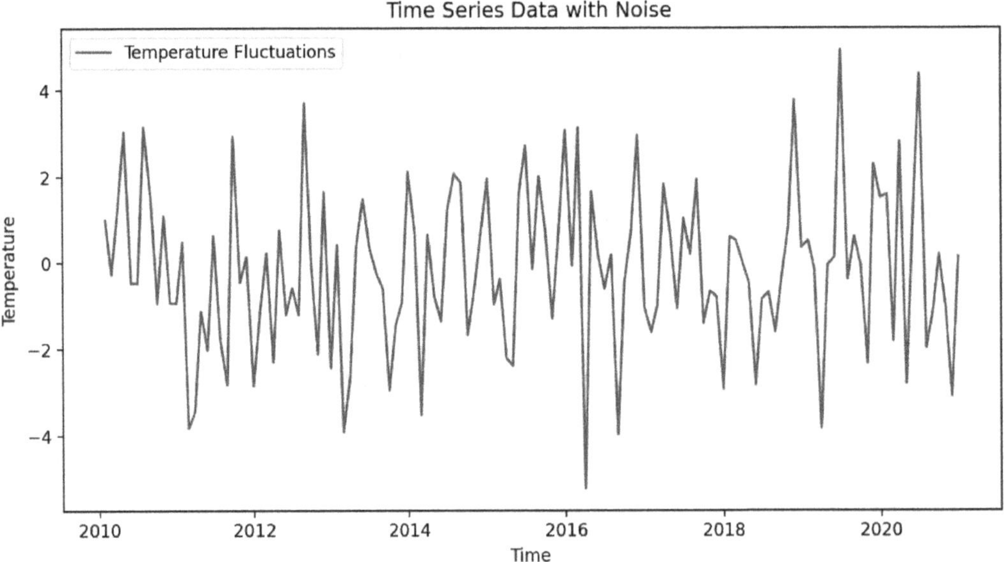

Figure 11.3 – Monthly sales data with noise

Figure 11.3 shows random, unpredictable variations that do not follow any specific pattern. These fluctuations occur over short timeframes, creating instability in the data and making it harder to see any patterns.

Now that we can identify the different time series components, let's have a look at the different types of time series.

Types of time series data

In this section, we will quickly revise the types of time series data – univariate and multivariate – while clarifying their distinctions and showcasing their applications.

Univariate time series data

Univariate time series data consists of a single variable or observation recorded over time. It is a one-dimensional time-ordered sequence, making it simpler to analyze compared to multivariate time series data.

Consider a univariate time series representing the monthly average temperature in a city over several years. You can find the full code here: `https://github.com/PacktPublishing/Python-Data-Cleaning-and-Preparation-Best-Practices/blob/main/chapter11/2.types/univariate.py`.

Let's generate our univariate time series data:

1. First, we will create the data range we want, in this case from `2010-01-01` to `2020-12-31`:

   ```
   date_rng = pd.date_range(start='2010-01-01', end='2020-12-31',
   freq='M')
   ```

2. Then, we must create the corresponding values for the temperatures by adding noise using a **normal distribution** (also known as a Gaussian distribution):

   ```
   temperature_data = pd.Series(np.random.normal(20, 5, len(date_
   rng)), index=date_rng)
   ```

   Let's understand the value parameters:

   - `20`: This is the **mean** of the normal distribution. The noise that's generated will have an average value of 20.
   - `5`: This is the **standard deviation** of the normal distribution. The noise values will typically vary by about ±5 units from the mean. A larger standard deviation means the noise will be more spread out, while a smaller standard deviation means the noise values are closer to the mean.
   - The date range we created previously is passed as an index to the DataFrame.

3. Now, let's plot the univariate time series data:

```
plt.figure(figsize=(10, 5))
plt.plot(temperature_data, label='Temperature Data')
plt.title('Univariate Time Series Data')
plt.xlabel('Time')
plt.ylabel('Temperature (°C)')
plt.legend()
plt.show()
```

This will output the following plot:

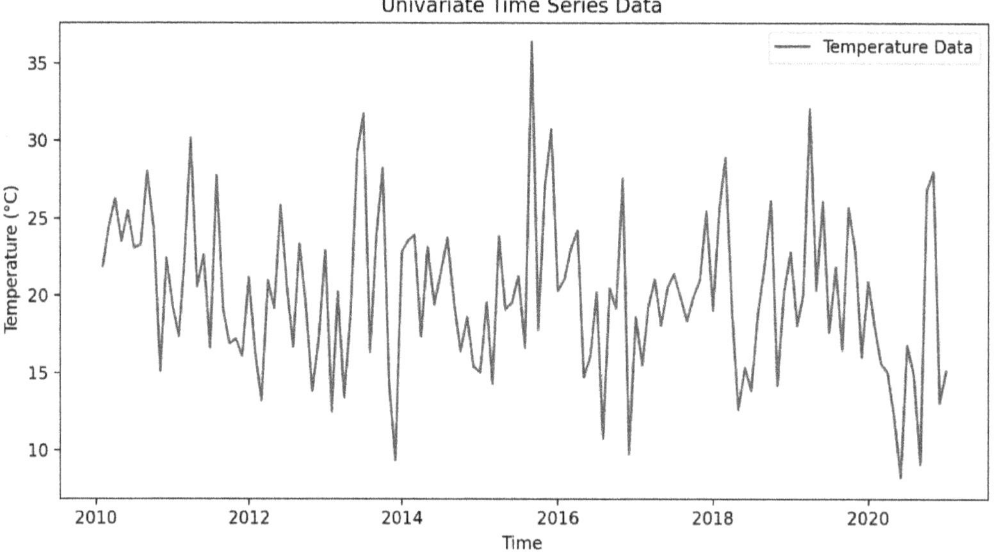

Figure 11.4 – Univariate temperature data

In this example, the univariate time series represents the monthly average temperature. Since the data is randomly generated with a mean of 20°C and some variation (a standard deviation of 5°C), the plot will exhibit random fluctuations around this average temperature.

Understanding the complexities of univariate time series data lays a solid foundation for delving into multivariate time series analysis. Unlike univariate data, which involves observing a single variable over time, multivariate time series data involves monitoring multiple interrelated variables simultaneously.

Multivariate time series data

Multivariate time series data involves multiple variables or observations recorded over time. Each variable is a time-ordered sequence, and the variables may be interdependent, capturing more complex relationships.

Consider a multivariate time series representing both the monthly average temperature and monthly rainfall in a city over several years. You can find the code for this example at `https://github.com/PacktPublishing/Python-Data-Cleaning-and-Preparation-Best-Practices/blob/main/chapter11/2.types/multivariate.py`. Let's get started:

1. Let's add the required libraries for this example:

```
import pandas as pd
import matplotlib.pyplot as plt
import numpy as np
```

2. Now, let's generate an example of multivariate time series data by using the temperature data we created previously and adding a new time series in the same DataFrame representing the rainfall data with different mean and standard deviation values:

```
date_rng = pd.date_range(start='2010-01-01', end='2020-12-31', freq='M')
temperature_data = pd.Series(np.random.normal(20, 5, len(date_rng)), index=date_rng)
rainfall_data = pd.Series(np.random.normal(50, 20, len(date_rng)), index=date_rng)
```

3. Combine all the time series into the same DataFrame, making sure to include both temperature and rainfall data:

```
multivariate_data = pd.DataFrame({'Temperature': temperature_data, 'Rainfall': rainfall_data})

print(multivariate_data.head())
```

4. The combined time series DataFrame is shown here:

```
            Temperature   Rainfall
2010-01-31    19.132623  56.621393
2010-02-28    18.551274  51.249927
2010-03-31    24.502358  65.679049
2010-04-30    27.069077  73.044307
2010-05-31    21.176376  41.317497
```

5. Finally, let's plot the multivariate time series data:

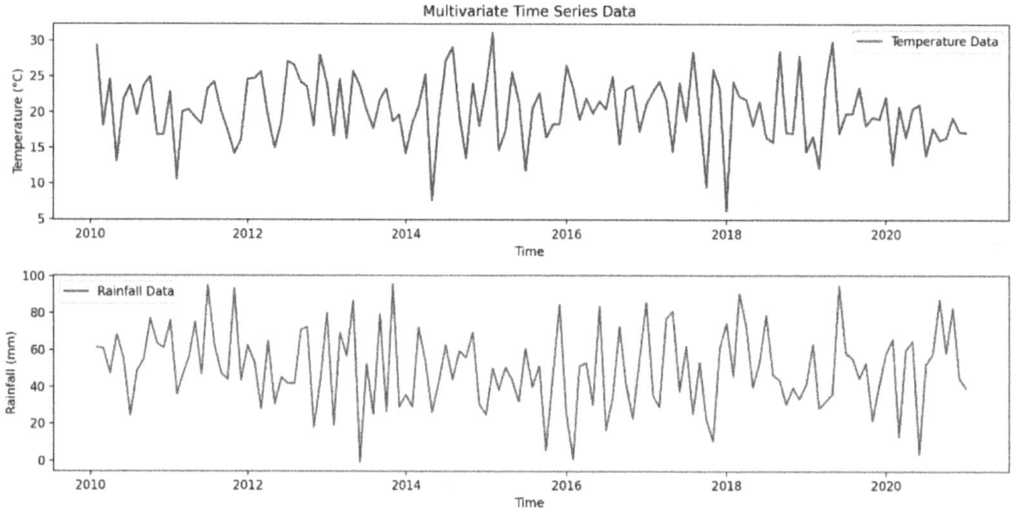

Figure 11.5 – Multivariate data

In this example, the multivariate time series includes both temperature and rainfall data, providing a more comprehensive view of environmental conditions.

Overall, univariate data is simpler to work with, while multivariate data allows us to capture more complex relationships and dependencies between variables over time. Multivariate analysis is essential for tackling real-world challenges in diverse fields such as economics, finance, environmental science, and healthcare, where understanding multifaceted relationships among variables is crucial.

Now that we have a strong understanding of time series data, we can explore methods for cleaning and managing this type of data effectively.

Identifying missing values in time series data

Identifying missing values in time series data is somewhat like identifying missing values in other types of data, but there are specific considerations due to the temporal nature of time series. Since we covered some of these techniques in *Chapter 8*, *Detecting and Handling Missing Values and Outliers*, let's summarize them here and highlight their specific adaptations for analyzing time series data using a stock market analysis use case.

Let's consider a use case where we have daily stock prices (open, high, low, and close) for a particular company over several years. Our goal is to identify missing data in this time series to ensure the integrity of the dataset. You can find the code for this example here: `https://github.com/PacktPublishing/Python-Data-Cleaning-and-Preparation-Best-Practices/blob/main/chapter11/3.missing_values/1.identify_missing_values.py`.

Let's start by generating the data:

1. First, we will generate a date range for business days from January 1, 2020, to December 31, 2023. Here, `freq='B'` is used to generate a range of dates that includes only *business days* (weekdays, excluding weekends):

```
date_range = pd.date_range(start='2020-01-01', end='2023-12-31',
freq='B')  # Business days
```

2. Next, we must generate random stock prices for the date range with a length of *n*:

```
n = len(date_range)
data = {
    'open': np.random.uniform(100, 200, n),
    'high': np.random.uniform(200, 300, n),
    'low': np.random.uniform(50, 100, n),
    'close': np.random.uniform(100, 200, n)
}
```

3. Next, we must create a DataFrame by passing all the separate data points that were created in the previous step:

```
df = pd.DataFrame(data, index=date_range)
```

4. Now, let's introduce random NaN values to simulate some missing values in the data:

```
nan_indices = np.random.choice(n, size=100, replace=False)
df.iloc[nan_indices] = np.nan
```

5. Next, drop random dates to simulate missing timestamps:

```
missing_dates = np.random.choice(date_range, size=50,
replace=False)
```

6. Finally, display the first few rows of the DataFrame:

```
                  open        high        low       close
2020-01-01  137.454012  262.589138  55.273685  183.849183
2020-01-02  195.071431  288.597775  82.839005  180.509032
2020-01-03  173.199394  261.586319  91.105158  182.298381
2020-01-06  159.865848  223.295947  69.021000  193.271051
2020-01-07         NaN         NaN        NaN         NaN
```

The key thing to notice here is that we have two kinds of missing data:

- Complete rows missing, so one full date index is not available
- Some observations in some of the columns are missing for the current date

We will mainly address the first case here as we covered the second case in *Chapter 8, Detecting and Handling Missing Values and Outliers*. Let's start with the simple but effective `isnull()` method.

Checking for NaNs or null values

Unlike regular datasets, time series data points are ordered in time. Missing values can break the continuity and affect the analysis of trends and seasonal patterns. Let's use the `isnull()` method to identify missing timestamps. Here, we are looking to find complete rows that are missing from the dataset:

1. To check which dates are missing from a time series DataFrame, we need to create a full date range (with no missing values) in the frequency of our current DataFrame index and compare it against the date range we have in our current DataFrame. Here, we are creating a complete date range for business days:

    ```
    complete_index = pd.date_range(start=df.index.min(), end=df.
    index.max(), freq='B')
    ```

2. To quickly see the missing index points, the DataFrame must be reindexed to this complete date range so that we can identify any missing timestamps:

    ```
    df_reindexed = df.reindex(complete_index)
    ```

3. Now, we can use the `isnull()` method to identify any missing timestamps:

    ```
    missing_timestamps = df_reindexed[df_reindexed.isnull().
    any(axis=1)]
    ```

Here, we can see that there are some missing timestamps in the data:

```
print(f"\nPercentage of Missing Timestamps: {missing_timestamps_
percentage:.2f}%")
Percentage of Missing Timestamps: 14.09%
```

The analysis so far tells us that we have complete dates missing from the dataset. Now, let's add some visual plots to help us better see the gaps in the data.

> **Note**
>
> As presented in *Chapter 8, Detecting and Handling Missing Values and Outliers*, you can use the `isnull()` method to see how many nulls we have in each column – for example, `missing_values = df.isnull().sum()`.

Visual inspection

Visualizing the data can help us identify missing values and patterns of missingness. Plots can reveal gaps in data that are not immediately obvious from a tabular inspection.

Continuing with the example from the previous section, let's plot our time series data and mark any missing values on our graph:

1. Plot the closing prices:

```
plt.figure(figsize=(14, 7))

plt.plot(df.index, df['close'], linestyle='-', label='Closing
Price', color='blue')
```

2. Mark missing timestamps with vertical lines:

```
for date in missing_dates:
    plt.axvline(x=date, color='red', linestyle='--',
linewidth=1)

plt.title('Daily Closing Prices with Missing Timestamps and NaN
Values Highlighted')
plt.xlabel('Date')
plt.ylabel('Closing Price')
plt.legend()
plt.grid(True)
plt.show()
```

This will result in the following plot:

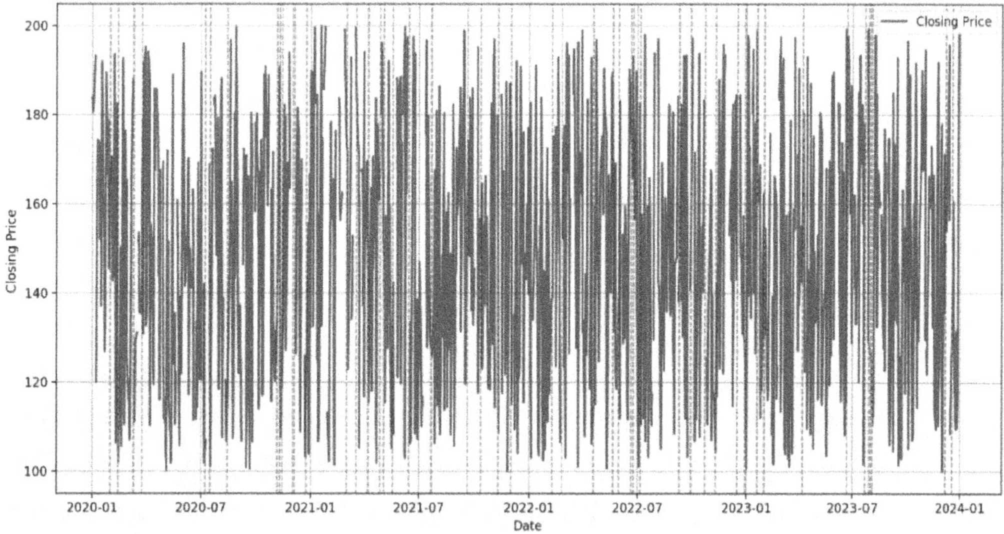

Figure 11.6 – Daily closing prices with missing timestamps highlighted

In *Figure 11.6*, the closing prices are plotted in blue with markers, while missing timestamps are highlighted with dotted lines, making it easy to see gaps in the data. Now, let's explore the final method, known as lagged analysis. In this method, we create a lagged version of the series and compare it with the original to detect inconsistencies.

> **Note**
>
> In *Chapter 3, Data Profiling – Understanding Data Structure, Quality, and Distribution* we demonstrated various data profiling methods. You can apply similar techniques to time series data by using the built-in gap analysis feature. Simply pass `tsmode=True` when creating the profile report – for example, `profile = ProfileReport(df, tsmode=True)`.

As we move forward, it's essential to explore effective strategies for handling missing data in time series.

Handling missing values in time series data

Missing values are a common challenge in time series data and can arise due to various reasons, such as sensor failures, data transmission issues, or simply the absence of recorded observations. As we've discussed, two main scenarios often arise:

- **Some null values in features**: Imagine a stock market analysis where daily trading data is collected. While all trading days are accounted for, the volume of shares traded on certain days may be missing due to reporting errors. This scenario presents a challenge: how do you maintain the integrity of the dataset while ensuring that analyses remain robust?
- **Complete rows are missing**: Conversely, consider a weather monitoring system that records daily temperatures. If entire days of data are missing – perhaps due to sensor failures – this poses a significant issue. Missing timestamps means you cannot simply fill in values; the absence of data for those days disrupts the entire time series.

In the next section, we will focus on solving the first scenario and consider the existence of null values in some features. Once we have done this, we can adjust it for the second one.

Removing missing data

Removing missing data is a straightforward approach, but it should be done with caution while considering the impact on the overall dataset. Here are some scenarios where removal might be appropriate:

- If the missing values constitute a small percentage of the dataset (for example, less than 5%), removing them might be feasible. This approach works well if the data loss does not significantly impact the analysis or the conclusions drawn from the dataset. For example, in a dataset with 10,000 time points, if 50 points are missing, removing these 50 points (0.5% of the data) might not significantly affect the overall analysis.

- If imputing missing values would introduce too much uncertainty, especially if the values are critical and cannot be accurately estimated. This scenario is common when the missing values are highly unpredictable data, making imputation unreliable.
- If missing values occur completely at random and do not follow any systematic pattern. An example of this is sensor data where occasional random failures cause missing readings, but there is no underlying pattern to these failures.

Let's revisit the stock market use case and see how we can drop the null values to see what effect this has on the dataset.

Removing missing data in the stock market use case

In our stock prices data scenario, we will add some NaN values and evaluate the impact of removing them. You can find the full code here: `https://github.com/PacktPublishing/Python-Data-Cleaning-and-Preparation-Best-Practices/blob/main/chapter11/3.missing_values/2.remove_missing_values.py`. Let's get started:

1. Continuing with the example from the previous section, we will create the stock data with the different features. Then, we will randomly select some indexes from specific features (for example, `close` and `open`) so that we can map the values for each feature of that index to a NaN value:

   ```
   nan_indices_close = np.random.choice(df.index, size=50,
   replace=False)
   nan_indices_open = np.random.choice(df.index, size=50,
   replace=False)
   ```

2. Then, we will map the indices that were randomly selected before to NaN values:

   ```
   df.loc[nan_indices_close, 'close'] = np.nan
   df.loc[nan_indices_open, 'open'] = np.nan
   ```

3. Let's check how many NaNs or null values are available in the data:

   ```
   missing_values = df.isnull().sum()
   Percentage of Missing Values in Each Column:
   open      4.793864
   high      0.000000
   low       0.000000
   close     4.793864
   ```

   As expected, some null values were introduced on the `open` and `close` features.

   Let's check how many rows we have before removing any based on the nulls we have in the dataset:

   ```
   print(f"\nNumber of rows before dropping NaN values: {len(df)}")
   Number of rows before dropping NaN values: 1043
   ```

4. At this stage, we will drop any rows that have NaN values in either the `close` or `low` columns:

```
df_cleaned = df.dropna()
print(f"\nNumber of rows after dropping NaN values: {len(df_
cleaned)}")
----
Number of rows after dropping NaN values: 945
```

5. Let's plot the time series data:

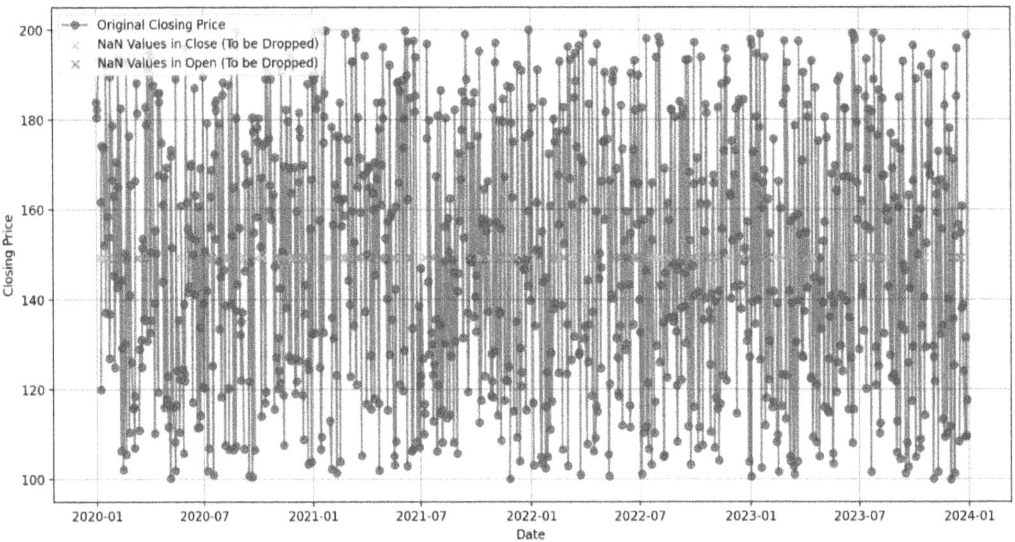

Figure 11.7 – Daily closing prices with missing data to be dropped/flagged

As shown in *Figure 11.7*, the original closing prices are plotted, and points that were dropped due to missing values are highlighted with red "x" markers. Remember that even with selective dropping, removing rows can lead to a loss of useful information as it reduces the sample size, which can decrease the statistical power of the analysis and affect the generalizability of the results.

In scenarios where retaining every timestamp is crucial but missing values within features need to be addressed, forward and backward filling offer practical solutions. These methods allow us to maintain the chronological integrity of time series data while efficiently filling in missing values based on adjacent observations. Let's explore how forward and backward filling can effectively handle missing data in time series analyses.

Forward and backward fill

Forward fill (ffill) and **backward fill (bfill)** are methods of imputing missing values by propagating the last known value forward or the next known value backward in the time series, respectively.

When dealing with time series backfilling, the choice between ffill and bfill depends on several factors and use cases. Here's an overview of when to use each approach and the thought process behind these decisions:

- **Ffill**: Forward filling, also known as **last observation carried forward** (LOCF), propagates the last known value forward to fill in missing data points.

  Here's when you should use it:

  - When you believe the most recent known value is the best predictor of missing future values
  - In financial time series, where carrying forward the last known price is often a reasonable assumption
  - When dealing with slowly changing variables where persistence is a good assumption
  - In scenarios where you want to maintain the most recent state until new information becomes available

  If you're still uncertain or find yourself pondering which method to use, answering "yes" to at least two of the following three questions can guide you in the right direction:

  - Is the variable likely to remain relatively stable over short periods?
  - Would using the last known value be a reasonable assumption for the missing data?
  - Is it more important to reflect the most recent known state rather than potential future changes?

- **Bfill**, on the other hand, propagates the next known value backward to fill in missing data points.

  Here's when you should use it:

  - When you have more confidence in future values than past values
  - In scenarios where you want to retroactively apply known outcomes to previous missing periods
  - When you're dealing with lagged effects where future events influence past missing data
  - In cases where you want to align data with the next known state rather than the previous one

  If you're still uncertain or find yourself pondering which method to use, answering "yes" to the following questions will guide you in the right direction:

  - Is the next known value likely to be more representative of the missing data than the previous known value?
  - Are you dealing with a scenario where future information should inform past missing values?
  - Would aligning with the next known state provide more meaningful insights for your analysis?

In practice, choosing between ffill and bfill often requires a combination of domain expertise, understanding of the data generation process, and consideration of the specific analytical goals. It's also worth experimenting with both methods and comparing the results to see which provides more meaningful and accurate insights for your particular use case.

As always, there are some important considerations when using ffill and bfill in time series data. Let's expand on those:

- **Sequential nature**: The sequential nature of time series data is indeed crucial for both ffill and bfill methods. Both methods rely on the assumption that adjacent data points are related, which is fundamental to time series analysis.
- **Ffill and increasing trends**: Ffill can be appropriate for increasing trends as it carries forward the last known value, potentially underestimating the true value in an upward trend. However, it may lead to a "staircase" effect in strongly increasing trends, potentially understating the rate of increase.
- **Bfill and decreasing trends**: Bfill can be suitable for decreasing trends as it pulls back future lower values, potentially overestimating the true value in a downward trend. It may create a similar "staircase" effect in strongly decreasing trends, potentially overstating the rate of decrease.
- The choice between ffill and bfill should consider not just the direction of the trend, but also its *strength* and the *length* of missing data periods. For subtle trends, either method might be appropriate, and the choice may depend more on other factors, such as the nature of the data or the specific analysis goals.
- Both methods can indeed propagate errors if the missing values are inconsistent with surrounding data points. This is particularly problematic for long stretches of missing data, where the filled values may significantly deviate from the true underlying pattern.
- **Handling outliers**: If an outlier precedes or follows a stretch of missing data, ffill or bfill can propagate this anomalous value, distorting the series.
- **Assumption of data continuity**: Both methods assume that the missing data can be **reasonably approximated** by adjacent known values, which may not always be true. For variables that can change abruptly or have discontinuities, these methods may be inappropriate.

Let's revisit the stock prices example and see how we can fill the missing values on the columns with nulls.

Filling nulls in the stock market use case

In this example, we will not be focusing on missing indexes, just on the missing data in some of the features available. Let's deep dive into the code – as always, you can find the full end-to-end code at `https://github.com/PacktPublishing/Python-Data-Cleaning-and-Preparation-Best-Practices/blob/main/chapter11/3.missing_values/3.back_forward_fill.py`:

1. This code introduces random missing values into the `close` and `open` columns of a DataFrame (`df`). It begins by randomly selecting 50 indices from the DataFrame's index using `np.random.`

choice. The selected indices are stored in two variables, nan_indices_close and nan_indices_open, which correspond to the rows where missing values will be inserted:

```
nan_indices_close = np.random.choice(df.index, size=50,
replace=False)
nan_indices_open = np.random.choice(df.index, size=50,
replace=False)
```

2. The following code uses the .loc accessor to assign NaN to the close column at the indices specified by nan_indices_close, and similarly to the open column at the indices specified by nan_indices_open. Effectively, this creates 50 random missing values in each of these columns, which can be useful for simulating real-world data scenarios or testing data handling techniques:

```
df.loc[nan_indices_close, 'close'] = np.nan
df.loc[nan_indices_open, 'open'] = np.nan
```

3. Fill NaN values using ffill and bfill:

```
df['close_ffill'] = df['close'].ffill() # Forward Fill
df['close_bfill'] = df['close'].bfill() # Backward Fill
```

4. Let's see the result:

```
print(df[['open', 'close', 'close_ffill', 'close_bfill']].
head(20)) # Show first 20 rows
```

This will display the following output:

	open	close	close_ffill	close_bfill
2020-01-01	137.454012	183.849183	183.849183	183.849183
2020-01-02	195.071431	180.509032	180.509032	180.509032
2020-01-03	173.199394	182.298381	182.298381	182.298381
2020-01-06	159.865848	193.271051	193.271051	193.271051
2020-01-07	**115.601864**	**NaN**	**193.271051**	**120.028202**
2020-01-08	115.599452	120.028202	120.028202	120.028202
2020-01-09	105.808361	161.678361	161.678361	161.678361
2020-01-10	186.617615	174.288149	174.288149	174.288149
2020-01-13	160.111501	173.791739	173.791739	173.791739
2020-01-14	170.807258	152.144902	152.144902	152.144902
2020-01-15	**102.058449**	**NaN**	**152.144902**	**137.111294**
2020-01-16	196.990985	137.111294	137.111294	137.111294

As we can see, on 2020-01-07 and 2020-01-15, there are missing values (NaN) in the close column. This indicates that the closing prices for these dates were not recorded or are otherwise unavailable.

As we've learned, the ffill method (`close_ffill`) fills missing values with the last valid observation:

- For `2020-01-07`, the closing price is filled with the last known value from `2020-01-06 (193.27)`
- For `2020-01-15`, the missing value is filled with the last valid price from `2020-01-14 (152.14)`

On the other hand, the bfill method (`close_bfill`) fills missing values with the next valid observation:

- For `2020-01-07`, since no subsequent valid price is recorded immediately, it takes the closing price from `2020-01-08 (120.03)`
- For `2020-01-15`, the value is filled with the next known price from `2020-01-16`

Let's take a closer look at what happened in the data after we performed the different filling methods:

- On `2020-01-07`, ffill overestimates the missing value compared to bfill, which aligns more closely with the next known value
- On `2020-01-15`, ffill and bfill provide different estimates, with ffill potentially overestimating the value compared to bfill

As a general recommendation, we need to investigate the pattern of missing values. If the missing values are *random and sparse*, either method might be appropriate. However, if there is a systematic pattern, more sophisticated imputation methods might be needed, such as *interpolation*. Interpolation allows us to estimate missing data points by leveraging the existing values in the dataset, providing a more nuanced approach that can capture trends and patterns over time. We'll discuss this in more detail next.

Interpolation

Interpolation is a method for estimating missing values by filling in the gaps based on the surrounding data points. Unlike ffill and bfill, which copy existing values, interpolation uses mathematical techniques to estimate missing values. There are different interpolation techniques and applications. So, let's have a look at the available options, along with their considerations:

- **Linear interpolation**: Linear interpolation connects two adjacent known data points with a straight line and estimates the missing values along this line. It is the simplest form of interpolation and assumes a linear relationship between the data points. It is suitable for datasets where changes between data points are expected to be linear or nearly linear. It is commonly used in financial data, temperature readings, and other environmental data where gradual changes are expected.
- **Polynomial interpolation**: Polynomial interpolation fits a polynomial function to the known data points and uses this function to estimate missing values. Higher-order polynomials can capture more complex relationships between data points. It is suitable for datasets with non-linear trends, and it is usually used in scientific and engineering applications, where data follows a polynomial trend.

- **Spline interpolation**: Spline interpolation uses piecewise polynomials, typically cubic splines, to fit the data points, ensuring smoothness at the data points and providing a smooth curve through the data. It is suitable for datasets requiring smooth transitions between data points and is commonly used in computer graphics, signal processing, and environmental data.

Let's use interpolation in our use case.

Interpolation in the stock market use case

Consider the same time series dataset presented previously with missing values. In this case, we want to impute these missing values using different interpolation methods. You can find the full code examples in this book's GitHub repository: `https://github.com/PacktPublishing/Python-Data-Cleaning-and-Preparation-Best-Practices/blob/main/chapter11/3.missing_values/4.interpolation.py`. Let's get started:

1. The following code introduces random missing values into the `close` and `open` columns of our DataFrame (`df`), as we did in the previous section:

```
nan_indices_close = np.random.choice(df.index, size=50,
replace=False)
nan_indices_open = np.random.choice(df.index, size=50,
replace=False)
df.loc[nan_indices_close, 'close'] = np.nan
df.loc[nan_indices_open, 'open'] = np.nan
```

2. The following line of code is used to fill in missing values in the `close` column of our DataFrame (`df`) using linear interpolation. The code specifically employs **linear interpolation**, where the missing values are estimated by drawing a straight line between the nearest known data points before and after the missing value:

```
df['close_linear'] = df['close'].interpolate(method='linear')
```

3. We can interpolate missing values using polynomial interpolation by changing the method argument to `method='polynomial'`. This specifies that the interpolation should be done using a polynomial function of `order=3`. The `order` argument indicates the degree of the polynomial to be used. In this case, a cubic polynomial (third degree) is used, which means the function that estimates the missing values will be a curve, potentially providing a better fit for more complex data trends compared to a simple straight line (as in linear interpolation):

```
df['close_poly'] = df['close'].interpolate(method='polynomial',
order=3)
```

4. We can interpolate missing values using spline interpolation by changing the method to `method='spline'`. This specifies that the interpolation should be done using spline interpolation, which is a piecewise polynomial function that ensures smoothness at the data points. The `order=3` argument indicates the degree of the polynomial used in each piece

of the spline. In this case, a cubic spline (third-degree polynomial) is used, meaning that the interpolation will involve fitting cubic polynomials to segments of the data:

```
df['close_spline'] = df['close'].interpolate(method='spline',
order=3)
```

5. Now, let's plot the interpolated data:

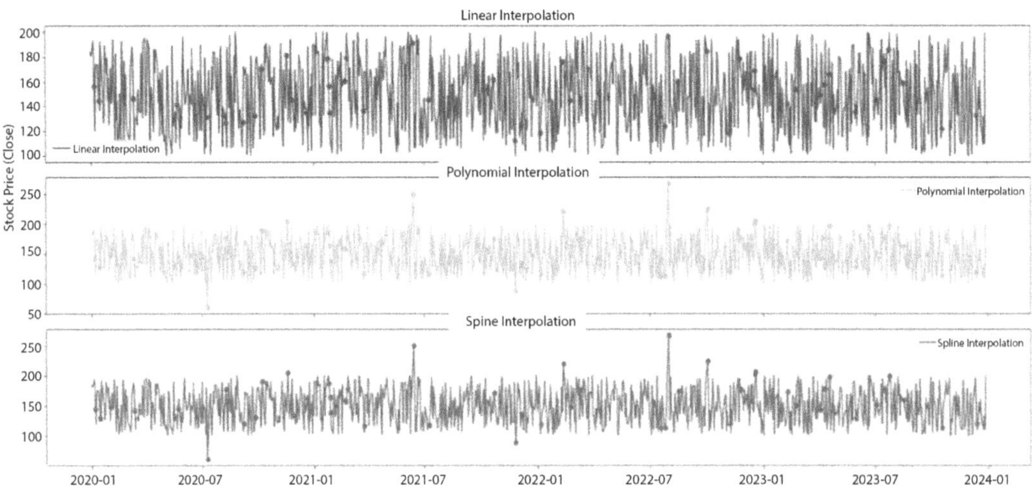

Figure 11.8 – Daily closing prices interpolated

In *Figure 11.8*, we can see how the data changes with the different interpolation methods. To better grasp the differences, let's have a look at the actual data after interpolation, as shown in *Figure 11.9*:

	open	high	low	close	close_linear	close_poly	close_spline
2020-01-01	137.454012	262.589138	55.273685	183.849183	183.849183	183.849183	183.849183
2020-01-02	195.071431	288.597775	82.839005	180.509032	180.509032	180.509032	180.509032
2020-01-03	173.199394	261.586319	91.105158	182.298381	182.298381	182.298381	182.298381
2020-01-06	159.865848	223.295947	69.021000	193.271051	193.271051	193.271051	193.271051
2020-01-07	115.601864	202.440078	88.780593	NaN	156.649626	142.704592	143.173016
2020-01-08	115.599452	287.009887	98.223833	120.028202	120.028202	120.028202	120.028202
2020-01-09	105.808361	202.126941	60.188323	161.678361	161.678361	161.678361	161.678361
2020-01-10	186.617615	287.470167	76.166482	174.288149	174.288149	174.288149	174.288149
2020-01-13	160.111501	252.893713	64.356898	173.791739	173.791739	173.791739	173.791739
2020-01-14	170.807258	293.906770	89.642721	152.144902	152.144902	152.144902	152.144902
2020-01-15	102.058449	279.878324	78.879668	NaN	144.628098	127.403857	128.666028
2020-01-16	196.990985	299.793411	81.729121	137.111294	137.111294	137.111294	137.111294
2020-01-17	183.244264	235.071182	89.895708	192.076661	192.076661	192.076661	192.076661
2020-01-20	121.233911	276.718829	69.798524	158.444891	158.444891	158.444891	158.444891
2020-01-21	118.182497	240.193091	95.754503	153.833194	153.833194	153.833194	153.833194
2020-01-22	118.340451	247.987562	76.651443	126.885351	126.885351	126.885351	126.885351
2020-01-23	130.424224	262.750546	57.897741	136.844892	136.844892	136.844892	136.844892
2020-01-24	152.475643	287.367711	84.794956	189.534609	189.534609	189.534609	189.534609
2020-01-27	143.194502	298.408347	89.663068	166.646305	166.646305	166.646305	166.646305

Figure 11.9 – Table of the daily closing prices interpolated

Let's compare the different interpolation methods and come to some conclusions:

On 2020-01-07, we have the following:

- **Linear interpolation**: 156.649626
- **Polynomial interpolation**: 142.704592
- **Spline interpolation**: 143.173016

On 2020-01-15, we have the following:

- **Linear interpolation**: 144.628098
- **Polynomial interpolation**: 127.403857
- **Spline interpolation**: 128.666028

Based on this data, linear interpolation seems to provide higher estimates compared to polynomial and spline interpolation. It assumes a linear trend between data points, which might not be accurate for non-linear data. Polynomial interpolation seems to provide lower estimates and capture more complex relationships but can be prone to overfitting. Finally, spline interpolation provides smooth estimates that are intermediate between linear and polynomial interpolation, offering a balance between simplicity and accuracy. In this specific case, we would go with spline interpolation as it provides a smooth curve that avoids abrupt changes, and the results are more realistic and closely aligned with the expected trends in the data. While spline interpolation is recommended based on the provided data, it is essential to validate the interpolated values against known data points or domain knowledge.

> **Note**
>
> Interpolation methods such as linear, polynomial, and spline interpolation can also be used to deal with *outliers* in time series data.

Choosing and tuning interpolation arguments for filling missing values involves understanding the characteristics of your data and the specific needs of your analysis. For straightforward data with a linear trend, linear interpolation is efficient and effective. However, if your data exhibits non-linear patterns, polynomial interpolation can provide a better fit, with the degree of the polynomial (`order`) influencing the complexity of the curve; lower orders work well for simpler trends, while higher orders may capture more detail but risk overfitting. Spline interpolation offers a smooth and flexible approach, with cubic splines (`order=3`) being commonly used for their balance of smoothness and flexibility. To tune these methods, start with simpler approaches and test more complex ones progressively while monitoring for overfitting and ensuring the fit aligns with the data's underlying trends. Employ cross-validation, visual inspection, and statistical metrics to evaluate and refine your interpolation choices.

Now that we have explored the various techniques for handling missing data in time series, it's essential to summarize the different filling methods to understand their unique applications and effectiveness.

Comparing the different methods for missing values

Handling missing values in time series data is an involved process that requires thoughtful consideration of the specific context and characteristics of the dataset. The decision to drop values, use bfill, or apply interpolation should be guided by a careful assessment of the impact on subsequent analyses and the preservation of critical information within the time series. The following table summarizes the different techniques and can be used as a guide:

Method	When to Use	Pros	Cons
Dropping missing values	A small percentage of missing values	- Simplicity - Avoids imputation uncertainty	- Information loss - Potential bias
Bfill	Missing values are expected to precede consistent values	- Preserves the general trend - Suitable for increasing trends	- Propagates errors if missing values are not similar to the following values
Ffill	Missing values are expected to follow consistent values	- Simple to implement - Maintains recent state until new data is available	- Can misrepresent data if trends change - Propagates errors if missing values differ from previous values
Linear interpolation	Missing values need to be estimated based on neighboring data points	- Simple and easy to implement - Preserves overall trend	- May not capture non-linear trends - Sensitive to outliers
Polynomial interpolation	Missing values need to be estimated with more complex relationships	- Captures complex relationships - Flexible with polynomial order	- Can lead to overfitting and oscillations - Computationally intensive
Spline interpolation	Missing values need to be estimated with smooth transitions	- Provides a smooth curve - Avoids oscillations of high-order polynomials	- More complex to implement - Computationally intensive

Table 11.1 – Comparison of the different methods to handle missing data in time series

Having examined the various methods for filling missing values in time series data, it is equally important to address another critical aspect: the correlation of a time series with its own lagged values.

Analyzing time series data

Autocorrelation and **partial autocorrelation** are crucial tools in time series analysis that provide insights into data patterns and guide model selection. For outlier detection, they help distinguish between genuine anomalies and expected variations, leading to more accurate and context-aware outlier identification.

Autocorrelation and partial autocorrelation

Autocorrelation refers to correlating a time series with its own lagged values. Simply put, it measures how each observation in a time series is related to its past observations. Autocorrelation is a crucial concept in understanding the temporal dependencies and patterns present in time series data.

Partial autocorrelation function (**PACF**), on the other hand, is a statistical tool that's used in time series analysis to measure the correlation *between a time series and its lagged values after removing the effects of intermediate lags*. It provides a more direct measure of the relationship between observations at different time points, excluding the indirect effects of shorter lags.

Both autocorrelation and partial autocorrelation help in the following cases:

- **Temporal patterns**: They help identify patterns that repeat over time. This is crucial for understanding the inherent structure of the time series data.
- **Stationarity assessment**: They help in assessing the stationarity of a time series. A lack of stationarity can impact the reliability of statistical analyses and model predictions.
- **Lag selection for models**: They guide the selection of appropriate lags for time series models, such as **autoregressive** (**AR**) components in **autoregressive moving average** (**ARIMA**) models.
- **Seasonality detection**: Significant peaks in the **autocorrelation function** (**ACF**) plot at specific lags indicate the presence of seasonality, providing insights for further analysis.
- **Anomaly detection**: Unusual patterns in the autocorrelation function may suggest anomalies or outliers in the data, prompting investigation and cleaning.

Now, let's perform an ACF and PACF analysis on the `close_filled` series from our stock price dataset. This analysis will help us determine the appropriate parameters (`p` and `q`) for the ARIMA modeling we will perform in the following section.

ACT and PACF in the stock market use case

We will continue with the example we've used so far and add the ACT and PACF charts. As always, you can have a look at the full code here: `https://github.com/PacktPublishing/Python-Data-Cleaning-and-Preparation-Best-Practices/blob/main/chapter11/4.analisis/autocorrelation.py`. Let's get started:

1. Create an autocorrelation plot:

```
plot_acf(df['close'].dropna(), lags=40, ax=plt.gca())
```

2. Create a partial autocorrelation plot:

```
plot_pacf(df['close'].dropna(), lags=40, ax=plt.gca())
```

The resultant plots are shown here:

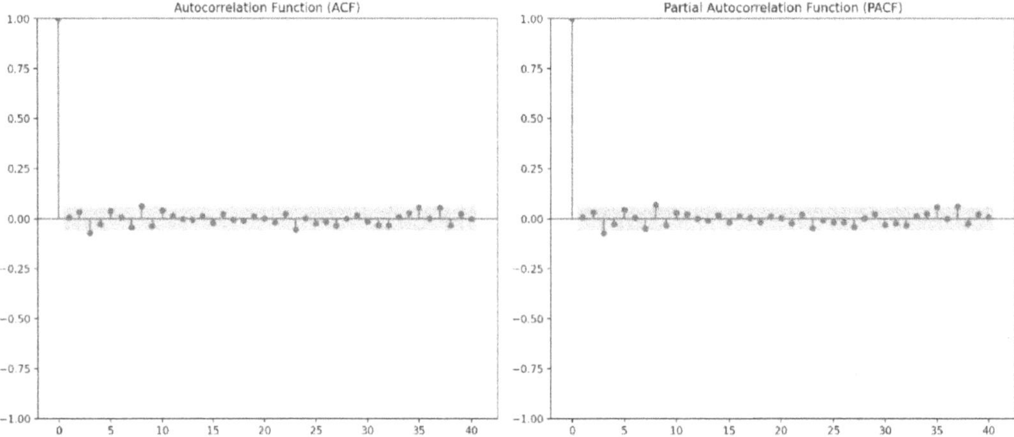

Figure 11.10 – ACF and PACF plots

Let's explain what we can see in the preceding charts. Here's what we can see for ACF:

- The ACF plot shows the correlation between the series and its lagged values at various lags (`lags=40`, in this example). The *X*-axis of the ACF plot represents the number of lags, indicating how many points back in time the correlation is being calculated.
- The *Y*-axis of the ACF plot represents the correlation coefficients between the original time series and its lagged values. The correlation values range from -1 to 1.

- The blue shaded area represents the confidence interval. Bars that extend beyond the shaded area are considered statistically significant and indicate strong autocorrelation at those lags, which suggest potential values for the *q parameter in the ARIMA model* (MA order), as we will see in the following section.
- Significant peaks at regular intervals indicate the presence of *seasonality* in the time series data.
- If there is a significant autocorrelation at lag 1 in the ACF plot (a spike that goes beyond the blue-shaded region, as in our case), it suggests that the series has a strong correlation *with its immediate previous value*. This might indicate that the series is *not stationary* and may need differencing (d > 0).

Here's what we can see for PACF:

- The PACF plot shows the correlation between the series and its lagged values after removing the effects explained by shorter lags.
- Significant spikes in the PACF plot indicate that lag 1 and potentially lag 2 could be good candidates for the *p parameter in the ARIMA model* (AR order).

> **Note**
>
> When we specify `lags=40` in the context of ACF and PACF plots, we are examining the autocorrelation and partial autocorrelation of the time series at 40 different lag intervals. This means we will see how the series is correlated with itself from *lag 1 up to lag 40*.

ACF and PACF plots are crucial for identifying the underlying structure in a time series. In the next section, we will link the ACF and PACF analysis to outlier detection and handling to ensure our time series model captures the underlying patterns accurately.

Dealing with outliers

Time series data often exhibit seasonal patterns (for example, sales spikes during holidays) and trends (for example, gradual growth over the years). An outlier in this context might not be an anomaly; rather, it could reflect a normal seasonal effect or a change in the underlying trend. For example, a sudden spike in retail sales during Black Friday is expected and should not be treated as an outlier. Techniques such as **seasonal decomposition of time series** (**STL**), autocorrelation, and seasonal indices can aid in understanding the expected behavior of the data, thus providing a clearer basis for identifying outliers.

Identifying outliers with seasonal decomposition

One way to identify outliers in time series is to decompose the series into trend, seasonality, and residual components, as outliers are often identified in the residual component. To decompose the series into trend, seasonality, and residual components, we can use the STL method. This method helps in identifying and handling outliers by analyzing the residual component, which ideally should be white noise. Let's see how we can do this using the stock market data. You can find the full code example at `https://github.com/PacktPublishing/Python-Data-Cleaning-and-Preparation-Best-Practices/blob/main/chapter11/5.outliers/1.seasonal_decomposition.py`:

```
result = seasonal_decompose(df['close'], model='additive', period=252)
```

In this code snippet, we decompose the time series while assuming there are 252 business days in a year. We will also calculate the Z-scores of residuals to identify outliers using the following code:

```
df['resid_z'] = zscore(df['residual'].dropna())
```

Finally, let's plot the decomposed series:

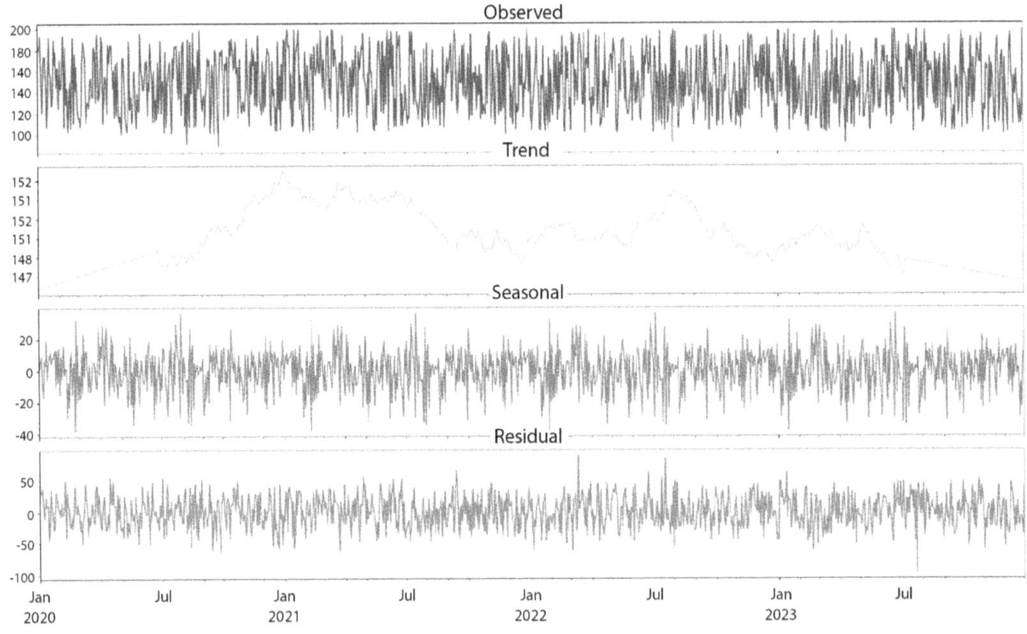

Figure 11.11 – Decomposed time series

Outliers can be detected by analyzing the residual component. Significant deviations from zero or sudden spikes in the residual component indicate potential outliers:

	close	close	trend	seasonal	residual	resid_z
2020-01-01	183.849183	183.849183	146.351742	9.229125	28.268317	1.088995
2020-01-02	180.509032	180.509032	146.367165	1.849790	32.292077	1.244582
2020-01-03	182.298381	182.298381	146.382589	6.832658	29.083133	1.120501
2020-01-06	193.271051	193.271051	146.398013	8.801893	38.071145	1.468043
2020-01-07	154.425397	154.425397	146.413436	-8.351888	16.363848	0.628682
2020-01-08	120.028202	120.028202	146.428860	-16.741557	-9.659102	-0.377554
2020-01-09	161.678361	161.678361	146.444284	6.465091	8.768986	0.335010
2020-01-10	174.288149	174.288149	146.459707	3.880661	23.947782	0.921932
2020-01-13	173.791739	173.791739	146.475131	19.387879	7.928730	0.302519
2020-01-14	152.144902	152.144902	146.490555	-5.486037	11.140384	0.426705
2020-01-15	106.845872	106.845872	146.505978	4.755037	-44.415144	-1.721474
2020-01-16	137.111294	137.111294	146.521402	9.459234	-18.869342	-0.733688
2020-01-17	192.076661	192.076661	146.536826	11.801938	33.737897	1.300488
2020-01-20	158.444891	158.444891	146.552249	4.146181	7.746461	0.295471
2020-01-21	153.833194	153.833194	146.567673	7.697386	-0.431865	-0.020762
2020-01-22	126.885351	126.885351	146.583096	8.193982	-27.891727	-1.082559
2020-01-23	136.844892	136.844892	146.598520	10.459132	-20.212761	-0.785634
2020-01-24	189.534609	189.534609	146.613944	13.148114	29.772552	1.147159
2020-01-27	166.646305	166.646305	146.629367	-10.411728	30.428665	1.172529
2020-01-28	178.738703	178.738703	146.644791	12.832703	19.261208	0.740715

Figure 11.12 – Table of decomposed values

Based on the decomposed time series in *Figure 11.11*, we can analyze the outliers by examining the `residual` and `resid_z` columns. Typically, Z-scores with an absolute value greater than 2 or 3 are considered potential outliers. In this dataset, the largest positive residuals are observed on `2020-01-06` (Z-score: 1.468043), `2020-01-17` (Z-score: 1.300488), and `2020-01-27` (Z-score: 1.172529), while the largest negative residuals are on `2020-01-15` (Z-score: -1.721474) and `2020-01-22` (Z-score: -1.082559). Although these values indicate some deviations from the trend and seasonal components, none of the Z-scores exceed the typical threshold of ±2 or ±3, suggesting that there are no extreme outliers in this dataset. The residuals appear to be relatively well-distributed around zero, indicating a good fit for the decomposition model. However, the dates with the largest deviations (`2020-01-06`, `2020-01-15`, and `2020-01-17`) might be worth investigating further for any unusual events or factors that could explain their deviation from the expected values.

On digging deeper into this data to understand the reasons behind the fluctuations and upon closer inspection, we can see that the deviations on these dates were due to specific events and system issues:

> **Disclaimer!**
>
> The following events correspond to made-up events!

- `2020-01-06`: A technical glitch in the stock exchange's trading system caused a temporary spike in prices
- `2020-01-15`: An erroneous trade input led to a sudden drop in prices, which was later corrected
- `2020-01-17`: A major economic announcement led to increased volatility and a brief surge in stock prices
- `2020-01-22`: A miscommunication about quarterly earnings results caused temporary panic selling
- `2020-01-27`: Rumors of a merger and acquisition led to speculative buying, temporarily inflating the prices

These findings helped us understand that the residuals' deviations were not random but were due to specific, identifiable events. While these events did not qualify as significant outliers statistically, they highlighted the inherent volatility and noise in stock price data. Given the noisy nature of stock prices, even without significant outliers, smoothing techniques become crucial!

Handling outliers – model-based approaches – ARIMA

ARIMA models are widely used for forecasting time series data. They predict future values based on past observations, making them effective tools for identifying outliers by comparing actual values against predicted values. The ARIMA model consists of three main components:

- **Autoregressive** (**AR**): Uses the dependency between an observation and several lagged observations (p)
- **Integrated** (**I**): Uses differencing of observations to make the time series stationary (d)
- **Moving average** (**MA**): Uses dependency between an observation and a residual error from a moving average model applied to lagged observations (q)

ARIMA models are effective in handling the following outliers:

- **Additive outliers** (**AO**): Sudden spikes or drops in the time series
- **Innovative outliers** (**IO**): Changes that affect the entire series from the point of occurrence onwards

Let's discuss how the ARIMA model can be used for outlier detection and smoothing in the context of the stock price data example we've been working with. You can find the full example at `https://github.com/PacktPublishing/Python-Data-Cleaning-and-Preparation-Best-Practices/blob/main/chapter11/5.outliers/3.arima.py`:

1. Fit the ARIMA model to the `close_filled` series:

   ```
   model = ARIMA(df['close_filled'], order=(2,1,1))
   results = model.fit()
   ```

2. Calculate the residuals and Z-scores:

```
df['residuals'] = results.resid
df['residuals_z'] = zscore(df['residuals'].dropna())
```

3. Identify any outliers based on the Z-score threshold (for example, ±3):

```
outliers_arima = df[np.abs(df['residuals_z']) > 3]
```

4. Visualize the original `close_filled` series and the smoothed series that was obtained from the ARIMA model:

```
df['arima_smooth'] = results.fittedvalues
```

Here's the output:

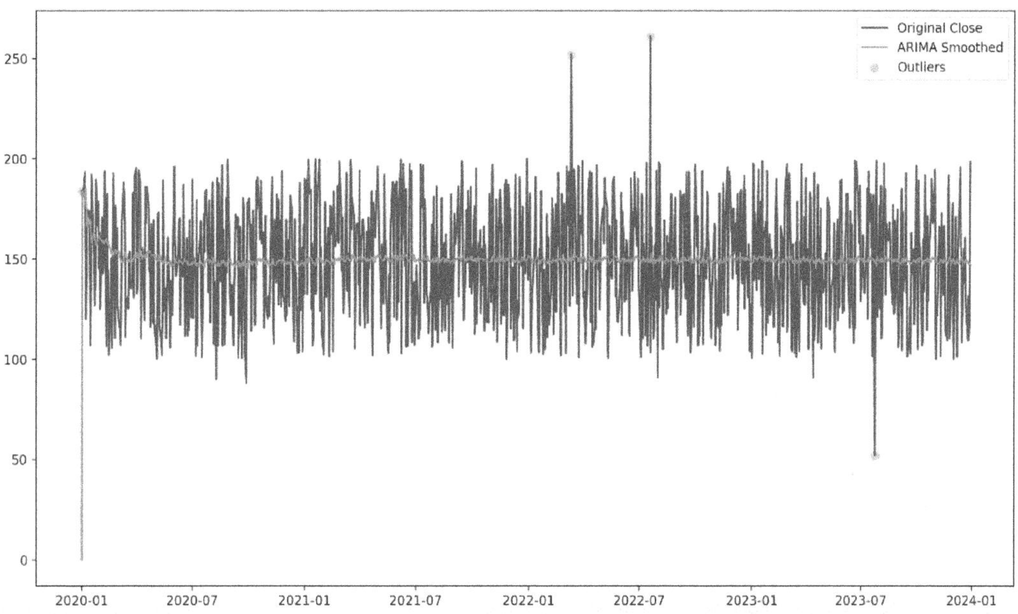

Figure 11.13 – ARIMA smoothing and outlier detection

5. Generate diagnostic plots to evaluate the model fit, including residual analysis, a **Quantile-Quantile** (**Q-Q**) plot, and standardized residuals:

```
results.plot_diagnostics(figsize=(14,8))
```

6. The resulting plots are shown here:

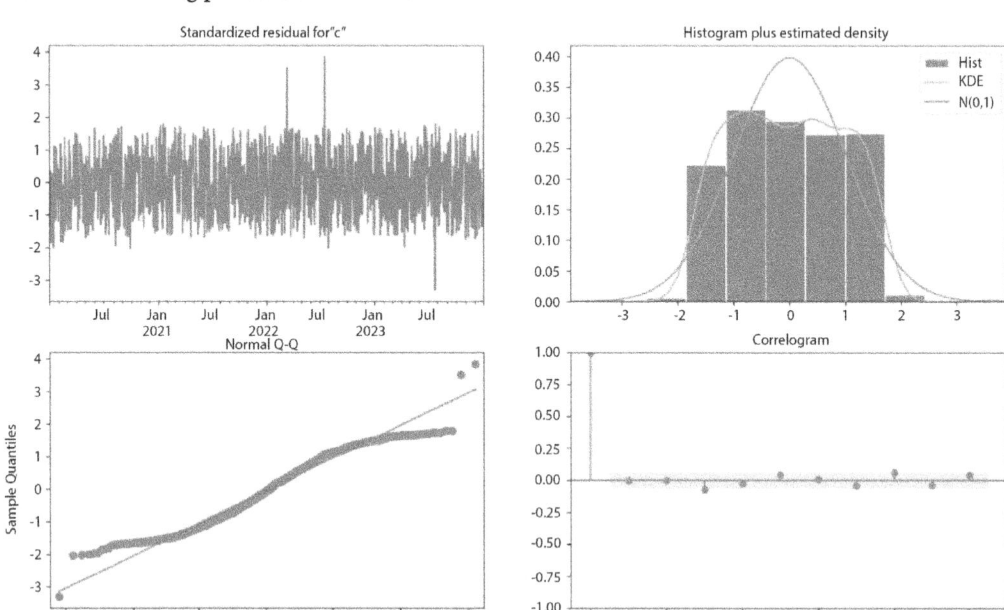

Figure 11.14 – Residual analysis, Q-Q plot, and standardized residuals

Let's dive a bit deeper into the diagnostic plots shown in *Figure 11.14*:

- **Standardized residuals**: Standardized residuals are the residuals from the ARIMA model scaled by their standard deviation. For the ARIMA model to be considered a good fit, the standardized residuals *should resemble white noise, meaning they should have no discernible pattern*. This implies that the residuals are randomly distributed with a mean of zero and constant variance. If a pattern is visible in the residuals, it suggests that the model has not captured some underlying structure in the data, and further refinement may be necessary. In our case, the *residuals* look like white noise.
- **Histogram plus KDE**: The histogram, combined with the **kernel density estimate** (**KDE**) plot of the residuals, provides a visual assessment of their distribution. For a well-fitted ARIMA model, the residuals should follow a normal distribution. The histogram should resemble the familiar bell curve, and the KDE plot should overlay a smooth curve that matches this shape. Deviations from the normal distribution, such as skewness or heavy tails, indicate that the residuals are not normally distributed, suggesting potential issues with the model. In our case, we don't see any significant skewness or tails in the residuals.
- **Normal Q-Q plot**: The Q-Q plot compares the quantiles of the residuals to the quantiles of a normal distribution. If the residuals are normally distributed, the points on the Q-Q plot

will lie along the 45-degree line. Significant deviations from this line indicate departures from normality. In our case, we don't see any significant deviations.

- **Correlogram** (**ACF of residuals**): The correlogram displays the ACF of the residuals. For a properly specified ARIMA model, the residuals should show no significant autocorrelation. This means that none of the lags should have statistically significant correlation coefficients. Significant spikes in the ACF plot indicate that the residuals are still correlated with their past values, suggesting that the model has not fully captured the time series' structure. This can guide further model refinement, such as increasing the order of the AR or MA components. In our case, everything looks good.

> **What is lag 0?**
>
> In the correlogram (ACF plot), the term **lag 0** refers to the autocorrelation of the time series with itself at lag 0, which is essentially the correlation of the time series with itself at the same time point. By definition, this correlation is always 1, because any time series is perfectly correlated with itself at lag 0. This means the autocorrelation value at lag 0 is always 1, which is why you see a spike at lag 0 in the ACF plot.

It is a good idea to go and play with the different settings and see their effect on the ARIMA model and the residuals.

Having explored the ARIMA method to detect and handle outliers in our stock price dataset, we have seen that outliers can significantly affect the accuracy and reliability of our time series model. While the ARIMA method helps in identifying and adjusting for these sudden changes, it's also valuable to consider other approaches to robust outlier detection and handling. One such approach involves using moving window techniques, as we will see in the next section.

Moving window techniques

Moving window techniques, also known as rolling or sliding window methods, involve analyzing a fixed-size subset or "window" of data that moves sequentially across a larger dataset. At each position of the window, a specific calculation or function is applied, such as computing the mean, median, sum, or more complex statistical measures. As the window shifts by one or more data points, the calculation is updated with the new subset of data. This method is particularly robust in time series analysis, where it is often used for smoothing data, identifying trends, or detecting anomalies over time.

The robustness of moving window techniques lies in their ability to provide localized analysis while maintaining a connection to the broader dataset. For example, when smoothing a time series, a moving average can reduce noise and highlight underlying trends without distorting the overall signal. Similarly, in financial data, moving windows can be used to compute rolling averages or volatilities, offering a real-time view of market conditions.

In this section, we will focus on two primary methods: **simple moving average** (**SMA**) and **exponential moving average** (**EMA**). Both can act as a basis that you can adjust with other statistics such as the median later.

SMA

The **SMA** is a commonly used statistical calculation that represents the average of a set of data points over a specified time. It is a type of moving average that smoothens out fluctuations in data to identify trends more easily. The SMA is calculated by summing up a set of values and dividing the sum by the number of data points. More advanced methods such as Kalman smoothing can estimate missing values by modeling the underlying process:

$$SMA_t = (X_t + X_{t-1} + X_{t-2} + \ldots + X_{t-n+1})/n$$

Here, we have the following:

- SMA_t is the SMA at time t
- $X_t + X_{t-1} + X_{t-2} + \ldots + X_{t-n+1}$ are the values for the time period
- n is the number of periods included in the calculations

Now, let's introduce the exponential moving average so that we can compare the two.

EMA

The **EMA** gives more weight to recent data points and less weight to older data points. It uses an exponential decay formula:

$$EMA_t = \alpha \cdot X_t + (1 - \alpha) \cdot EMA_{t-1}$$

Here, α is the smoothing factor.

Now, let's discuss how the SMA and EMA can be used for outlier detection and smoothing in the context of the stock price data example we've been working with.

Smoothing with SMA and EMA on the stock price use case

Continuing with the stock price data example we presented previously, let's see the effect that SMA and EMA have on the data:

First, let's calculate the SMA with a window of 12 months:

1. Define the `window` size for SMA and the `span` size for EMA:

```
window_size = 20
span = 20
```

2. Calculate the SMA:

```
df['SMA'] = df['close'].rolling(window=window_size, min_
periods=1).mean()
```

3. Calculate the EMA:

```
df['EMA'] = df['close'].ewm(span=span, adjust=False).mean()
```

4. Calculate the residuals for the SMA and EMA:

```
df['SMA_residuals'] = df['close'] - df['SMA'] df['EMA_
residuals'] = df['close'] - df['EMA'] sma_window = 12
data['SMA'] = data['Passengers'].rolling(window=sma_window).
mean()
```

5. Plot the original time series and the SMA:

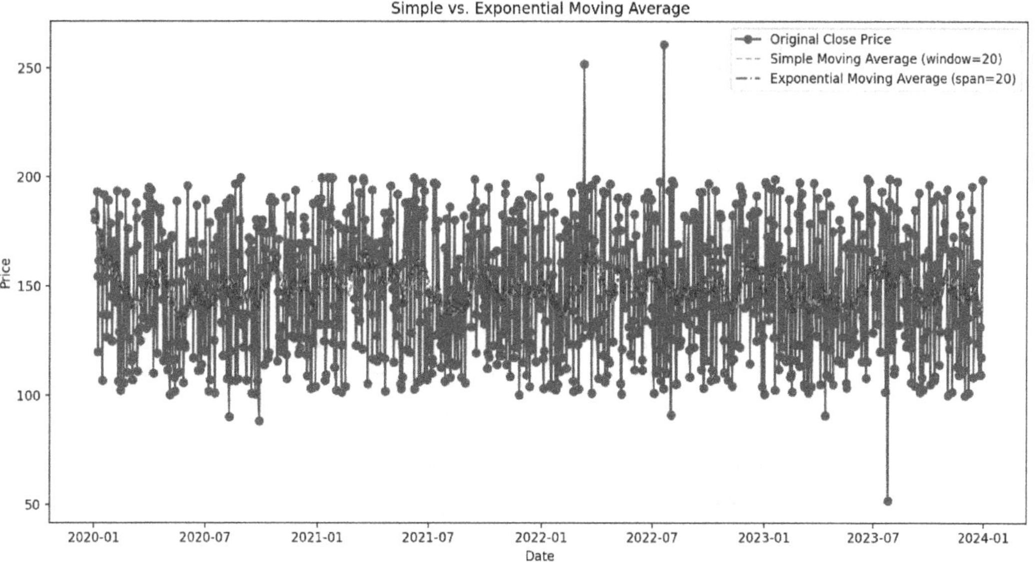

Figure 11.15 – SMA and EMA

In this example, we calculated the SMA and EMA using a window size of 20 and a span of 20, respectively. The window size for SMA determines how many previous data points are included in calculating the average at each point in time. Just like SMA, the frequency of your data points influences the choice of span. If your data is daily, a span zone of 20 might represent roughly 20 days of historical data.

Let's discuss the generated plot a little more:

- **SMA**:
    - **Smoothing effect**: The SMA smooths the time series data by averaging the values within the window, reducing noise, and highlighting the underlying trend

    - **Outlier impact**: While SMA reduces the impact of outliers, it can still be influenced by them since it considers all values within the window equally

- EMA:

    - **Smoothing effect**: The EMA also smooths the data but gives more weight to recent observations, making it more responsive to recent changes
    - **Outlier impact**: EMA is less influenced by older outliers but can be more affected by recent ones due to its weighting scheme

> **Finding a balance between smoothness and responsiveness**
>
> Larger window sizes result in smoother moving averages but may lag behind changes in the data. Smaller window sizes make the moving average more responsive to short-term fluctuations but might introduce more noise.

Remember the autocorrelation plot we created in *Figure 11.10*? We can use the analysis to adjust the span or window size based on the observed autocorrelation patterns. The following points will help you guide the selection of the window size and span:

- Consider the frequency of your data points (daily, weekly, monthly).
- If the autocorrelation plot shows significant autocorrelation at shorter lags, a smaller span in EMA or a smaller window size in SMA can help maintain responsiveness to recent changes while mitigating the influence of short-term noise.
- If your data exhibits seasonal patterns, you might choose a window size or span that aligns with the seasonal cycle. For example, if there's a weekly seasonality, you might consider a window size of 5 or 7. Use the autocorrelation chart to figure this out.

To evaluate how well the window models perform, we can use the **mean absolute error** (**MAE**), as well as the **mean squared error** (**MSE**) and **root mean squared error** (**RMSE**). We can compare the errors between the original data and the smoothed values generated by these models, as shown in the following figure:

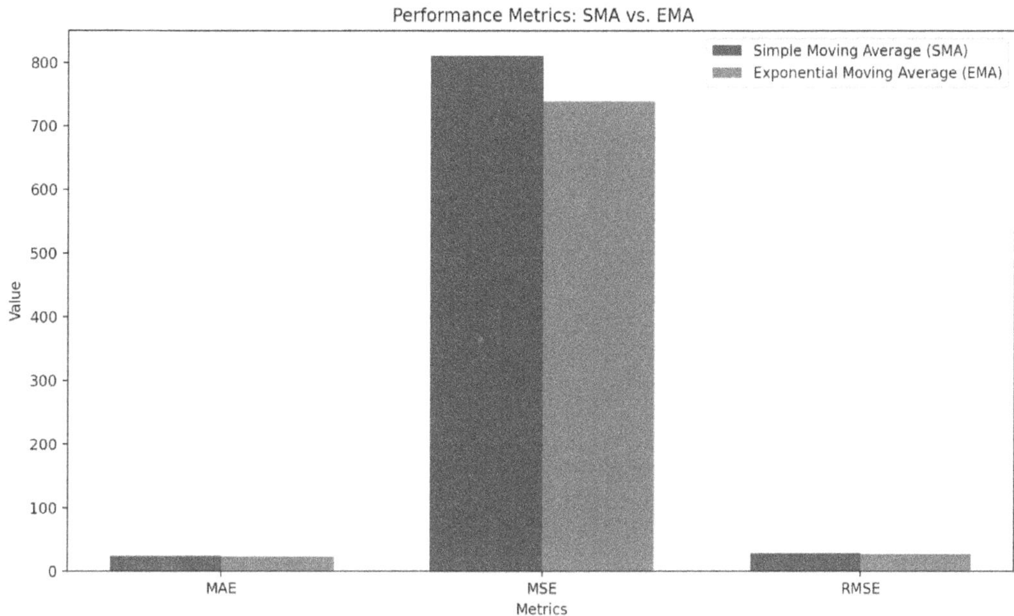

Figure 11.16 – Performance metrics for SMA and EMA

To make sure we have a clear understanding of the different metrics presented in *Figure 11.16*, let's look at this in more detail:

- **MAE**: This represents the average magnitude of the errors in a set of predictions, providing a simple average of the absolute differences between predicted and actual values
- **MSE**: This measures the average squared differences between predicted and actual values, penalizing larger errors more heavily than MAE
- **RMSE**: RMSE is the square root of MSE, offering an interpretable measure of the average magnitude of error, aligning with the scale of the original data

Now that we know what these terms represent, let's unpick what they show for our stock prices use case. Lower MAE, MSE, and RMSE values indicate better performance of the smoothing method. While MAE and RMSE are very close for SMA and EMA, the MSE is lower for the exponential method.

The following table compares and summarizes when to use the SMA and EMA:

Criteria	SMA	EMA
Type of smoothing	Simple and uniform smoothing of data points over a window	More responsive and adaptable, giving more weight to recent data points
Weighing data points	Equal weight to all data points in the window	More weight to recent observations; older observations receive exponentially decreasing weights
Responsiveness to changes	Lagging indicator; slower to respond to recent changes	More responsive to recent changes; adapts quickly to shifts in the data
Suitability for stability	Suitable for stable and less volatile time series	Suitable for volatile or rapidly changing time series
Adaptability to trends	Smooths out long-term trends, suitable for identifying overall patterns	Adapts quickly to shifting trends, suitable for capturing recent changes
Use case example	Analyzing long-term trends and identifying seasonality patterns	Capturing short-term fluctuations and reacting to market volatility
Calculation complexity	Simpler calculation and easier to understand and implement	More complex calculations involve a smoothing factor

Table 11.2 – Comparison between SMA and EMA

Beyond moving average techniques, exploring advanced feature engineering steps such as lags and differencing can significantly enrich our understanding and predictive capabilities. We'll explore those in the next section.

Feature engineering for time series data

Effective feature engineering is essential in time series analysis to uncover meaningful patterns and enhance predictive accuracy. It involves transforming raw data into informative features that capture temporal dependencies, seasonal variations, and other relevant aspects of the time series. The first technique we are going to explore is creating lags of features.

Lag features and their importance

Lag features are a crucial aspect of time series feature engineering as they allow us to transform time series data into a format suitable for supervised learning models. Lag features involve creating new variables that represent past observations of the target variable:

- **Lag 1**: The value from the previous time step
- **Lag 2**: The value from two time steps ago
- **Lag k**: The value from k time steps ago

By shifting the time series data by a specified number of time steps (referred to as the lag), these past values are included as features in the model at the current timestamp. As we know, time series data often exhibits temporal dependencies, where the current value is related to past observations. Lag features help capture these dependencies, allowing the model to learn from historical patterns.

Now, let's discuss how the lag features can be used in the context of the stock price data example we've been working with.

Creating lag features in the stock price use case

Continuing with the stock price data example we presented previously, let's see the effect lag features have on the data:

1. First, introduce more aggressive outliers in the `close` column:

```
outlier_indices = np.random.choice(df.index, size=10,
replace=False)
df.loc[outlier_indices[:5], 'close'] = df['close'] * 1.5 #
Increase by 50%
df.loc[outlier_indices[5:], 'close'] = df['close'] * 0.5 #
Decrease by 50%
```

2. Use the following function to create lagged features:

```
def create_lagged_features(df, column, lags):
    for lag in lags:
        df[f'{column}_lag_{lag}'] = df[column].shift(lag)
    return df # Define the lags to create lags = [1, 5, 10, 20]
```

3. Create lagged features for the `close` column:

```
df = create_lagged_features(df, 'close', lags)
```

4. Plot the original time series and lagged datasets:

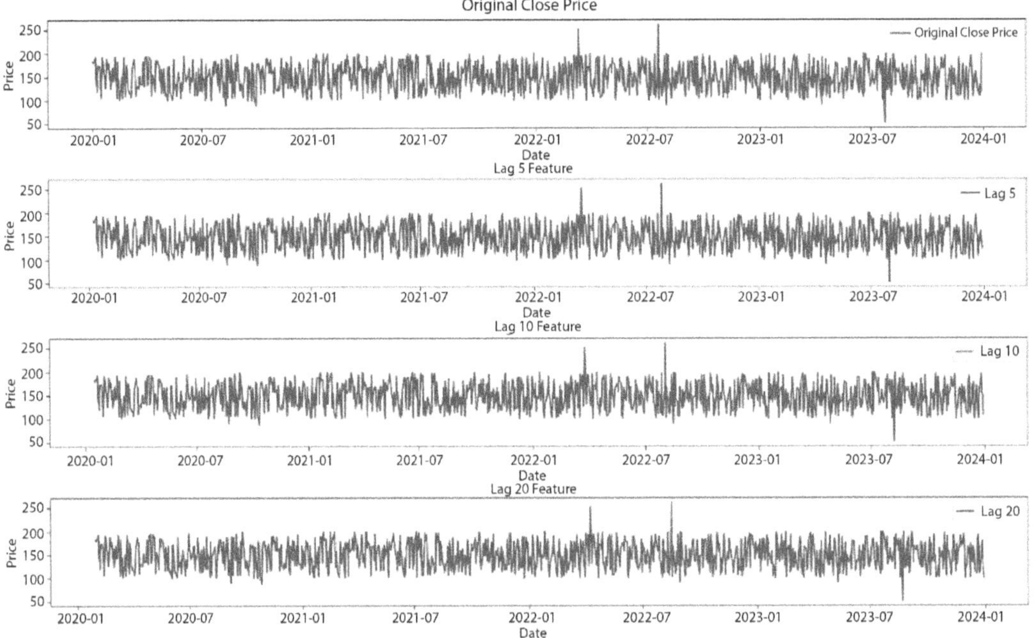

Figure 11.17 – Original versus lagged features

As we can see in *Figure 11.17*, lag 1 (`close_lag_1`) represents the closing price from the previous day, lag 5 (`close_lag_5`) represents the closing price from 5 days ago, and so on. You can observe how each lag captures the historical values of the target variable. When adding lagged features to a time series, the start date of the data shifts forward because the first few data points cannot be used until the specified lag period is complete. This shift means that if you add more lags, the number of initial data points that lack complete lagged data increases, effectively pushing the start date forward.

Feel free to experiment with different lag values and see the effect on the dataset. Adjusting the lag values allows you to capture different degrees of temporal dependencies and trends in the data.

Differencing time series

In *Chapter 4, Cleaning Messy Data and Data Manipulation*, we discussed how calculating the time difference between two datetime objects using the `diff()` function can help us measure the time elapsed between consecutive events. This technique is useful for understanding the temporal gaps in a sequence of timestamps. Similarly, in time series analysis, differencing is a powerful technique that's used to stabilize the mean of a time series by removing changes in the level of a time series, thus eliminating trend and seasonality. Just as we calculated the time elapsed in the previous chapter, we can apply differencing to our stock market data to highlight changes over time. However, we will also introduce a new term – seasonal differencing.

Seasonal differencing

Seasonal differencing is a technique that's used to remove seasonal patterns from time series data, making it more stationary and suitable for analysis and forecasting. Seasonal differencing involves subtracting the value of an observation from a previous observation at a lag equal to the **seasonal** period. So, we need to identify the seasonal period with all the tools we provided previously and then take the seasonal period and difference the data on that.

For monthly data with an annual seasonal pattern, we can use the following formula:

$$y'_t = y_t - y_{t-12}$$

For quarterly data, we can use the following formula:

$$y'_t = y_t - y_{t-4}$$

Here, is the seasonally differenced series and is the original series.

Now, let's discuss how the difference can be used in the context of the stock price data example we've been working with.

Differencing the stock price data

To showcase the seasonal differencing, we will introduce some seasonality in the stock market data. Based on the analysis we've done so far, there is not a big seasonal component in the data. Let's get started:

1. Create a seasonal component (weekly seasonality with higher amplitude):

```
seasonal_component = 50 * np.sin(2 * np.pi * np.arange(n) / 5) #
5-day seasonality
```

2. Generate random stock prices with added seasonality:

```
data = {
'open': np.random.uniform(100, 200, n) + seasonal_component,
'high': np.random.uniform(200, 300, n) + seasonal_component,
'low': np.random.uniform(50, 100, n) + seasonal_component,
'close': np.random.uniform(100, 200, n) + seasonal_component
}
df = pd.DataFrame(data, index=date_range)
```

3. Calculate the first difference:

```
df['First Difference'] = df['close'].diff()
```

4. Calculate the second difference:

```
df['Second Difference'] = df['First Difference'].diff()
```

5. Finally, calculate the seasonal difference (weekly seasonality):

```
df['Seasonal Difference'] = df['close'].diff(5)
```

Let's demonstrate differencing by plotting the first, second, and seasonal differences:

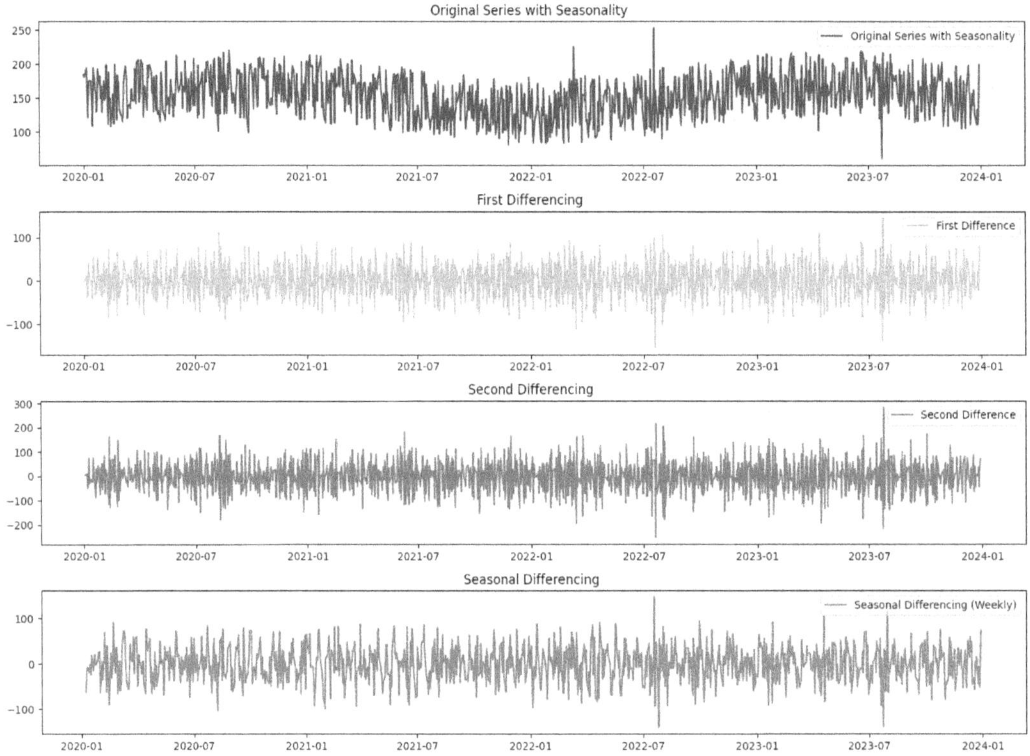

Figure 11.18 – Original versus differenced series

In *Figure 11.18*, we can observe the first, second, and seasonal differencing. We can see that in the original plot, there is some seasonality, but after the first difference, we can see that the seasonal component is minimized. But how can we evaluate this more statistically? Let's perform some statistical tests to check for stationarity in the time series.

The Augmented Dickey-Fuller (ADF) test

The **ADF** test is a statistical test that's used to determine whether a time series is stationary or not. The ADF test examines the null hypothesis that a unit root is present in a time series sample. The presence of a unit root indicates that the time series is non-stationary. The alternative hypothesis is that the time series is stationary. For the ADF test, a more negative value indicates stronger evidence against the null hypothesis.

The p-value represents the probability of obtaining test results at least as extreme as the observed results, assuming that the null hypothesis is true. In the case of the ADF test, we want to see *a small p-value to reject the null hypothesis of non-stationarity.*

To conclude that a time series is stationary, we typically want to see the following:

- **p-value < 0.05**: This is the most common threshold that's used in statistical testing. If $p < 0.05$, we reject the null hypothesis at the 5% significance level. This means we have strong evidence to conclude the series is stationary.
- **Even smaller p-values**: $p < 0.01$ (1% significance level) and $p < 0.001$ (0.1% significance level) provide even stronger evidence of stationarity.

Let's code up this test:

```
def adf_test(series, title=''):
    result = adfuller(series.dropna(), autolag='AIC')
    print(f'Augmented Dickey-Fuller Test: {title}')
    print(f'ADF Statistic: {result[0]}')
    print(f'p-value: {result[1]}')
    for key, value in result[4].items():
        print(f' {key}: {value}')
    print('\n')
```

Now, it's time for the results! We will run the test for the original time series (to check if it is stationary or not) and then for each of the differenced time series. Let's explain the results:

```
Augmented Dickey-Fuller Test: Original Series
ADF Statistic: -3.5898552445987595
p-value: 0.005957961883734467
   1%: -3.4367333690404767
   5%: -2.8643583648001925
   10%: -2.568270618452702
```

The ADF statistic of -3.5899 is less than the 5% critical value of -2.8644, and the p-value is below 0.05. This indicates that we can reject the null hypothesis of the presence of a unit root, suggesting that *the original series is likely stationary*. However, the result is relatively close to the critical value, indicating *borderline stationarity*:

```
Augmented Dickey-Fuller Test: First Difference
ADF Statistic: -11.786384523171499
p-value: 1.0064914317100746e-21
   1%: -3.4367709764382024
   5%: -2.8643749513463637
   10%: -2.568279452717228
```

The ADF statistic of -11.7864 is well below the 5% critical value of -2.8644, and the p-value is extremely small. This strongly suggests that the first-differenced series is stationary. The significant drop in the

ADF statistic compared to the original series indicates that first differencing has effectively removed any remaining trends or unit roots:

```
Augmented Dickey-Fuller Test: Second Difference
ADF Statistic: -14.95687341689794
p-value: 1.2562905072914351e-27
   1%: -3.4367899468008916
   5%: -2.8643833180472744
   10%: -2.5682839089705536
```

The ADF statistic of -14.9569 is much lower than the 5% critical value, and the p-value is extremely small. This result suggests that the second-differenced series is also stationary. However, *over-differencing can lead to loss of meaningful patterns and increase noise*, so it's essential to balance between achieving stationarity and maintaining the integrity of the series:

```
Augmented Dickey-Fuller Test: Seasonal Differencing
ADF Statistic: -11.48334880444129
p-value: 4.933051350797084e-21
   1%: -3.4367899468008916
   5%: -2.8643833180472744
   10%: -2.5682839089705536
```

Finally, the ADF statistic of -11.4833 is well below the 5% critical value, and the p-value is very small. This indicates that seasonal differencing has successfully made the series stationary. Seasonal differencing is particularly useful if the series exhibits periodic patterns at specific intervals.

Given these results, the first difference appears to be the most appropriate choice for the following reasons:

- The original series is already stationary at the 1% level, but first differencing significantly improves stationarity
- The first difference yields a highly significant result (p-value: 1.006e-21) without risking over-differencing
- While the second difference shows an even more significant result, it may lead to over-differencing, which can introduce unnecessary complexity and potentially remove important information from the series
- Seasonal differencing also shows strong results, but unless there's a clear seasonal pattern in your data, the simpler first difference method is generally preferred

In conclusion, first difference strikes a good balance between achieving stationarity and avoiding over-differencing. Now, it's time to discuss some of the most common use cases in the time series space.

Applying time series techniques in different industries

The ability to analyze temporal patterns provides a competitive advantage in today's data-driven world across various industries. Here are some popular use cases across various industries:

Sector	Use Case	Explanation
Finance	Stock market analysis	Analyzing historical stock prices and trading volumes to make informed investment decisions
	Portfolio management	Assessing the performance of investment portfolios over time to optimize asset allocation
	Risk assessment	Modeling and forecasting financial risks such as market volatility and credit defaults
Healthcare	Patient monitoring	Continuously tracking vital signs and health metrics for early detection of abnormalities
	Epidemiology	Analyzing temporal patterns of disease spread and predicting outbreaks
	Treatment effectiveness	Assessing the impact of medical interventions over time
Meteorology	Weather forecasting	Analyzing historical weather patterns to predict future conditions
	Climate change studies	Monitoring long-term trends and variations in climate data
	Natural disaster prediction	Early detection of potential disasters such as hurricanes, floods, and droughts
Manufacturing	Production planning	Forecasting demand and optimizing production schedules
	Quality control	Monitoring and ensuring product quality over time
	Equipment maintenance	Predictive maintenance based on the performance history of machinery
Marketing	Sales forecasting	Predicting future sales based on historical data
	Customer engagement	Analyzing patterns of customer interaction with products and services
	Campaign optimization	Evaluating the impact of marketing initiatives over time

Sector	Use Case	Explanation
Transportation	Traffic flow analysis	Monitoring and optimizing traffic patterns in urban areas
	Vehicle tracking	Tracking the movement and efficiency of transportation fleets
	Supply chain optimization	Forecasting demand and optimizing the movement of goods over time

Table 11.3 – Time series techniques use cases

With that, we can summarize this chapter.

Summary

Time series analysis plays a pivotal role in extracting meaningful insights and making informed decisions in a wide range of industries. As technology advances, sophisticated time series techniques will become increasingly integral to understanding complex temporal patterns and trends. Whether in finance, healthcare, or transportation, the ability to analyze and forecast time-dependent data empowers organizations to adapt, optimize, and make strategic decisions in an ever-evolving landscape.

In this chapter, we covered techniques for handling missing values and outliers, differencing methods, and feature engineering in time series analysis. We learned how to use ffill and bfill for missing values and compared their effects on stock price data. Differencing techniques, including first, second, and seasonal differencing, were applied to achieve stationarity and were evaluated using ADF tests. We also explored lagged features for capturing temporal dependencies and assessed model performance using metrics such as MAE, MSE, and RMSE. These skills will prepare you so that you can manage and analyze time series data effectively.

In the next chapter, we will pivot to a different type of data – text. Analyzing text data involves unique challenges and methodologies distinct from those used for numerical time series. We will deep dive into text preprocessing and cover text cleaning techniques, tokenization strategies, and spelling correction approaches, all of which are essential for any **natural language processing** (**NLP**) task.

Part 3:
Downstream Data Cleaning – Consuming Unstructured Data

This part focuses on the challenges and techniques involved in processing unstructured data, such as text, images, and audio, in the context of modern machine learning, particularly **large language models** (**LLMs**). It provides a comprehensive overview of how to prepare unstructured data types for machine learning applications, ensuring that the data is properly preprocessed for analysis and model training. The chapters cover essential preprocessing methods for text, as well as image and audio data, offering readers the tools to work with more complex and varied datasets in today's AI-driven landscape.

This part has the following chapters:

- *Chapter 12, Text Preprocessing in the Era of LLMs*
- *Chapter 13, Image and Audio Preprocessing with LLMs*

12

Text Preprocessing in the Era of LLMs

In the era of **Large Language Models** (**LLMs**), mastering text preprocessing is more crucial than ever. As LLMs grow in complexity and capability, the foundation of successful **Natural Language Processing** (**NLP**) tasks still lies in how well the text data is prepared. In this chapter, we will discuss text preprocessing, the foundation for any NLP Task. We will also explore essential preprocessing techniques, focusing on adapting them to maximize the potential of LLMs.

In this chapter, we'll cover the following topics:

- Relearning text preprocessing in the era of LLMs
- Text cleaning techniques
- Handling rare words and spelling variations
- Chunking
- Tokenization strategies
- Turning tokens into embeddings

Technical requirements

The complete code for this chapter can be found in the following GitHub repository:

`https://github.com/PacktPublishing/Python-Data-Cleaning-and-Preparation-Best-Practices/tree/main/chapter12`

Let's install the necessary libraries we will use in this chapter:

```
pip install transformers==4.42.4
pip install beautifulsoup4==4.12.3
pip install langchain-text-splitters==0.2.2
pip install tiktoken==0.7.0
pip install langchain==0.2.10
pip install langchain-experimental==0.0.62
pip install langchain-huggingface==0.0.3
pip install presidio_analyzer==2.2.355
pip install presidio_anonymizer==2.2.355
pip install rapidfuzz-3.9.4 thefuzz-0.22.1
pip install stanza==1.8.2
pip install tf-keras-2.17.0
```

Relearning text preprocessing in the era of LLMs

Text preprocessing involves the application of various techniques to raw textual data with the aim of cleaning, organizing, and transforming it into a format suitable for analysis or modeling. The primary goal is to enhance the quality of the data by addressing common challenges associated with unstructured text. This entails tasks such as cleaning irrelevant characters, handling variations, and preparing the data for downstream NLP tasks.

With the rapid advancements in LLMs, the landscape of NLP has evolved significantly. However, fundamental preprocessing techniques such as text cleaning and tokenization remain crucial, albeit with some shifts in approach and importance.

Staring with text cleaning, while LLMs have shown remarkable robustness to noise in input text, clean data still yields better results and is especially important for fine-tuning tasks. Basic cleaning techniques such as removing HTML tags, handling special characters, and normalizing text are still relevant. However, more advanced techniques such as spelling correction may be less critical for LLMs, as they can often handle minor spelling errors. Domain-specific cleaning remains important, especially when dealing with specialized vocabulary or jargon.

Tokenization has evolved with the advent of subword tokenization methods used by most modern LLMs such as **Byte-Pair Encoding** (**BPE**) or WordPiece. Traditional word-level tokenization is less common in LLM contexts. Some traditional NLP preprocessing steps such as stopword removal, stemming, and lemmatization have become less critical. Stopword removal, which involves eliminating common words such as "and" or "the," is less necessary because LLMs can understand their contextual importance and how they contribute to the meaning of a sentence. Similarly, stemming and lemmatization, which reduce words to their base forms (e.g., "running" to "run"), are less frequently used because LLMs can interpret different word forms accurately and understand their relationships within the text. This shift allows for a more nuanced understanding of language, capturing subtleties that rigid preprocessing might miss.

The key message is that while LLMs can handle raw text impressively, preprocessing remains crucial in certain scenarios as it can improve model performance on specific tasks. Remember: **garbage in, garbage out**. Cleaning and standardizing text can also reduce the number of tokens processed by an LLM, potentially lowering computational costs. New approaches are emerging that blend traditional preprocessing with LLM capabilities, using LLMs themselves for data cleaning and preprocessing tasks.

In conclusion, while LLMs have reduced the need for extensive preprocessing in many NLP tasks, understanding and judiciously applying these fundamental techniques remains valuable. In the following sections, we will focus on the text preprocessing techniques that remain relevant.

Text cleaning

The primary goal of text cleaning is to transform unstructured textual information into a standardized and more manageable form. While cleaning text, several operations are commonly performed, such as the removal of HTML tags, special characters, and numerical values, as well as the standardization of letter cases and the handling of whitespaces and formatting issues. These operations collectively contribute to refining the quality of textual data and reducing its ambiguity. Let's deep dive into these techniques.

Removing HTML tags and special characters

HTML tags are often present due to the extraction of content from web pages. These tags, such as `<p>`, `<a>`, or `<div>`, carry *no semantic meaning* in the context of NLP and must be removed. The cleaning process involves the identification and stripping of HTML tags, leaving behind only the actual words.

For this example, let's consider a scenario where we have a dataset of user reviews for a product and want to prepare the text data for sentiment analysis. You can find the code for this section in the GitHub repository at `https://github.com/PacktPublishing/Python-Data-Cleaning-and-Preparation-Best-Practices/blob/main/chapter12/1.text_cleaning.py`. In this script, the data generation is also available for you, and you can follow the example step by step.

> **Important note**
>
> Throughout this chapter, we've included key code snippets to illustrate the most important concepts. However, to see the complete code, including the libraries used, and to run the full end-to-end examples, please visit the repository.

The first text preprocessing step that we will execute is the removal of HTML tags. Let's have a look at the code step by step:

1. Let's import the libraries for this example:

```
from bs4 import BeautifulSoup
from transformers import BertTokenizer
```

2. The sample user reviews are shown here:

```
reviews = [
  "<html>This product is <b>amazing!</b></html>",
  "The product is good, but it could be better!!!",
  "I've never seen such a terrible product. 0/10",
  "The product is AWESOME!!! Highly recommended!",
]
```

3. Next, we create a function that uses `BeautifulSoup` to parse the HTML content and extract only the text, removing any HTML tags:

```
def clean_html_tags(text):
    soup = BeautifulSoup(text, "html.parser")
    return soup.get_text()
```

4. Then we preprocess all the reviews:

```
def preprocess_text(text):
    text = clean_html_tags(text)
    return text

preprocessed_reviews = [preprocess_text(review) for review in
reviews]
```

5. Finally, we get the preprocessed reviews as follows:

```
- This product is amazing!
- The product is good, but it could be better!!!
- I've never seen such a terrible product. 0/10
- The product is AWESOME!!! Highly recommended!
```

As we can see, all the HTML tags have been removed and the text is clean. We will continue enhancing this example by adding another common preprocessing step: handling the capitalization of text.

Handling capitalization and letter case

Text data often comes in various cases—uppercase, lowercase, or a mix of both. Inconsistent capitalization can lead to ambiguity in language processing tasks. Therefore, one common text-cleaning practice is to standardize the letter case throughout the corpus. This not only aids in maintaining consistency but also ensures that the model generalizes well across different cases.

Building on the previous example, we are going to expand the preprocessing function to add one extra step: that of letter standardization:

1. Let's first remind ourselves what the reviews looked like after the removal of HTML tags from the previous preprocessing step:

```
- This product is amazing!
- The product is good, but it could be better!!!
- I've never seen such a terrible product. 0/10
- The product is AWESOME!!! Highly recommended!
```

2. The following function will convert all characters into lowercase:

```
def standardize_case(text):
    return text.lower()
```

3. We will expand the `preprocess_text` function we presented in the previous example to convert all characters in the text to lowercase, making the text case insensitive:

```
def preprocess_text(text):
    text = clean_html_tags(text)
    text = standardize_case(text)
    return text
```

4. Let's print the preprocessed reviews:

```
for preprocessed_review in preprocessed_reviews:
    print(f"- {preprocessed_review}")
```

The lower-cased reviews are presented here:

```
- this product is amazing!
- the product is good, but it could be better!!!
- i've never seen such a terrible product. 0/10
- the product is awesome!!! highly recommended!
```

Notice how all the letters have turned to lower case! Go ahead and update the capitalization function as follows to turn everything to upper case:

```
def standardize_case(text):
    return text.upper()
```

The upper case reviews are presented here:

```
- THIS PRODUCT IS AMAZING!
- THE PRODUCT IS GOOD, BUT IT COULD BE BETTER!!!
- I'VE NEVER SEEN SUCH A TERRIBLE PRODUCT. 0/10
- THE PRODUCT IS AWESOME!!! HIGHLY RECOMMENDED!
```

In case you are wondering whether you should use lower or upper case, we've got you covered.

Lower or upper case?

The choice between using lowercase or uppercase text depends on the specific requirements of the NLP task. For instance, tasks such as sentiment analysis typically benefit from lowercasing, as it simplifies the text and reduces variability. Conversely, tasks such as **Named Entity Recognition** (**NER**) may require preserving case information to accurately identify and differentiate entities.

For example, in German, all nouns are capitalized, so maintaining the case is crucial for correct language representation. In contrast, English typically does not use capitalization to convey meaning, so lowercasing might be more appropriate for general text analysis.

When dealing with text data from user inputs, such as social media posts or reviews, it's important to consider the role of case variations. For instance, a tweet may use mixed case for emphasis or tone, which could be relevant for sentiment analysis.

Modern LLMs such as **Bidirectional Encoder Representations from Transformers** (**BERT**) and GPT-3 are trained on mixed-case text and handle both uppercase and lowercase effectively. These models utilize case information to enhance context and understanding. Their tokenizers are designed to manage case sensitivity inherently, processing text without needing explicit conversion.

If your task requires distinguishing between different cases (e.g., recognizing proper nouns or acronyms), it is better to preserve the original casing. However, always consult the documentation and best practices for the specific model you are using. Some models might be optimized for lowercased input and could perform better if the text is converted to lowercase.

The next step is to learn how we can deal with numerical values and symbols in the text.

Dealing with numerical values and symbols

Numerical values, symbols, and mathematical expressions may be present in text data but may not always contribute meaningfully to the context. Cleaning them involves deciding whether to retain, replace, or remove these elements based on the specific requirements of the task.

For instance, in sentiment analysis, numerical values might be less relevant, and their presence could be distracting. In contrast, for tasks related to quantitative analysis or financial sentiment, preserving numerical information becomes crucial.

Building on the previous example, we are going to remove all the numbers and symbols in the text:

1. Let's review how the data looked like after the previous preprocessing step:

```
- This product is amazing!
- The product is good, but it could be better!!!
- I've never seen such a terrible product. 0/10
- The product is AWESOME!!! Highly recommended!
```

2. Now let's add a function that removes all characters from the text *except alphabetic characters and spaces*:

```
def remove_numbers_and_symbols(text):
    return ''.join(e for e in text if e.isalpha() or
e.isspace())
```

3. Apply the text preprocessing pipeline:

```
def preprocess_text(text):
    text = clean_html_tags(text)
    text = standardize_case(text)
    text = remove_numbers_and_symbols(text)
    return text
```

4. Let's have a look at the preprocessed reviews:

```
- this product is amazing
- the product is good but it could be better
- ive never seen such a terrible product
- the product is awesome highly recommended
```

As you can see, after this preprocessing step, all the punctuations and symbols have been removed from the text. The decision to retain, replace, or remove symbols and punctuation during text preprocessing depends on the specific goals of your NLP task and the characteristics of your dataset.

Retaining symbols and punctuation

With the advancements in LLMs, the approach to handling punctuation and symbols during text preprocessing has evolved significantly. Modern LLMs benefit from retaining punctuation and symbols due to their extensive training on diverse datasets. This retention helps these models understand context more accurately by capturing nuances such as emotions, emphasis, and sentence boundaries. For instance, punctuation marks such as exclamation points and question marks play a crucial role in sentiment analysis by conveying strong emotions, which improves the model's performance. Similarly, in tasks such as text generation, punctuation maintains readability and structure, while in NER and translation, it aids in identifying proper nouns and sentence boundaries.

On the other hand, there are scenarios where removing punctuation and symbols can be advantageous. Modern LLMs are robust enough to handle noisy data, but in certain applications, simplifying text by removing punctuation can streamline preprocessing and *reduce the number of unique tokens*. This approach is beneficial for tasks such as topic modeling and clustering, where the focus is on content rather than structural elements. For example, removing punctuation can help identify core topics by eliminating distractions from sentence structure, and in text classification, it can standardize input data when punctuation does not add significant value.

Another approach is replacing punctuation and symbols with spaces or specific tokens, which helps in normalizing text while preserving some level of separation between tokens. This method can be particularly useful for custom tokenization strategies. In specialized NLP pipelines, replacing punctuation with specific tokens can retain important distinctions without adding unnecessary clutter to the text, facilitating more effective tokenization and preprocessing for downstream tasks.

Let's see a quick example on how to remove or replace symbols and punctuation. You can find the code for this section at `https://github.com/PacktPublishing/Python-Data-Cleaning-and-Preparation-Best-Practices/blob/main/chapter12/2.punctuation.py`:

- Create the sample text:

```
text = "I love this product!!! It's amazing!!!"
```

- Option 1: replace symbols and punctuation with spaces:

```
replaced_text = text.translate(str.maketrans(string.punctuation,
" " * len(string.punctuation)))
print("Replaced Text:", replaced_text)
```

This will print the following output:

```
I love this product    It s amazing
```

- Option 2: remove symbols and punctuation:

```
removed_text = "".join(char for char in text if char.isalnum()
or char.isspace())
print("Removed Text:", removed_text)
```

This will print the following output:

```
I love this product Its amazing
```

Removing symbols and numbers is a crucial preprocessing step in text analysis that simplifies text by eliminating non-alphanumeric characters. The last thing we will discuss in this section is addressing whitespace issues to enhance text readability and ensure consistent formatting.

Addressing whitespace and formatting issues

Whitespaces and formatting inconsistencies can be prevalent in text data, especially when it originates from diverse sources. Cleaning involves addressing issues such as multiple consecutive spaces, leading or trailing whitespaces, and variations in formatting styles. Regularization of whitespace ensures a standardized text representation, reducing the risk of misinterpretation by downstream models.

Addressing whitespace and formatting issues remains crucial in the world of LLMs. Although modern LLMs exhibit robustness to various formatting inconsistencies, managing whitespace and formatting effectively can still enhance model performance and ensure data consistency.

Standardizing whitespace and formatting creates a uniform dataset, which facilitates model training and analysis by minimizing noise and focusing attention on the content rather than formatting discrepancies. Enhanced readability, achieved through proper whitespace management, aids both human and machine learning interpretation by clearly delineating text elements. Furthermore, consistent whitespace handling is essential for accurate tokenization—a fundamental process in many NLP tasks—as it ensures precise identification and processing of words and phrases.

So, let's go back to the review example and add another step in the pipeline to remove whitespaces:

1. Let's start by addressing whitespace and formatting issues. This function removes extra spaces and ensures that there is only one space between words:

    ```
    def remove_extra_whitespace(text):
        return ' '.join(text.split())
    ```

2. Next, we'll add this to our text preprocessing pipeline:

    ```
    def preprocess_text(text):
        text = clean_html_tags(text)
        text = standardize_case(text)
        text = remove_numbers_and_symbols(text)
        text = remove_extra_whitespace(text)
        return text
    ```

Let's have a look at the reviews before applying the new step and focus on the whitespaces marked here:

```
- this product     is amazing
- the product is good but it could be better
- ive never seen such a terrible      product
- the product is awesome highly recommended
```

Finally, let's review the clean dataset, after having applied the whitespace removal:

```
- this product is amazing
- the product is good but it could be better
- ive never seen such a terrible product
- the product is awesome highly recommended
```

Let's move from pure text cleaning to focusing on safeguarding the data.

Removing personally identifiable information

When preprocessing text data, removing **Personally Identifiable Information** (**PII**) is crucial for maintaining privacy, ensuring compliance with regulations, and improving data quality. For instance, consider a dataset of user reviews that includes names, email addresses, and phone numbers. If this sensitive information is not anonymized or removed, it poses significant risks such as privacy violations and potential misuse. Regulations such as the **General Data Protection Regulation** (**GDPR**), **California Consumer Privacy Act** (**CCPA**), and **Health Insurance Portability and Accountability Act** (**HIPAA**) mandate that personal data must be handled carefully. Failing to remove PII can lead to legal penalties and loss of trust. Moreover, including identifiable details can introduce bias into machine learning models and compromise their generalization. Removing PII is essential for responsible AI development, as it allows for the creation and use of datasets that maintain individual privacy while still providing valuable insights for research and analysis

The following code snippet demonstrates how to use the presidio-analyzer and presidio-anonymizer libraries to detect and anonymize PII. Let's have a look at the code step by step. The full code can be accessed at `https://github.com/PacktPublishing/Python-Data-Cleaning-and-Preparation-Best-Practices/blob/main/chapter12/3.pii_detection.py`:

1. Let's start by importing the required libraries for this example:

   ```
   import pandas as pd
   from presidio_analyzer import AnalyzerEngine
   from presidio_anonymizer import AnonymizerEngine
   from presidio_anonymizer.entities import OperatorConfig
   ```

2. We create a sample DataFrame with one column named `text` containing sentences with different types of PII (e.g., names, email addresses, and phone numbers):

   ```
   data = {
       'text': [
           "Hello, my name is John Doe. My email is john.doe@
   example.com",
           "Contact Jane Smith at jane.smith@work.com",
           "Call her at 987-654-3210.",
           "This is a test message without PII."
       ]
   }
   df = pd.DataFrame(data)
   ```

3. We initialize `AnalyzerEngine` for *detecting* PII entities and `AnonymizerEngine` for *anonymizing* the detected PII entities:

   ```
   analyzer = AnalyzerEngine()
   anonymizer = AnonymizerEngine()
   ```

4. Next, we'll define an anonymization function that detects PII in the text and applies masking rules based on the entity type:

   ```
   def anonymize_text(text):
       analyzer_results = analyzer.analyze(text=text,
   entities=["PERSON", "EMAIL_ADDRESS", "PHONE_NUMBER"],
   language="en")
   ```

```
    operators = {
        "PERSON": OperatorConfig("mask", {"masking_char": "*",
"chars_to_mask": 4, "from_end": True}),
        "EMAIL_ADDRESS": OperatorConfig("mask", {"masking_char":
"*", "chars_to_mask": 5, "from_end": True}),
        "PHONE_NUMBER": OperatorConfig("mask", {"masking_char":
"*", "chars_to_mask": 6, "from_end": True})
    }
    anonymized_result = anonymizer.anonymize(
        text=text, analyzer_results=analyzer_results,
        operators=operators)
    return anonymized_result.text
```

The `anonymize_text` function is designed to protect sensitive information within a given text by anonymizing specific types of entities. It first analyzes the text to identify entities such as names (`PERSON`), email addresses (`EMAIL_ADDRESS`), and phone numbers (`PHONE_NUMBER`) using an analyzer. For each entity type, it then applies a masking operation to conceal part of the information. Specifically, it masks the last four characters of a person's name, the last five characters of an email address, and the last six characters of a phone number. The function returns the text with these sensitive entities anonymized, ensuring that personal information is obscured while retaining the overall structure of the text.

5. Apply the anonymization function to the DataFrame:

```
df['anonymized_text'] = df['text'].apply(anonymize_text)
```

6. Display the DataFrame:

```
0    Hello, my name is John. My email is john.d...
1             Contact Jane S at jane.smith@wor*
2                           Call her at 987-65.
3                    This is a test message without PII.
```

By using these configurations, you can tailor the anonymization process to meet specific requirements and ensure that sensitive information is properly protected. This approach helps you comply with privacy regulations and protect sensitive information in your datasets.

While removing PII is essential for protecting privacy and ensuring data compliance, another critical aspect of text preprocessing is handling rare words and spelling variations.

Handling rare words and spelling variations

The rise of LLMs has revolutionized how we interact with technology and process information, particularly in the world of handling spelling variations and rare words. Before the emergence of LLMs, managing these linguistic challenges required extensive manual effort, often involving specialized knowledge and painstakingly crafted algorithms. Traditional spell-checkers and language processors struggled with rare words and variations, leading to frequent errors and inefficiencies. Today, LLMs such as GPT-4, Lllama3, and others have transformed this landscape by leveraging vast datasets and sophisticated machine-learning techniques to understand and generate text that accommodates a wide range of spelling variations and uncommon terminology. These models can recognize and correct misspellings, provide contextually appropriate suggestions, and accurately interpret rare words, enhancing the precision and reliability of text processing.

Dealing with rare words

In the era of LLMs such as GPT-3 and GPT-4, handling rare words has become less of a challenge compared to traditional NLP methods. These models have been trained on vast and diverse datasets, enabling them to understand and generate text with rare or even unseen words. However, there are still some considerations for text preprocessing and handling rare words effectively.

So, how can we handle rare words with LLMs? There are some key concepts we need to understand, starting with tokenization. We won't explore tokenization in detail here as we have a dedicated section later on; for now, let's say that LLMs use **subword tokenization** methods that break down rare words into more common subword units. This helps in managing **Out-of-Vocabulary** (**OOV**) words by decomposing them into familiar components. The other interesting thing about LLMs is that even if they don't know the word per se, they have contextual understanding capabilities, meaning that LLMs leverage context to infer the meaning of rare words.

In the following code example, we will test GPT-2 to see if it can handle rare words. You can find the code in the repository at `https://github.com/PacktPublishing/Python-Data-Cleaning-and-Preparation-Best-Practices/blob/main/chapter12/4.rare_words.py`:

1. Let's import the required libraries:

```
from transformers import GPT2LMHeadModel, GPT2Tokenizer
```

2. Initialize the GPT-2 tokenizer and model:

```
tokenizer = GPT2Tokenizer.from_pretrained("gpt2")
model = GPT2LMHeadModel.from_pretrained("gpt2")
```

3. Define a text prompt with a rare word:

```
text = "The quokka, a rare marsupial,"
```

4. Encode the input text to tokens:

```
indexed_tokens = tokenizer.encode(text, return_tensors='pt')
```

5. Generate text until the output length reaches 50 tokens. The model generates text based on the input prompt, leveraging its understanding of the context to handle the rare word:

```
output_text = model.generate(indexed_tokens, max_length=50, num_
beams=5, no_repeat_ngram_size=2, early_stopping=True)
```

The `generate` function in the given code snippet is used to produce text output from a model based on the input tokens provided. The parameters used in this function call control various aspects of the text generation process:

- `indexed_tokens`: This represents the input sequence that the model will use to start generating text. It consists of tokenized text that serves as the starting point for generation.
- `max_length=50`: This parameter sets the maximum length of the generated text. The model will generate up to 50 tokens, including the input tokens, ensuring that the output doesn't exceed this length.
- `num_beams=5`: This controls the beam search process, where the model keeps track of the top five most likely sequences during generation. Beam search helps improve the quality of the generated text by exploring multiple possible outcomes simultaneously and selecting the most likely one.
- `no_repeat_ngram_size=2`: This prevents the model from repeating any sequence of two tokens (bigrams) within the generated text. It helps produce more coherent and less repetitive output by ensuring that the same phrases don't appear multiple times.
- `early_stopping=True`: This parameter allows the generation process to stop early if all beams have reached the end of the text sequence (e.g., a sentence-ending token). This can make the generation process more efficient by avoiding unnecessary continuation when a complete and sensible output has already been produced.

These parameters can be adjusted depending on the desired output. For instance, increasing `max_length` generates longer text, while modifying `num_beams` can balance quality and computational cost. Adjusting `no_repeat_ngram_size` changes the strictness of repetition prevention, and toggling `early_stopping` can affect the efficiency and length of the generated text. *I would advise that you go and play with these configurations to see how their output is affected*:

1. The generated tokens are decoded back into human-readable text:

```
output_text_decoded = tokenizer.decode(output_text[0], skip_
special_tokens=True)
```

2. Print the decoded text:

```
The quokka, a rare marsupial, is one of the world's most
endangered species.
```

As we can see, the model understood the meaning of *quokka* and created a sequence of words, additional text that continues from the prompt, showcasing the language generation capabilities of LLMs. This is possible because LLMs turn the tokens into a numerical representation called **embeddings**, as we will see later on, that capture the meaning of words.

We discussed the use of rare words in text preprocessing. Let's now move to another challenge—spelling errors and typos.

Addressing spelling variations and typos

The challenge with spelling variations and typos is that it can lead to *different tokenizations for similar words*. In the era of LLMs, handling spelling and typos has become more sophisticated. LLMs can understand contexts and generate text that often corrects such errors implicitly. However, explicit preprocessing to correct spelling mistakes can still enhance the performance of these models, especially in applications where accuracy is critical. There are different ways to address spelling variations and mistakes, as we will see in the following section, starting with spelling correction.

Spelling correction

Let's create an example of fixing spelling mistakes using an LLM with Hugging Face Transformers. We'll use the experimental `oliverguhr/spelling-correction-english-base` spelling correction model for this demonstration. You can find the full code at `https://github.com/PacktPublishing/Python-Data-Cleaning-and-Preparation-Best-Practices/blob/main/chapter12/5.spelling_checker.py`:

1. Define the spelling function pipeline. Inside this function, we initialize the spelling correction pipeline using the `oliverguhr/spelling-correction-english-base` model. This model is specifically trained for spelling correction tasks:

```
def fix_spelling(text):
    spell_check = pipeline("text2text-generation",
model="oliverguhr/spelling-correction-english-base")
```

2. We use the pipeline to generate the corrected text. The `max_length` parameter is set to `2048` to allow for longer input texts:

```
    corrected = spell_check(text, max_length=2048)[0]
['generated_text']
    return corrected
```

3. Test the function with some sample text containing spelling mistakes:

```
sample_text = "My name si from Grece."
corrected_text = fix_spelling(sample_text)
Corrected text: My name is from Greece.
```

It's important to note that this is an experimental model, and its performance may vary depending on the complexity and context of the input text. For more robust spelling and grammar correction, you might consider using more advanced models; however, some of them need authentication to download or sign agreements. So, for simplicity, we used an experimental model here. You can replace it with any model you have access to, from Llama3 to GPT4 and others.

The significance of spelling correction in text preprocessing tasks takes us nicely to the concept of fuzzy matching, a technique that further enhances the accuracy and relevance of generated content by accommodating minor errors and variations in input text.

Fuzzy matching

Fuzzy matching is a technique used to compare strings for similarity, even when they are not exactly the same. It's like finding words that are "kind of similar" or "close enough." So, we can use fuzzy matching algorithms to identify and map similar words, as well as to solve for variations and minor misspellings. We can enhance the spelling correction function by adding fuzzy matching using the `TheFuzz` library.

Let's go through the code that you can find at `https://github.com/PacktPublishing/Python-Data-Cleaning-and-Preparation-Best-Practices/blob/main/chapter12/6.fuzzy_matching.py`:

1. We'll start by installing the library:

   ```
   pip install thefuzz==0.22.1
   ```

2. Let's import the required libraries:

   ```
   from transformers import pipeline
   from thefuzz import process, fuzz
   ```

3. Initialize the spelling correction pipeline:

   ```
   def fix_spelling(text, threshold=80):
       spell_check = pipeline("text2text-generation",
   model="oliverguhr/spelling-correction-english-base")
   ```

   The `oliverguhr/spelling-correction-english-base` model is specifically fine-tuned for the task of spelling correction, making it a highly effective and efficient tool for spelling correction. This model has been trained to recognize and correct common spelling errors in English text, leading to greater accuracy. It is optimized for text-to-text generation, which allows it to efficiently generate corrected versions of input text with minimal computational overhead. Additionally, its training likely involved exposure to datasets containing spelling errors and their corrections, enabling it to make informed and contextually appropriate corrections.

4. Generate the corrected text as in the previous section:

```
    corrected = spell_check(text, max_length=2048)[0]
['generated_text']
```

5. Split the original and corrected texts into words:

```
    original_words = text.split()
    corrected_words = corrected.split()
```

6. Create a dictionary of common English words (you can expand this list):

```
    common_words = set(['the', 'be', 'to', 'of', 'and', 'a',
'in', 'that', 'have', 'I', 'it', 'for', 'not', 'on', 'with',
'he', 'as', 'you', 'do', 'at'])
```

7. Fuzzy match each word:

```
    final_words = []
    for orig, corr in zip(original_words, corrected_words):
        if orig.lower() in common_words:
            final_words.append(orig)
        else:
            matches = process.extractOne(orig, [corr],
scorer=fuzz.ratio)
            if matches[1] >= threshold:
                final_words.append(matches[0])
            else:
                final_words.append(orig)
    return ' '.join(final_words)
```

8. Test the function with some sample text containing spelling mistakes:

```
sample_text = "Lets do a copmarsion of speling mistaks in this
sentense."
corrected_text = fix_spelling(sample_text)
```

9. Print the results:

```
Original text: Lets do a copmarsion of speling mistaks in this
sentense.
Corrected text: Let's do a comparison of speling mistaks in this
sentence.
```

Now, as you can see, not all the spelling mistakes are corrected. We could get some better performance by fine-tuning the model on the examples it usually misses. However, there is good news! The rise of LLMs has made it less critical to correct spelling mistakes because these models are designed to understand and process text contextually. Even when words are misspelled, LLMs can infer the intended meaning by analyzing the surrounding words and overall sentence structure. This ability reduces the need for perfect spelling, as the primary focus shifts to conveying the message rather than ensuring every word is spelled correctly.

After completing the initial text preprocessing steps, the next critical phase is **chunking**. This process involves breaking the cleaned text into smaller, meaningful units. Let's discuss that in the following section.

Chunking

Chunking is an essential preprocessing step in NLP that involves breaking down text into smaller, manageable units, or "chunks." This process is crucial for various applications, including text summarization, sentiment analysis, information extraction, and more.

Why is chunking becoming more and more important? By breaking down large documents, chunking enhances manageability and efficiency, particularly for models with *token limits*, preventing overload and enabling smoother processing. It also improves accuracy by allowing models to *focus on smaller, coherent segments of text*, which reduces noise and complexity compared to analyzing entire documents. Additionally, chunking helps maintain context within each segment, which is essential for tasks such as machine translation and text generation, ensuring that the model comprehends and processes the text effectively.

Chunking can be implemented in many different ways; for instance, summarization may benefit from paragraph-level chunks, whereas sentiment analysis might use sentence-level chunks to capture nuanced emotional tones. In the following sections, we will focus on fixed-length, recursive, and semantic chunking as we see them more often in the data world.

Implementing fixed-length chunking

Fixed-length chunking involves breaking text into chunks of a *predefined length*, either by character count or token count. It is usually preferred because it is very simple to implement and ensures uniform chunk sizes. However, as the split is random, it may split sentences or semantic units, leading to a loss of context. It is suitable for tasks where uniform chunk sizes are needed, such as certain types of text classification.

To showcase fixed-length chunking, we are going to work with review data again, but we will include a few lengthier reviews. You can see the full example at `https://github.com/PacktPublishing/Python-Data-Cleaning-and-Preparation-Best-Practices/blob/main/chapter12/7.fixed_chunking.py`:

1. Let's start by loading the example data:

```
reviews = [
    "This smartphone has an excellent camera. The photos are
sharp and the colors are vibrant. Overall, very satisfied with
my purchase.",
```

```
    "I was disappointed with the laptop's performance. It
frequently lags and the battery life is shorter than expected.",
    "The blender works great for making smoothies. It's powerful
and easy to clean. Definitely worth the price.",
    "Customer support was unresponsive. I had to wait
a long time for a reply, and my issue was not resolved
satisfactorily.",
    "The book is a fascinating read. The storyline is engaging
and the characters are well-developed. Highly recommend to all
readers."
]
```

2. Import the `TokenTextSplitter` class:

```
from langchain_text_splitters import TokenTextSplitter
```

3. Initialize the `TokenTextSplitter` class with a chunk size of `50` tokens and no overlap:

```
text_splitter = TokenTextSplitter(chunk_size=50, chunk_
overlap=0)
```

4. Combine the reviews into a single text block for chunking:

```
text_block = " ".join(reviews)
```

5. Split the text into token-based chunks:

```
chunks = text_splitter.split_text(text_block)
```

6. Print the chunks:

```
Chunk 1:
This smartphone has an excellent camera. The photos are sharp
and the colors are vibrant. Overall, very satisfied with my
purchase. I was disappointed with the laptop's performance. It
frequently lags and the battery life is shorter than expected.
The blender works
Chunk 2:
great for making smoothies. It's powerful and easy to clean.
Definitely worth the price. Customer support was unresponsive.
I had to wait a long time for a reply, and my issue was not
resolved satisfactorily. The book is a
Chunk 3:
fascinating read. The storyline is engaging and the characters
are well-developed. Highly recommend to all readers.
```

To understand how varying chunk sizes affect the output, you can modify the `chunk_size` parameter. For instance, you might try chunk sizes of `20`, `70`, and `150` tokens. Here, you can see how you can adapt the code to test different chunk sizes:

```
chunk_sizes = [20, 70, 150]
for size in chunk_sizes:
    print(f"Chunk Size: {size}")
    text_splitter = TokenTextSplitter(chunk_size=size, chunk_
overlap=0)
    chunks = text_splitter.split_text(text_block)
    for i, chunk in enumerate(chunks):
        print(f"Chunk {i + 1}:")
        print(chunk)
        print("\n")
```

We successfully divided our review into the required chunks, but before moving forward, it's crucial to understand the significance of the `chunk_overlap=0` parameter.

Chunk overlap

Chunk overlap refers to the number of characters or tokens that are *shared* between adjacent chunks when splitting a text. It's the amount of text that "overlaps" between one chunk and the next.

Chunk overlap is crucial as it helps preserve context and enhance the coherence of the text. By ensuring that adjacent chunks share some common content, overlap *maintains continuity* and prevents important information from being lost at the boundaries. For instance, if a document is divided into chunks without overlap, a critical piece of information could be split between two chunks, potentially rendering it inaccessible or causing a loss of meaning. In retrieval tasks, such as searching or question-answering, overlap ensures that relevant details are captured even if they fall across chunk boundaries, thereby improving the effectiveness of the retrieval process. For example, if a chunk ends mid-sentence, the overlap ensures that the entire sentence is considered, which is essential for accurate comprehension and response generation.

Let's consider a simple example to illustrate chunk overlap:

```
Original text:
One of the most important things I didn't understand about the world
when I was a child is the degree to which the returns for performance
are superlinear.
```

With a chunk size of five words and an overlap of one word, we'll get the following results:

```
Chunk 1: "One of the most important"
Chunk 2: "important things I didn't understand"
```

```
Chunk 3: "understand about the world when"
Chunk 4: "when I was a child"
Chunk 5: "child is the degree to"
Chunk 6: "to which the returns for"
Chunk 7: "for performance are superlinear."
```

As you can see, each chunk overlaps with the next by *two words*, helping to maintain context and prevent loss of meaning at chunk boundaries. Fixed-length chunking divides text into segments of a uniform size, but this method can sometimes fail to capture meaningful units of text, especially when dealing with natural language's inherent variability. Transitioning to paragraph chunking, on the other hand, allows for a more contextually coherent approach by segmenting text based on its natural structure.

Implementing RecursiveCharacter chunking

`RecursiveCharacterTextSplitter` is a sophisticated text-splitting tool designed to handle more complex text segmentation tasks, especially when dealing with lengthy documents that need to be broken down into smaller, meaningful chunks. Unlike basic text splitters that simply cut text into fixed or variable-sized chunks, `RecursiveCharacterTextSplitter` uses a recursive approach to divide text, ensuring that each chunk is both contextually coherent and appropriately sized for processing by natural language models. Continuing from the review example, we will now demonstrate how to split a document into paragraphs using `RecursiveCharacterTextSplitter`. You can find the full code at `https://github.com/PacktPublishing/Python-Data-Cleaning-and-Preparation-Best-Practices/blob/main/chapter12/8.paragraph_chunking.py`:

1. We create a `RecursiveCharacterTextSplitter` instance:

   ```
   text_splitter = RecursiveCharacterTextSplitter(
       separators=["\n\n", "\n", " ", ""],
       chunk_size=200,
       chunk_overlap=0,
       length_function=len
       )
   ```

   The `RecursiveCharacterTextSplitter` instance is instantiated with specific parameters:

   - `separators`: This is a list of separators used to split the text. Here, it includes double newlines (`\n\n`), single newlines (`\n`), spaces ( ), and empty strings (`""`). This helps the splitter to use natural text boundaries and whitespace for chunking.
   - `chunk_size`: This is the maximum size of each chunk, set to 200 characters. This means each chunk will be *up to* 200 characters long.
   - `chunk_overlap`: This is the number of characters overlapping between adjacent chunks, set to 0. This means there is no overlap between chunks.

   - `length_function`: This is a function used to measure the length of the text, set to `len`, which calculates the number of characters in a string.

2. Split the text into chunks:

   ```
   chunks = text_splitter.split_text(text_block)
   ```

3. Print the chunks. In this first one, the user is very satisfied with the smartphone camera, praising the sharpness and vibrant colors of the photos. However, the user is disappointed with the laptop's performance, citing frequent lags:

   ```
   Chunk 1:
   This smartphone has an excellent camera. The photos are sharp
   and the colors are vibrant. Overall, very satisfied with my
   purchase. I was disappointed with the laptop's performance. It
   frequently lags
   ```

   The user is pleased with the blender, noting its effectiveness in making smoothies, its power, and its ease of cleaning. They consider it a good value for the price:

   ```
   Chunk 2:
   and the battery life is shorter than expected. The blender works
   great for making smoothies. It's powerful and easy to clean.
   Definitely worth the price. Customer support was unresponsive. I
   had to
   ```

   The user had a negative experience with customer support, mentioning long wait times and unresolved issues. The user finds the book to be a fascinating read with an engaging storyline and well-developed characters, and they highly recommend it to readers:

   ```
   Chunk 3:
   wait a long time for a reply, and my issue was not resolved
   satisfactorily. The book is a fascinating read. The storyline is
   engaging and the characters are well-developed. Highly recommend
   to all
   ```

   We are left with one remaining word:

   ```
   Chunk 4:
   Readers.
   ```

Now, these chunks are not perfect, but let's understand how `RecursiveCharacterTextSplitter` works so that you can adjust it to your use case:

- **Chunk size target**: The splitter aims for chunks of about 200 characters, but this is a maximum rather than a strict requirement. It will try to create chunks as close to 200 characters as possible without exceeding this limit.
- **Recursive approach**: The recursive nature means it will apply these rules repeatedly, working its way through the separator list until it finds an appropriate split point.

- **Preserving semantic meaning**: By using this approach, the splitter attempts to keep semantically related content together. For example, it will try to avoid splitting in the middle of a paragraph or sentence if possible.
- **No overlap**: With `chunk_overlap` set to `0`, there's no repetition of content between chunks. Each chunk is distinct.
- **Length function**: The `len` function is used to measure chunk size, meaning it's counting characters rather than tokens.

> **The length_function parameter**
>
> The `length_function` parameter in `RecursiveCharacterTextSplitter` is a flexible option that allows you to define *how the length of text chunks is measured*. While `len` is the default and most common choice, there are many other options, from token-based to word-based to custom implementations.

While recursive chunking focuses on creating chunks based on fixed sizes and natural separators, semantic chunking takes this a step further by grouping text based on its meaning and context. This method ensures that chunks are not only coherent in length but also semantically meaningful, improving the relevance and accuracy of downstream NLP tasks.

Implementing semantic chunking

Semantic chunking involves breaking text into chunks based on semantic meaning rather than just syntactic rules or fixed lengths. Behind the scenes, they use *embeddings* to group related sentences together (we will deep dive into embeddings in *Chapter 13, Image and Audio Preprocessing with LLMs*). We usually use semantic chunking for tasks requiring a deep understanding of context, such as question answering and thematic analysis. Let's deep dive into the process behind semantic chunking:

1. **Text input**: The process begins with a text input, which could be a document, a collection of sentences, or any textual data that needs to be processed.
2. **Embedding generation**: Each segment of the text (typically sentences or small groups of sentences (chunks)) is converted into a high-dimensional vector representation using embeddings. These embeddings are generated by pre-trained language models and the key to understand here is that these embeddings capture the semantic meaning of the text. In other words, we convert text into a numerical representation *that encodes its meaning*!
3. **Similarity measurement**: The embeddings are then compared to measure the semantic similarity between different parts of the text. Techniques such as cosine similarity are often used to quantify how closely related different segments are.
4. **Clustering**: Based on the similarity scores, sentences or text segments are clustered together. The clustering algorithm groups sentences that are semantically similar into the same chunk. This ensures that each chunk maintains semantic coherence and context.

5. **Chunk creation**: The clustered sentences are then combined to form chunks. These chunks are designed to be semantically meaningful units of text, which can be more effectively processed by NLP models.

Let's go back to the product review example and see what chunk we generate with semantic chunking. You can find the code at `https://github.com/PacktPublishing/Python-Data-Cleaning-and-Preparation-Best-Practices/blob/main/chapter12/9.semantic_chunking.py`:

1. Initialize `SemanticChunker` with `HuggingFaceEmbeddings`:

   ```
   text_splitter = SemanticChunker(HuggingFaceEmbeddings())
   ```

2. Split the text into chunks:

   ```
   docs = text_splitter.create_documents([text_block])
   ```

3. Print the chunks:

   ```
   Chunk 1:
   This smartphone has an excellent camera. The photos are sharp
   and the colors are vibrant. Overall, very satisfied with my
   purchase. I was disappointed with the laptop's performance. It
   frequently lags and the battery life is shorter than expected.
   The blender works great for making smoothies. It's powerful and
   easy to clean.
   Chunk 2:
   Definitely worth the price. Customer support was unresponsive.
   I had to wait a long time for a reply, and my issue was not
   resolved satisfactorily. The book is a fascinating read. The
   storyline is engaging and the characters are well-developed.
   Highly recommend to all readers.
   ```

Each chunk contains related sentences that form a coherent segment. For example, chunk 1 discusses various product performances, while chunk 2 includes customer support experience and a book review. The chunks also maintain context within each segment, ensuring that related information is grouped together. A point for improvement is that chunk 1 includes reviews of different products (smartphone, laptop, and blender), and chunk 2 mixes a customer support experience with a book review, which could be seen as semantically unrelated. In this case, we could further split the text into smaller, more focused chunks to make it more coherent or/and tweak the parameters of the sematic chunker:

```
text_splitter = SemanticChunker(
    embeddings=embedding_model,
    buffer_size=200,
    add_start_index=True,
    breakpoint_threshold_type='percentile',
    breakpoint_threshold_amount=0.9,
    number_of_chunks=4,
```

```
    sentence_split_regex=r'\.|\n|\s'
)
```

You can find more details about these parameters in the documentation:

`https://api.python.langchain.com/en/latest/text_splitter/langchain_experimental.text_splitter.SemanticChunker.html`

However, the steps to improve chunking in our case could look something like this:

- Use different embedding models to see which provides the best embeddings for your text
- Tweak the buffer size to find the right balance between chunk size and coherence
- Adjust the threshold type and amount to optimize where chunks are split based on semantic breaks
- Customize the regular expression for sentence splitting to better fit the structure of your text

Transitioning from chunking to tokenization involves moving from a process where text is divided into larger, often syntactically significant segments (chunks) to a process where text is divided into smaller, more granular units (tokens). Let's take a look at how **tokenization** works.

Tokenization

Tokenization is the process of breaking down a sequence of text into smaller units, or tokens, which can be words, subwords, or characters. This process is essential for converting text into a format suitable for *computational processing*, enabling models to learn patterns at a finer granularity.

Some key terms in the Tokenization phase are **vocabulary** and **Unique Identifiers** (**IDs**). The vocabulary is a fixed set of tokens *that a model knows*. It can include words, subwords, punctuation, and special tokens (such as `[CLS]` for classification, `[SEP]` for separation, etc.). Each token in the vocabulary is assigned an ID, which the model uses to represent the token internally. These IDs are integers and typically range from 0 to the size of the vocabulary minus one.

Can all the words in the world fit into a vocabulary? The answer is *no*! OOV words are words that are not present in the model's vocabulary.

Now that we know the main terms that are used, let's explore the different types of tokenization and the challenges associated with them.

Word tokenization

Word tokenization involves splitting text into individual words.

For example, the sentence "Tokenization is crucial in NLP!" would be tokenized into `["Tokenization", "is", "crucial", "in", "NLP", "!"]`.

Word tokenization preserves whole words, which can be beneficial for tasks requiring word-level understanding. It works well for languages with clear word boundaries. It is a simple solution but can lead to problems with OOV words, especially in specialized domains such as medical texts and texts with a lot of misspellings. You can find the full code at `https://github.com/PacktPublishing/Python-Data-Cleaning-and-Preparation-Best-Practices/blob/main/chapter12/10.word_tokenisation.py`.

Let's see a code example:

1. Download the necessary NLTK data (run this once):

```
nltk.download('punkt')
```

2. Take the following as sample text:

```
text = "The quick brown fox jumps over the lazy dog. It's
unaffordable!"
```

3. Perform word tokenization:

```
word_tokens = word_tokenize(text)
```

4. Print the output:

```
Tokens:
['The', 'quick', 'brown', 'fox', 'jumps', 'over', 'the', 'lazy',
'dog', '.', 'It', "'s", 'unaffordable', '!']
```

This type of word tokenization is useful when dealing with simple, well-formed text where each word is clearly separated by spaces and punctuation. It's an easy method that aligns well with how humans perceive words. However, different forms of the same word (e.g., "run", "running", "ran") are treated as separate tokens, which can dilute the model's understanding. It can also result in a large vocabulary, especially for languages with rich morphology or many unique words. Finally, these models struggle with words not seen during training, leading to OOV tokens.

Given the limitations of word tokenization, subword tokenization methods have become popular. Subword tokenization strikes a balance between word-level and character-level tokenization, addressing many of the shortcomings of both.

Subword tokenization

Subword tokenization splits text into smaller units than words, typically subwords. By breaking the words into known subwords, it can handle OOV words. It reduces the vocabulary size and parameter count significantly. Let's see the different options in the subword tokenization in the following sections.

Byte Pair Encoding (BPE)

BPE starts with individual characters and iteratively merges the most frequent pairs of tokens to create subwords. It was originally developed as a data compression algorithm but has been adapted for tokenization in NLP tasks. The process looks like this:

1. Start with a vocabulary of individual characters.
2. Calculate the frequency of all character pairs in the text.
3. Merge the most frequent pair of characters to form a new token.
4. Repeat the process until the desired vocabulary size is reached.

This frequency-based merging strategy can be useful for languages with simpler morphological structures (e.g., English) or when a straightforward yet robust tokenization is needed. It is simple and computationally efficient due to the frequency-based merging. Let's demonstrate an example of how to implement BPE for tokenization. You can find the full example at `https://github.com/PacktPublishing/Python-Data-Cleaning-and-Preparation-Best-Practices/blob/main/chapter12/11.bpe_tokeniser.py`:

1. Load the pre-trained tokenizer:

```
tokenizer = GPT2Tokenizer.from_pretrained('gpt2')
```

2. Load the sample text:

```
text = "Tokenization in medical texts can include words like
hyperlipidemia."
```

3. Tokenize the text:

```
tokens = tokenizer.tokenize(text)
```

4. Convert tokens to input IDs:

```
input_ids = tokenizer.convert_tokens_to_ids(tokens)
```

5. Print the tokens and the IDs, as shown here:

```
Tokens: ['Token', 'ization', 'Ġin', 'Ġmedical', 'Ġtexts',
'Ġcan', 'Ġinclude', 'Ġwords', 'Ġlike', 'Ġhyper', 'lip', 'idem',
'ia', '.']
Input IDs: [21920, 3666, 287, 1400, 1562, 649, 4551, 3545, 588,
20424, 3182, 1069, 257, 13]
```

The special character "Ġ" in the tokenization output has a specific meaning in the context of BPE tokenization. It indicates that the token following it originally had a preceding space or was at the beginning of the text, so it allows for preserving information about word boundaries and spacing in the original text.

Let's explain the output we see:

- `Token` and `ization`: These are subwords of "Tokenization", split without a "Ġ" because *they're part of the same word.*
- `Ġin`, `Ġmedical`, `Ġtexts`, and so on: These tokens start with Ġ, indicating they were *separate* words in the original text.
- `hyper`, `lip`, `id`, and `emia`: These are subwords of "hyperlipidemia", a medical term. The Ġ before `hyper` shows it's a new word, while the subsequent subwords don't have Ġ as they're part of the same word. In medical terminology, many words are compounds or have specific prefixes or suffixes. The BPE tokenization helps break these down into meaningful subunits. For example, `hyperlipidemia` is broken into `hyper` (prefix meaning *excessive*), `lip` (related to fats), `id` (connecting element), and `emia` (suffix meaning *blood condition*).

Having explored BPE tokenization and its impact on text processing, we now turn our attention to WordPiece tokenization, another powerful method that further refines the handling of subword units in NLP tasks.

WordPiece tokenization

WordPiece, used by BERT, starts with a base vocabulary of characters, and iteratively adds the most frequent subword units. The process looks like this:

1. Start with a base vocabulary of individual characters and a special token for unknown words.
2. Iteratively merge the most frequent pairs of tokens (starting with characters) to form new tokens until a predefined vocabulary size is reached.
3. For any given word, the longest matching subword units from the vocabulary are used. This process is known as **maximum matching**.

WordPiece tokenization is effective for languages with complex word structures (e.g., Korean and Japanese) and when handling a diverse vocabulary efficiently is crucial. Its effectiveness comes from the fact that the merges are chosen based on maximizing the likelihood, leading to potentially more meaningful subwords. However, everything comes with a cost, and in this case, it is more computationally intensive due to the likelihood maximization step. Let's have a look at a code example of how to perform WordPiece tokenization using the BERT tokenizer. You can find the full code at `https://github.com/PacktPublishing/Python-Data-Cleaning-and-Preparation-Best-Practices/blob/main/chapter12/12.tokenisation_wordpiece.py`:

1. Load the pre-trained tokenizer:

```
tokenizer = BertTokenizer.from_pretrained('bert-base-uncased')
```

2. Take some sample text:

```
text = "Tokenization in medical texts can include words like
hyperlipidemia."
```

3. Tokenize the text. This method splits the input text into WordPiece tokens. For example, `unaffordable` is broken down into un, `##afford`, `##able`:

```
tokens = tokenizer.tokenize(text)
```

4. Convert tokens to input IDs:

```
input_ids = tokenizer.convert_tokens_to_ids(tokens)
Tokens:
['token', '##ization', 'in', 'medical', 'texts', 'can',
'include', 'words', 'like', 'hyper', '##lip', '##idem', '##ia']
Input IDs:
[19204, 10859, 1999, 2966, 4524, 2064, 2421, 2540, 2066, 15088,
17750, 28285, 3676]
```

The ## prefix is used to denote that the token is a continuation of the previous token. Thus, it helps in reconstructing the original word by indicating that the token should be appended to the preceding token without a space.

After examining tokenization methods such as BPE and WordPiece, it is crucial to consider how tokenizers can be tailored to handle specialized data, such as medical texts, to ensure precise and contextually relevant processing in these specific domains.

Domain-specific data

When working with domain-specific data such as medical texts, it's crucial to ensure that the tokenizer can handle specialized vocabulary effectively. When a domain has a high frequency of unique terms or specialized vocabulary, standard tokenizers may not perform optimally. In this case, domain-specific tokenizers can better capture the nuances and terminology of the field, leading to improved model performance. When you are faced with this challenge, there are some options available:

- Train a tokenizer on a corpus of domain-specific texts to create a vocabulary that includes specialized terms
- Consider extending existing tokenizers with domain-specific tokens instead of training from scratch

However, how can you know that you need to go the extra mile and tune the tokenizer on the dataset? Let's find out.

Evaluating the need for specialized tokenizers

As we explained, when working with domain-specific data, such as medical texts, it's essential to evaluate whether a specialized tokenizer is needed to ensure accurate and contextually relevant processing. Let's have a look at several key factors to consider:

- **Analyze OOV rate**: Determine the percentage of words in your domain-specific corpus that are not included in the standard tokenizer's vocabulary. A high OOV rate suggests that many important terms in your domain are not being recognized, highlighting the need for a specialized tokenizer to better handle the unique vocabulary.
- **Examine tokenization quality**: Check how standard tokenizers split domain-specific terms by manually reviewing sample tokenizations. If crucial terms, such as medical terminology, are frequently broken into meaningless subwords, this indicates that the tokenizer is not well-suited for the domain and may require customization.
- **Compression ratio**: Measure the average number of tokens per sentence using both standard and domain-specific tokenizers. A significantly lower ratio with a domain-specific tokenizer suggests it is more efficient at compressing and representing domain knowledge, reducing redundancy, and improving performance.

For instance, in a medical corpus, terms such as *myocardial infarction* might be tokenized as `myo`, `cardial`, and `infarction` by a standard tokenizer, leading to a loss of meaningful context. A specialized medical tokenizer, however, might recognize *myocardial infarction* as a single term, preserving its meaning and improving the quality of downstream tasks such as entity recognition and text generation. Similarly, if a standard tokenizer results in an OOV rate of 15% compared to just 3% with a specialized tokenizer, it clearly indicates the need for a tailored approach. Lastly, if the compression ratio using a standard tokenizer is 1.8 tokens per sentence versus 1.2 tokens with a specialized tokenizer, it shows that the specialized tokenizer is more efficient and effective in capturing the domain-specific nuances.

Let's implement a small application to evaluate the tokenizers for different medical data. The code for the example is available at `https://github.com/PacktPublishing/Python-Data-Cleaning-and-Preparation-Best-Practices/blob/main/chapter12/13.specialised_tokenisers.py`:

1. Initialize Stanza for biomedical text:

```
stanza.download('en', package='mimic', processors='tokenize')
nlp = stanza.Pipeline('en', package='mimic',
processors='tokenize')
```

2. Initialize the standard GPT-2 tokenizer:

```
standard_tokenizer = GPT2Tokenizer.from_pretrained("gpt2")
```

3. Set `pad_token` to `eos_token`:

```
standard_tokenizer.pad_token = standard_tokenizer.eos_token
model = GPT2LMHeadModel.from_pretrained("gpt2")
```

4. Set `pad_token_id` for the model:

```
model.config.pad_token_id = model.config.eos_token_id
```

5. A sample medical corpus consisting of sentences related to myocardial infarction and heart conditions is defined:

```
corpus = [
  "The patient suffered a myocardial infarction.",
  "Early detection of heart attack is crucial.",
  "Treatment for myocardial infarction includes medication.",
  "Patients with heart conditions require regular check-ups.",
  "Myocardial infarction can lead to severe complications."
]
```

6. The following `stanza_tokenize` function uses the Stanza pipeline to tokenize the text and return a list of tokens:

```
def stanza_tokenize(text):
    doc = nlp(text)
    tokens = [word.text for sent in doc.sentences for word in
sent.words]
    return tokens
```

7. The `calculate_oov_and_compression` function tokenizes each sentence in the corpus and calculates the OOV rate, as well as the average tokens per sentence, and returns all tokens. For standard tokenizers, it checks whether tokens are in the vocabulary, while for Stanza, it does not check OOV tokens explicitly:

```
def calculate_oov_and_compression(corpus, tokenizer):
    oov_count = 0
    total_tokens = 0
    all_tokens = []

    for sentence in corpus:
        tokens = tokenizer.tokenize(sentence) if
hasattr(tokenizer, 'tokenize') else stanza_tokenize(sentence)
        all_tokens.extend(tokens)
        total_tokens += len(tokens)
        oov_count += tokens.count(tokenizer.oov_token) if
hasattr(tokenizer, 'oov_token') else 0
```

```
    oov_rate = (oov_count / total_tokens) * 100 if total_tokens
> 0 else 0
    avg_tokens_per_sentence = total_tokens / len(corpus)

return oov_rate, avg_tokens_per_sentence, all_tokens
```

8. The `analyze_token_utilization` function calculates the frequency of each token in the corpus and returns a dictionary of token utilization percentages:

```
def analyze_token_utilization(tokens):
    token_counts = Counter(tokens)
    total_tokens = len(tokens)
    utilization = {token: count / total_tokens for token, count
in token_counts.items()}
    return utilization
```

9. The `calculate_perplexity` function calculates the perplexity of the model on the given text, which is a measure of how well the model predicts the sample:

```
def calculate_perplexity(tokenizer, model, text):
    inputs = tokenizer(text, return_tensors="pt", padding=True,
truncation=True)
    with torch.no_grad():
        outputs = model(inputs, labels=inputs["input_ids"])
    return torch.exp(outputs.loss).item()
```

10. The following script evaluates both the standard GPT-2 tokenizer and the Stanza medical tokenizer by calculating the OOV rate, average tokens per sentence, token utilization, and perplexity. Finally, it prints the results for each tokenizer and compares their performance on the `myocardial infarction` term:

```
for tokenizer_name, tokenizer in [("Standard GPT-2", standard_
tokenizer), ("Stanza Medical", stanza_tokenize)]:
    oov_rate, avg_tokens, all_tokens = calculate_oov_and_
compression(corpus, tokenizer)
    utilization = analyze_token_utilization(all_tokens)
    print(f"\n{tokenizer_name} Tokenizer:")
    print(f"OOV Rate: {oov_rate:.2f}%")
    print(f"Average Tokens per Sentence: {avg_tokens:.2f}")
    print("Top 5 Most Used Tokens:")
    for token, freq in sorted(utilization.items(), key=lambda x:
x[1], reverse=True)[:5]:
        print(f" {token}: {freq:.2%}")
```

Let's see the results of the two tokenizers presented in the following table:

Metric	Standard GPT-2 tokenizer	Stanza Medical tokenizer
OOV rate	0.00%	0.00%
Average tokens per sentence	10.80	7.60
Top five most used tokens	. : 9.26%	. : 13.16%
	ocard : 5.56%	infarction : 7.89%
	ial : 5.56%	myocardial : 5.26%
	Ġinf : 5.56%	heart : 5.26%
	ar : 5.56%	The : 2.63%

Table 12.1 – Comparison of GPT-2 tokenizer and specialized medical tokenizer

As we see in the table, both tokenizers show an OOV rate of 0.00%, indicating that all tokens in the corpus are recognized by both tokenizers. The Stanza Medical tokenizer has a lower average rate of tokens per sentence (7.60) compared to the Standard GPT-2 tokenizer (10.80). This suggests that the Stanza Medical tokenizer is more efficient at compressing domain-specific terms into fewer tokens. The Standard GPT-2 tokenizer splits meaningful medical terms into smaller subwords, leading to less meaningful token utilization (e.g., `ocard`, `ial`, `Ġinf`, and `ar`). However, the Stanza Medical tokenizer maintains the integrity of medical terms (e.g., `infarction` and `myocardial`), making the tokens more meaningful and contextually relevant. Based on the analysis, the Stanza Medical tokenizer should be preferred for medical text processing because of the following reasons:

- It efficiently tokenizes domain-specific terms into fewer tokens
- It preserves the integrity and meaning of medical terms
- It provides more meaningful and contextually relevant tokens, which is crucial for tasks such as entity recognition and text generation in the medical domain

The Standard GPT-2 tokenizer, while useful for general text, splits medical terms into subwords, leading to potential loss of context and meaning, making it less suitable for specialized medical text.

Vocabulary size trade-offs

Larger vocabularies can capture more domain-specific terms but increase the model size and computational requirements. Find a balance that adequately covers domain terminology without excessive growth.

Having evaluated the performance of different tokenization methods, including their handling of OOV terms and their efficiency in compressing domain-specific knowledge, the next logical step is to explore how these tokenized outputs are transformed into meaningful numerical representations through embedding techniques. This transition is crucial, as embeddings form the foundation of how models understand and process the tokenized text.

Turning tokens into embeddings

Embeddings are numerical representations of words, phrases, or entire documents in a high-dimensional vector space. Essentially, we represent words as arrays of numbers to capture their semantic meaning. These numerical arrays aim to encode the underlying significance of words and sentences, allowing models to understand and process text in a meaningful way. Let's explore the process from tokenization to embedding.

The process starts with tokenization, whereby text is split into manageable units called tokens. For instance, the sentence "The cat sat on the mat" might be tokenized into individual words or subword units such as ["The", "cat", "sat", "on", "the", "mat"]. Once the text is tokenized, each token is mapped to an embedding vector using an embedding layer or lookup table. This table is often initialized with random values and then trained to capture meaningful relationships between words. For instance, `cat` might be represented as a 300-dimensional vector.

Advanced models such as transformers (e.g., BERT or GPT) generate contextual embeddings, where the vector representation of a word is influenced by its surrounding words. This allows the model to understand nuances and context, such as distinguishing between "bank" in "river bank" versus "financial bank."

Let's go and have a look at these models in more detail.

BERT – Contextualized Embedding Models

BERT is a powerful NLP model developed by Google. It belongs to the family of transformer-based models and is pre-trained on massive amounts of text data to learn contextualized representations of words.

The BERT embedding model is a *component* of the BERT architecture that generates contextualized word embeddings. Unlike traditional word embeddings that assign a fixed vector representation to each word, BERT embeddings are context-dependent, capturing the meaning of words in the context of the entire sentence. Here's an explanation of how to use the BERT embedding model `https://github.com/PacktPublishing/Python-Data-Cleaning-and-Preparation-Best-Practices/blob/main/chapter12/14.embedding_bert.py`:

1. Load a pre-trained BERT model and tokenizer:

```
tokenizer = BertTokenizer.from_pretrained("bert-base-uncased")
model = BertModel.from_pretrained("bert-base-uncased")
```

2. Encode the input text:

```
input_text = "BERT embeddings capture contextual information."
inputs= tokenizer(input_text, return_tensors="pt")
```

3. Obtain BERT embeddings:

```
with torch.no_grad():
    outputs = model(inputs)
```

4. Print the embeddings:

```
print("Shape of the embeddings tensor:", last_hidden_states.
shape)
Shape of the embeddings tensor: torch.Size([1, 14, 768])
```

The shape of the embeddings tensor will be (`1`, `sequence_length`, `hidden_size`). Here, `sequence_length` is the number of tokens in the input sentence, and `hidden_size` is the size of the hidden states in the BERT model (`768` for `bert-base-uncased`).

The `[CLS]` token embedding represents the entire input sentence and is often used for classification tasks. It's the first token in the output tensor:

```
CLS token embedding: [ 0.23148441 -0.32737488 ...  0.02315655]
```

The embedding for the first actual word in the sentence represents the contextualized embedding for that specific word. The embedding associated with the first actual word in a sentence is not just a static or isolated representation of that word. Instead, it's a "context-aware" or "contextualized" embedding, meaning it reflects how the word's meaning is influenced by the surrounding words in the sentence. In simpler terms, this embedding captures not only the word's intrinsic meaning but also how that meaning changes based on the context provided by the other words around it. This is a key feature of models such as BERT, which generate different embeddings for the same word depending on its usage in different contexts:

```
First word embedding: [ 0.00773875  0.24699381 ... -0.09120814]
```

The key thing to understand here is that we started with text and we have vectors or embeddings as outputs. The tokenization step is happening behind the scenes when we use the tokenizer provided by the `transformers` library. The tokenizer converts the input sentence into tokens and their corresponding token IDs, which are then passed to the BERT model. Remember that each word in the sentence has its own embedding that reflects its meaning within the context of the sentence.

BERT's versatility has enabled it to perform exceptionally well across a wide array of NLP tasks. However, the need for more efficient and task-specific embeddings has led to the development of models such as **BAAI General Embedding** (**BGE**). BGE is designed to be smaller and faster, providing high-quality embeddings optimized for tasks such as semantic similarity and information retrieval.

BGE

The BAAI/bge-small-en model is part of a series of BGE models developed by the **Beijing Academy of Artificial Intelligence** (**BAAI**). These models are designed for generating high-quality embeddings for text, typically used in various NLP tasks such as text classification, semantic search, and more.

These models generate embeddings (vector representations) for text. The embeddings capture the semantic meaning of the text, making them useful for tasks such as similarity search, clustering, and classification. The `bge-small-en` model is a smaller, English-specific model in this series. Let's see an example. The full code for this example is available at `https://github.com/PacktPublishing/Python-Data-Cleaning-and-Preparation-Best-Practices/blob/main/chapter12/15.embedding_bge.py`:

1. Define the model name and parameters:

```
model_name = "BAAI/bge-small-en"
model_kwargs = {"device": "cpu"}
encode_kwargs = {"normalize_embeddings": True}
```

2. Initialize the embeddings model:

```
bge_embeddings = HuggingFaceBgeEmbeddings(
    model_name=model_name,
    model_kwargs=model_kwargs,
    encode_kwargs=encode_kwargs
)
```

3. We sample a few sentences to embed:

```
sentences = [
    "The quick brown fox jumps over the lazy dog.",
    "I love machine learning and natural language processing."
]
```

4. Generate embeddings for each sentence:

```
embeddings = [bge_embeddings.embed_query(sentence) for sentence
in sentences]
```

5. Print the embeddings:

```
[-0.07455343008041382, -0.004580824635922909,
0.021685084328055382, 0.06458176672458649,
0.020278634503483772]...
Length of embedding: 384
Embedding for sentence 2: [-0.025911744683980942,
0.0050039878115057945, -0.011821565218269825,
-0.020849423483014107, 0.06114110350608826]...
```

BGE models, such as bge-small-en, are designed to be smaller and more efficient for embedding generation tasks compared to larger, more general models such as BERT. This efficiency translates to reduced memory usage and faster inference times, making BGE models particularly suitable for applications where computational resources are limited or where real-time processing is crucial. While BERT is a versatile, general-purpose model capable of handling a wide range of NLP tasks, BGE models are specifically optimized for generating high-quality embeddings. This optimization allows BGE models to provide comparable or even superior performance for specific tasks, such as semantic search and information retrieval, where the quality of embeddings is paramount. By focusing on the precision and semantic richness of embeddings, BGE models leverage advanced techniques such as learned sparse embeddings, which combine the benefits of both dense and sparse representations. This targeted optimization enables BGE models to excel in scenarios that demand nuanced text representation and efficient processing, making them a better choice for embedding-centric applications compared to the more generalized BERT model.

Building on the success of both BERT and BGE, the introduction of **General Text Embeddings** (**GTEs**) marks another significant step forward. GTE models are specifically fine-tuned to deliver robust and efficient embeddings tailored for various text-related applications.

GTE

GTEs represent the next generation of embedding models, designed to address the growing demand for specialized and efficient text representations. GTE models excel in providing high-quality embeddings for specific tasks such as semantic similarity, clustering, and information retrieval. Let's see them in action. The full code is available at `https://github.com/PacktPublishing/Python-Data-Cleaning-and-Preparation-Best-Practices/blob/main/chapter12/16.embedding_gte.py`:

1. Load the GTE-base model:

```
model = SentenceTransformer('thenlper/gte-base')
```

2. We sample a few random texts to embed:

```
texts = [
    "The quick brown fox jumps over the lazy dog.",
    "I love machine learning and natural language processing.",
    "Embeddings are useful for many NLP tasks."
]
```

3. Generate the embeddings:

```
embeddings = model.encode(texts
```

4. Print the shape of the embeddings:

```
print(f"Shape of embeddings: {embeddings.shape}")
Shape of embeddings: (3, 768)
```

5. Print the first few values of the first embedding:

```
[-0.02376037 -0.04635307  0.02570779  0.01606994  0.05594607]
```

One of the standout features of GTE is its efficiency. By maintaining a smaller model size and faster inference times, GTE is well-suited for real-time applications and environments with constrained computational resources. This efficiency does not come at the cost of performance; GTE models continue to deliver exceptional results across various text-processing tasks. However, their reduced complexity handling can be a limitation, as the smaller model size may impede their ability to process highly intricate or nuanced texts effectively. This could result in the less accurate capture of subtle contextual details, affecting performance in more complex scenarios. Additionally, a GTE's focus on efficiency might lead to diminished generalization capabilities; although it excels in specific tasks, it may struggle to adapt to a wide variety of diverse or less common language inputs. Furthermore, the model's smaller size may constrain its fine-tuning flexibility, potentially limiting its ability to adapt to specialized tasks or domains due to a reduced capacity for learning and storing intricate patterns specific to niche applications.

Selecting the right embedding model

When selecting a model for your application, start by identifying your specific use case and domain. Whether you need a model for classification, clustering, retrieval, or summarization, and whether your domain is legal, medical, or general text, will significantly influence your choice.

Next, evaluate the model's *size and memory usage*. Larger models generally provide better performance but come with increased computational requirements and higher latency. Begin with a smaller model for initial prototyping and consider transitioning to a larger one if your needs evolve. Pay attention to the embedding dimensions, as larger dimensions offer a richer representation of the data but are also more computationally intensive. Striking a balance between capturing detailed information and maintaining operational efficiency is key.

Assess *inference time carefully*, particularly if you have real-time application requirements; models with higher latency might necessitate GPU acceleration to meet performance standards. Finally, evaluate the model's performance using benchmarks such as the **Massive Text Embedding Benchmark (MTEB)** to compare across various metrics. Consider both intrinsic evaluations, which examine the model's understanding of semantic and syntactic relationships, and extrinsic evaluations, which assess performance on specific downstream tasks.

Solving real problems with embeddings

With the advancements in embedding models such as BERT, BGE, and GTE, we can tackle a wide range of challenges across various domains. These models enable us to solve different problems, as presented here:

- **Semantic search**: Embeddings improve search relevance by capturing the contextual meaning of queries and documents, enhancing the accuracy of search results.

- **Recommendation systems**: They facilitate personalized content suggestions based on user preferences and behaviors, tailoring recommendations to individual needs.
- **Text classification**: Embeddings enable accurate categorization of documents into predefined classes, such as for sentiment analysis or topic identification.
- **Information retrieval**: They enhance the accuracy of retrieving relevant documents from extensive datasets, improving the efficiency of information retrieval systems.
- **Natural language understanding**: Embeddings support tasks such as NER, helping systems identify and classify key entities within text.
- **Clustering techniques**: They improve the organization of similar documents or topics in large datasets, aiding in better clustering and data management.
- **Multimodal data processing**: Embeddings are essential for integrating and analyzing text, image, and audio data, leading to more comprehensive insights and enhanced decision-making capabilities.

Let's summarize the learnings from this chapter.

Summary

In this chapter, we had a look at text preprocessing, which is an essential step in NLP. We saw different text cleaning techniques, from handling HTML tags and capitalization to addressing numerical values and whitespace challenges. We deep-dived into tokenization, examining word and subword tokenization, with practical Python examples. Finally, we explored various methods for embedding documents and introduced some of the most popular embedding models available today.

In the next chapter, we will continue our journey with unstructured data, delving into image and audio preprocessing.

13
Image and Audio Preprocessing with LLMs

In this chapter, we delve into the preprocessing of unstructured data, specifically focusing on images and audio. We explore various techniques and models designed to extract meaningful information from these types of media. The discussion includes a detailed examination of image preprocessing methods, the use of **optical character recognition** (**OCR**) for extracting text from images, the capabilities of the BLIP model for generating image captions, and the application of the Whisper model for converting audio into text.

In this chapter, we'll cover the following topics:

- The current era of image preprocessing
- Extracting text from images
- Handling audio data

Technical requirements

The complete code for this chapter can be found in the following GitHub repository: `https://github.com/PacktPublishing/Python-Data-Cleaning-and-Preparation-Best-Practices/tree/main/chapter13`.

Let's install the necessary libraries we will use in this chapter:

```
pip install torchvision
pip install keras==3.4.1
pip install tensorflow==2.17.0
pip install opencv-python==4.10.0.84
pip install opencv-python==4.10.0.84
pip install paddleocr==2.8.1
pip install paddlepaddle==2.6.1
```

The current era of image preprocessing

In the era of advanced visual models, such as diffusion models, and models such as OpenAI's CLIP, preprocessing has become crucial to ensure the quality, consistency, and suitability of images for training and inference. These models require images to be in a format that maximizes their ability to learn intricate patterns and generate high-quality results. In this section, we will go through all the preprocessing steps to make your images ready for the subsequent tasks.

Across this section, we will use a common use case, which is to prepare images for training a diffusion model. You can find the code for this exercise in the GitHub repository: `https://github.com/PacktPublishing/Python-Data-Cleaning-and-Preparation-Best-Practices/blob/main/chapter13/1.image_prerpocessing.py`.

Let's start by loading some images.

Loading the images

Perform the following steps to load the images:

1. First, we load the required packages for this exercise:

   ```
   from PIL import Image
   import numpy as np
   import cv2
   import requests
   from io import BytesIO
   import matplotlib.pyplot as plt
   import tensorflow as tf
   from tensorflow.keras.preprocessing.image import
   ImageDataGenerator
   ```

   Then, we load the images into our environment. We'll use the Python Pillow library to handle loading the images.

2. Next, we create a function to load an image from a URL. This function fetches the image from the given URL and loads it into a `PIL` image object using `BytesIO` to handle the byte data:

   ```
   def load_image_from_url(url):
       response = requests.get(url)
       img = Image.open(BytesIO(response.content))
       return img
   ```

3. Then, we'll create a helper function to display our images. We will be using this function across the chapter:

   ```
   def show_image(image, title="Image"):
       plt.imshow(image)
   ```

```
        plt.title(title)
        plt.axis('off')
        plt.show()
```

4. Now, we'll pass the image URL to our `load_image_from_url` function. Here, we are using a random image URL from Unsplash, but you can use any image you have access to:

```
image_url = "https://images.unsplash.com/photo-1593642532871-8b12e02d091c"
image = load_image_from_url(image_url)
```

5. Let's display the original image that we just loaded using the function we created:

```
show_image(image, "Original Image")
```

This will display the following output image:

Figure 13.1 – Original image before any preprocessing

Image preprocessing is crucial for preparing visual data for ingestion by **machine learning** (**ML**) models. Let's delve deeper into each technique, explaining the concepts and demonstrating their application with Python code.

Resizing and cropping

Effective preprocessing can significantly enhance the performance of AI and ML models by ensuring that the most relevant features are highlighted and easily detectable in the images. **Cropping** is a technique that can help the model focus on relevant features. The main idea is to trim or cut away the outer edges of an image to improve framing, focus on the main subject, or eliminate unwanted elements. The size of the crop depends on the specific requirements of the task. For example, in object detection, the crop should focus on the object of interest, while in image classification, the crop should ensure that the main subject is centered and occupies most of the frame.

There are many different techniques for cropping images from a simple fixed-size cropping to more involved object-aware cropping. **Fixed-size cropping** involves adjusting all images to a predetermined size, ensuring uniformity across the dataset, which is useful for applications that require standardized input sizes, such as training certain types of neural networks. However, it may result in the loss of important information if the main subject is not centered. **Aspect ratio preservation** avoids distortion by maintaining the original image's aspect ratio while cropping, which is achieved through padding (adding borders to the image to reach the desired dimensions) or scaling (resizing the image while maintaining its aspect ratio, followed by cropping to the target size). **Center cropping** involves cropping the image around its center, assuming the main subject is generally located in the middle, and is commonly used in image classification tasks where the main subject should occupy most of the frame. **Object-aware cropping** uses algorithms to detect the main subject within the image and crop around it, ensuring that the main subject is always emphasized, regardless of its position within the original image. This technique is particularly useful in object detection and recognition tasks.

Resizing is a fundamental step in image preprocessing for AI and ML tasks, focusing on adjusting the dimensions of an image to a standard size required by the model. This process is crucial for ensuring that the input data is consistent and suitable for the specific requirements of various AI and ML algorithms.

Let's add some steps to the image preprocessing pipeline we started in the previous section to see the effects of cropping and resizing. The following function resizes the image to a specified target size (256x256 pixels in this case). We expect the image to appear uniformly sized down to fit within the target dimensions:

```
def resize_and_crop(image, target_size):
    image = image.resize((target_size, target_size),
    Image.LANCZOS)
    return image

target_size = 256
processed_image = resize_and_crop(image, target_size)
```

Let's print the resulting image using the following code:

```
show_image(processed_image, "Resized and Cropped Image")
```

This will display the following output:

Figure 13.2 – Image after resizing and cropping

As we can see from *Figure 13.2*, the image is resized to a square of 256x256 pixels, altering the aspect ratio of the original image that was not square. Thus, resizing ensures a uniform input size for all data, which facilitates the batch processing and the passing of data to models for training.

Next, we will discuss the normalization of images, which is not far from the normalization of features discussed in previous chapters.

Normalizing and standardizing the dataset

To ensure data consistency and help the training of models converge faster, we can force the input data to a common range of values. This adjustment involves scaling the input data between `0` and `1`, also known as **standardization** or **normalizing** using the mean and standard deviation of the dataset.

For most deep learning models, forcing pixel values to the range `[0, 1]` or `[-1, 1]` is standard practice. This can be achieved by dividing pixel values by 255 (for `[0, 1]`) or by subtracting the mean and dividing by the standard deviation (for `[-1, 1]`). In image classification tasks, this tactic ensures that the input images have consistent pixel values. For example, in a dataset of handwritten digits (such as MNIST), normalizing or standardizing the pixel values helps the model learn the patterns of the digits more effectively. In object detection tasks, it helps in accurately detecting and classifying objects within an image. However, normalization and standardization are not limited to image preprocessing; they are a fundamental step in preparing data for any ML problem.

Let's expand the previous example by adding the normalization and the standardization step. The first function performs the normalization to ensure that the pixel values are in a common scale, in this case, between the range `[0, 1]` and we do that by dividing by 255:

```
def normalize(image):
    image_array = np.array(image)
    normalized_array = image_array / 255.0
    return normalized_array

normalized_image = normalize(processed_image)
```

The normalized image can be seen in the following figure:

Figure 13.3 – Picture after normalization

As we can see from *Figure 13.3*, visually, the image remains the same, at least to the human eye. Normalization does not alter the relative intensity of the pixels; it only scales them to a different range so the content and details should remain unchanged.

Let's move on to the standardization exercise. Before standardization, pixel values are in the range `[0, 255]` and follow the natural distribution of image intensities. The idea with standardization is that all the pixel values will be transformed to have a mean of `0` and a standard deviation of `1`. Let's see how we can do that in the following code:

```
def standardize(image):
    image_array = np.array(image)
    mean = np.mean(image_array, axis=(0, 1), keepdims=True)
    std = np.std(image_array, axis=(0, 1), keepdims=True)
    standardized_array = (image_array - mean) / std
    return standardized_array

standardized_image = standardize(processed_image)
```

In this case, the appearance of the image might change since standardization shifts the mean to `0` and scales the values. This can make the image look different, possibly more contrasted, or with changed brightness. However, the image content should still be recognizable.

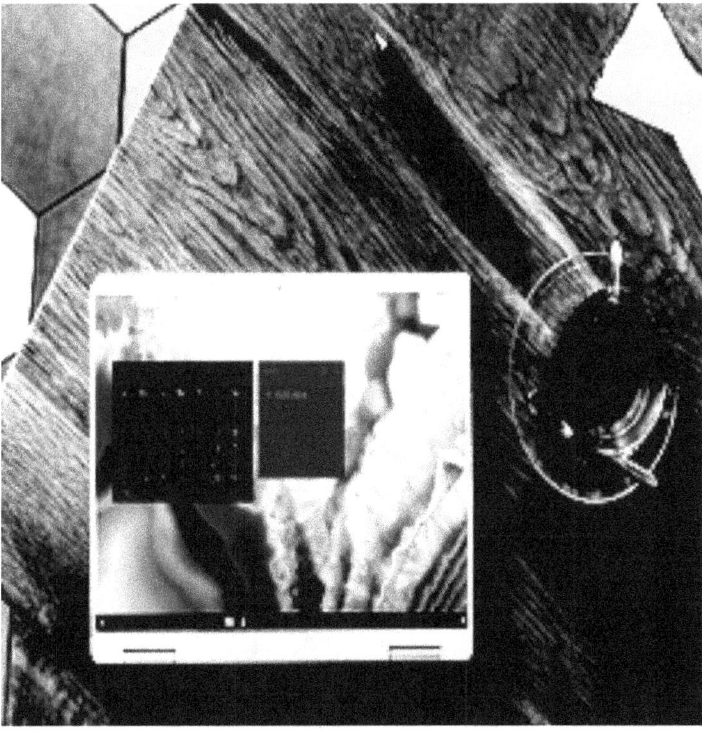

Figure 13.4 – Picture after standardization

Apart from the transformed image shown in *Figure 13.4*, the mean and standard deviation for the values are printed here:

```
Mean after standardization: 0.0
Standard deviation after standardization: 1.000000000000416
```

This confirms that the standardization has correctly scaled the pixel values. Let's now move on to the data augmentation part.

Data augmentation

Data augmentation aims to create more variability in the dataset by applying random transformations, such as rotation, flipping, translation, color jittering, and contrast adjustment. This artificially expands the dataset with modified versions of existing images, which helps with model generalization and performance, especially when working with limited data.

Common augmentation techniques include geometric transformations, such as rotation, flipping, and scaling, which change the spatial orientation and size of the images. For example, rotating an image by 15 degrees or flipping it horizontally can create new perspectives for the model to learn from. Color space alterations, such as adjusting brightness, contrast, or hue, can simulate different lighting conditions and improve the model's ability to recognize objects in varying environments. Adding noise or blur can help the model become more resilient to imperfections and distortions in real-world data.

Let's go back to our example to see how we can create image variations:

1. First, we will define the transformations that we will apply to the image:
   - **Rotation range**: Randomly rotate the image within a range of 40 degrees.
   - **Width shift range**: Randomly shift the image horizontally by 20% of the width.
   - **Height shift range**: Randomly shift the image vertically by 20% of the height.
   - **Shear range**: Randomly apply shearing transformations.
   - **Zoom range**: Randomly zoom in or out by 20%.
   - **Horizontal flip**: Randomly flip the image horizontally.
   - **Fill mode**: Define how to fill in newly created pixels after a transformation. (Here, using "nearest" pixel values.)
2. Let's create a function to apply these transformations:

```
datagen = ImageDataGenerator(
    rotation_range=40,
    width_shift_range=0.2,
    height_shift_range=0.2,
```

```
    shear_range=0.2,
    zoom_range=0.2,
    horizontal_flip=True,
    fill_mode='nearest'
)
```

3. Then, we will apply the transformations we just defined to the image:

```
def augment_image(image):
    image = np.expand_dims(image, axis=0) # Add batch dimension
    augmented_iter = datagen.flow(image, batch_size=1)
    augmented_image = next(augmented_iter)[0]
    return augmented_image
augmented_image = augment_image(normalized_image)
```

Let's print the augmented image:

```
show_image(augmented_image, "Augmented Image")
```

This will display the following image:

Figure 13.5 – Picture augmentation

As we can see from *Figure 13.5*, the image has some significant changes; however, the image still remains recognizable and the concept in the picture remains the same.

> **Note**
>
> As we are using some random parameters in the data augmentation phase, you may produce a slightly different image at this stage.

Data augmentation's importance lies in its ability to increase dataset diversity, which by extension helps prevent overfitting, as the model learns to recognize patterns and features from a wider range of examples rather than memorizing the training data. Let's move on to the next part and dive deep into the noise reduction options.

Noise reduction

Noise in images refers to the random variations in pixel values that can distort the visual quality of an image and by extension affect the performance of models during training. These variations often appear as tiny, irregular spots or textures, such as random dots, patches, or a gritty texture, disrupting the smoothness and clarity of the image. They often make the image look less sharp and can obscure important details, which can be problematic for both visual interpretation and for models that rely on clear, accurate data for training.

Noise reduction attempts to reduce the random variations and make the data simpler. The minimization of these random variations in pixel values helps improve image quality and model accuracy as they can mislead models during training. In the following subsections, we expand on some common denoising techniques used in the data field, including Gaussian smoothing, non-local means denoising, and wavelet denoising.

Gaussian smoothing

Gaussian blur (or **Gaussian smoothing**) applies a Gaussian filter to the image, which works by taking the pixel values within a specified neighborhood around each pixel and averaging them. The filter assigns higher weights to the pixels closer to the center of the neighborhood and lower weights to those farther away, following the Gaussian distribution. The denoised image will appear smoother but with slightly blurred edges, making it useful in applications where slight blurring is acceptable or desired, such as artistic effects or before edge detection algorithms to reduce noise. Let's see the code for applying Gaussian smoothing to the image:

```
def gaussian_blur(image):
    blurred_image = cv2.GaussianBlur(image, (5, 5), 0)
    return blurred_image
```

Let's display the denoised image:

```
blurred_image = gaussian_blur(noisy_image)
show_image(blurred_image, "Gaussian Blur")
```

The denoised image can be seen in *Figure 13.6*:

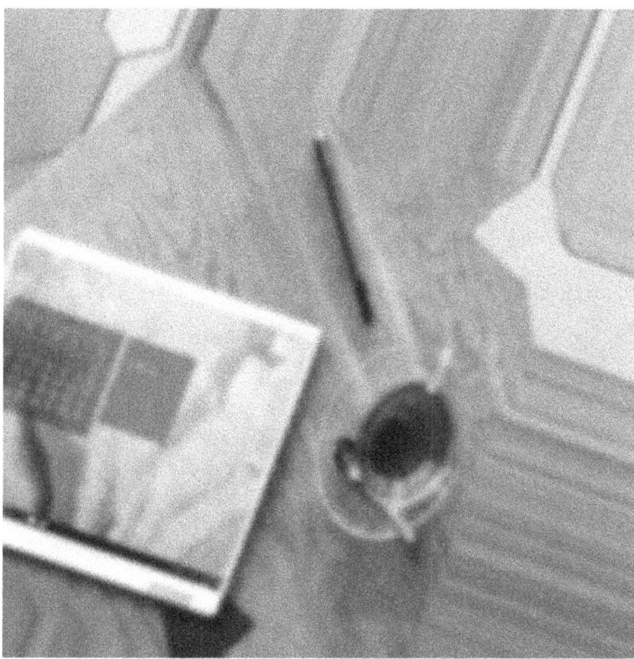

Figure 13.6 – Denoised images – Gaussian blur on the median blur on the right

In the next section, we will discuss the bilateral filter.

The bilateral filter

The **bilateral filter** smoothens images by considering both spatial and intensity differences. It averages the pixel values based on their spatial closeness and color similarity. Let's have a look at the code:

```
def bilateral_filter(image):
    image_uint8 = (image * 255).astype(np.uint8)
    filtered_image = cv2.bilateralFilter(
        image_uint8, 9, 75, 75)
    filtered_image = filtered_image / 255.0
    return filtered_image
```

The `bilateralFilter` function takes some arguments that we need to explain:

- `9`: This is the diameter of each pixel neighborhood used during filtering. A larger value means that more pixels will be considered in the computation, resulting in a stronger smoothing effect.
- `75`: This is the filter sigma in the color space. A larger value means that farther colors within the pixel neighborhood will be mixed, resulting in larger areas of semi-equal color.
- `75`: This is the filter sigma in the coordinate space. A larger value means farther pixels will influence each other if their colors are close enough. It controls the amount of smoothing.

Let's use the function and see the resulting output:

```
bilateral_filtered_image = bilateral_filter(noisy_image)
show_image(bilateral_filtered_image, "Bilateral Filter")
```

The denoised image can be seen in *Figure 13.7*.

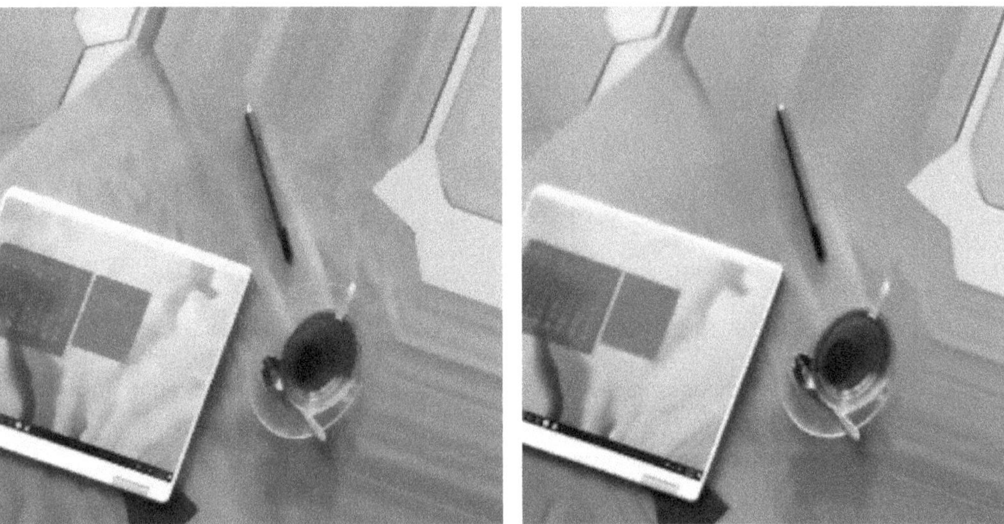

Figure 13.7 – Denoised images – left bilateral filter, right non-local mean denoising

In the next section, we will discuss the non-local means denoising.

Non-local means denoising

Non-local means denoising reduces noise by comparing patches of the image and averaging similar patches, even if they are far apart. This method works by comparing small patches of pixels across the entire image, rather than just neighboring pixels. Unlike simpler methods that only consider nearby pixels, non-local means denoising searches the image for patches that are similar, even if they are located far apart. When a match is found, the method averages these similar patches together to determine the final pixel value.

This approach is particularly effective at preserving fine details and textures because it can recognize and retain patterns that are consistent throughout the image, rather than just smoothing over everything indiscriminately. By averaging only the patches that are truly similar, it reduces noise while maintaining the integrity of important image features, making it an excellent choice for applications where detail preservation is crucial.

Let's have a look at the code:

```
def remove_noise(image):
    image_uint8 = (image * 255).astype(np.uint8)
    denoised_image = cv2.fastNlMeansDenoisingColored(
        image_uint8, None, h=10, templateWindowSize=7,
        searchWindowSize=21)
    denoised_image = denoised_image / 255.0
    return denoised_image
denoised_image = remove_noise(noisy_image)
show_image(denoised_image, "Non-Local Means Denoising")
```

The `fastNlMeansDenoisingColored` function applies the non-local means denoising algorithm to the image. The `h=10` argument reflects the filtering strength. A higher value removes more noise but may also remove some image details. The size in pixels of the template patch used to compute weights is reflected in the `templateWindowSize` variable. This value should be an odd number. A greater value means more smoothing. Finally, `searchWindowSize=21` means the size of the window in pixels used to compute a weighted average for a given pixel should be odd. A greater value means more smoothing.

Why use an odd number for window sizes, such as `templateWindowSize` and `searchWindowSize`?

The primary reason for using an odd number is to ensure that there is a clear center pixel within the window. For example, in a 3x3 window (where 3 is an odd number), the center pixel is the one at position "(2,2)". This center pixel is crucial because the algorithm often calculates how similar the surrounding pixels are in comparison to this central pixel. If an even-sized window were used, there would be no single, central pixel, as shown in *Figure 13.8*.

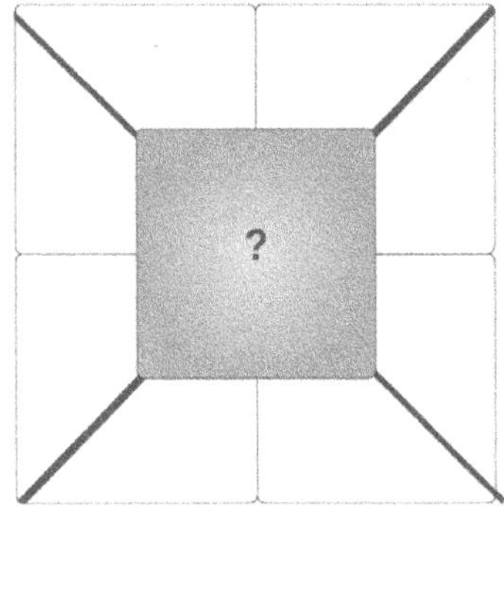

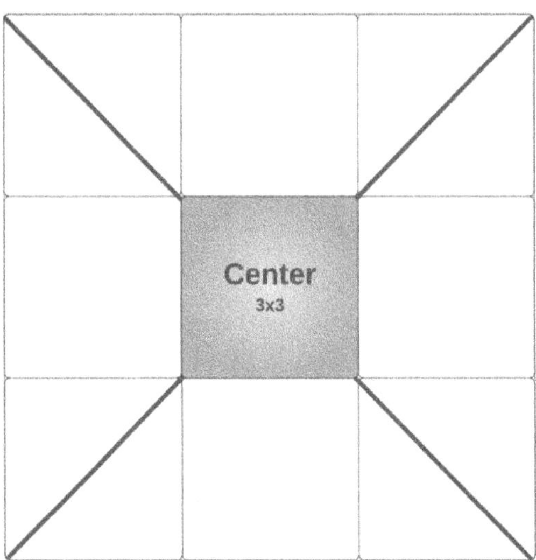

Figure 13.8 – Use an odd number for window sizes

Using an odd number simplifies the computation of weights and distances between the central pixel and its neighbors. This simplification is essential in algorithms such as non-local means, where the distances between pixels influence the weight given to each pixel in the averaging process. An odd-sized window naturally allows for straightforward indexing and less complex calculations.

Regarding the `searchWindowSize` parameter, this defines the area within which the algorithm looks for similar patches to the one currently being processed. Having an odd-sized window for this search area ensures that there is a central pixel around which the search is centered. This helps in accurately identifying similar patches and applying the denoising effect uniformly across the image.

The denoised image can be seen in the previous section in *Figure 13.7*.

In the next section, we will discuss the last method, the median blur.

Median blur

Median blur replaces each pixel's value with the median value of the neighboring pixels. This method is particularly effective for removing "salt-and-pepper" noise, where pixels are randomly set to black or white, as we will see later. Let's first denoise the image with the median blur method and then we will see how this method solves the salt-and-pepper effect.

The following function performs the **median blur** technique. It calls the `medianBlur` function, which requires the input image to be in an 8-bit unsigned integer format (`uint8`), where pixel values range from `0` to `255`. By multiplying the image by 255, the pixel values are scaled to the range `[0, 255]`:

```
def perform_median_blur(image):
    image_uint8 = (image * 255).astype(np.uint8)
```

```
    #The parameter below specifies the size of the kernel (5x5).
    blurred_image = cv2.medianBlur(image_uint8, 5)
    blurred_image = blurred_image / 255.0
    return blurred_image
```

Let's display the denoised image using the following code:

```
median_blurred_image = median_blur(noisy_image)
show_image(median_blurred_image, "Median Blur")
```

The denoised image can be seen in *Figure 13.9*:

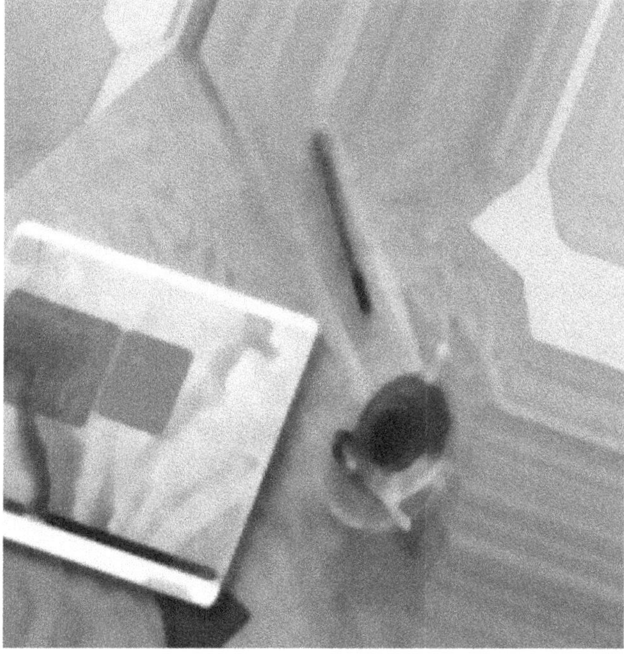

Figure 13.9 – Denoised images – median blur

As promised, let's now discuss the salt-and-pepper noise effect.

Salt-and-pepper noise

Salt-and-pepper noise is a type of impulse noise characterized by the presence of randomly distributed black-and-white pixels in an image. This noise can be caused by various factors, such as errors in data transmission, malfunctioning camera sensors, or environmental conditions during image acquisition. The black pixels are referred to as "pepper noise," while the white pixels are known as "salt noise." This noise type is particularly detrimental to image quality as it can obscure important details and make edge detection and image restoration challenging.

To showcase this, we have created a function that adds this noise effect to the original image so that we can then denoise it:

```
def add_salt_and_pepper_noise(image, salt_prob=0.02, pepper_
prob=0.02):
    noisy_image = np.copy(image)
    num_salt = np.ceil(salt_prob * image.size)
    coords = [np.random.randint(0, i - 1, int(num_salt)) for i in
image.shape]
     noisy_image[coords[0], coords[1], :] = 1
     num_pepper = np.ceil(pepper_prob * image.size)
    coords = [np.random.randint(0, i - 1, int(num_pepper)) for i in
image.shape]
     noisy_image[coords[0], coords[1], :] = 0
    return noisy_image
```

This function takes three arguments:

- `image`: The input image to which noise will be added
- `salt_prob`: The probability of a pixel being turned into salt noise (white)
- `pepper_prob`: The probability of a pixel being turned into pepper noise (black)

This function adds salt-and-pepper noise to an image. It starts by creating a copy of the input image to avoid altering the original. To introduce salt noise (white pixels), it calculates the number of pixels to be affected based on the `salt_prob` parameter, generates random coordinates for these pixels, and sets them to white. Similarly, for pepper noise (black pixels), it calculates the number of affected pixels using the `pepper_prob` parameter, generates random coordinates, and sets these pixels to black. The noisy image is then returned.

To apply this effect on the data you need to set the following flag to `True`. The flag can be found in the code after the `add_salt_and_pepper_noise` function definition:

```
use_salt_and_pepper_noise = True

if use_salt_and_pepper_noise:
noisy_image = add_salt_and_pepper_noise(tensor_to_image(tensor_image))
show_image(noisy_image, "Salt-and-Pepper Noisy Image")
```

The image with the noise can be seen in *Figure 13.10*:

Figure 13.10 – Salt-and-pepper noise

Now, let's apply the different denoising techniques we've learned so far to the preceding image. The different denoising effects can be seen in *Figure 13.11* and *Figure 13.12*.

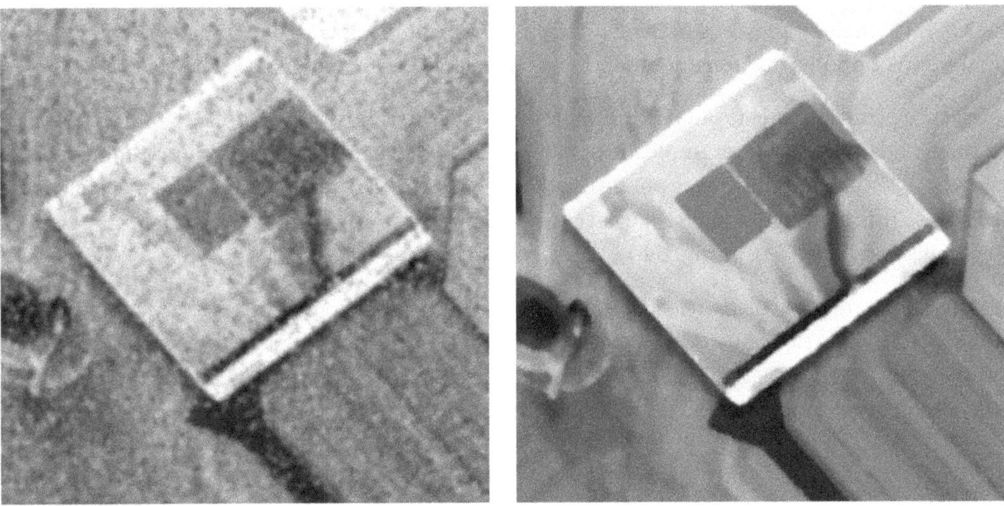

Figure 13.11 – Left: Gaussian blur, right: median blur

Figure 13.12 – Left: bilateral filter, right: non-local means denoising

As we can see, the median blur method really excels at removing this kind of noise, whereas all the other methods really struggle to remove it. In the next part of the chapter, we will discuss some image use cases that are becoming more popular in the data world, such as creating image captions and extracting text from images.

Extracting text from images

When discussing ways to extract text from images, the OCR technology is the one that comes to mind. The OCR technology allows us to handle textual information embedded in images, allowing for the digitization of printed documents, automating data entry, and enhancing accessibility.

One of the primary benefits of OCR technology today is its ability to significantly reduce the need for manual data entry. For example, businesses can convert paper documents into digital formats using OCR, which not only saves physical storage space but also enhances document management processes. This conversion makes it easier to search, retrieve, and share documents, streamlining operations and improving productivity.

In transportation, particularly with self-driving cars, OCR technology is used to read road signs and number plates. This capability is vital for navigation and ensuring compliance with traffic regulations. By accurately interpreting signage and vehicle identification, OCR contributes to the safe and efficient functioning of autonomous vehicles.

Moreover, OCR technology is employed in social media monitoring to detect brand logos and text in images. This application is particularly beneficial for marketing and brand management, as it enables companies to track brand visibility and engagement across social platforms. For instance, brands can use OCR to identify unauthorized use of their logos or monitor the spread of promotional materials, thereby enhancing their marketing strategies and protecting their brand identity.

Let's see how we can apply OCR in the data world with an open source solution.

PaddleOCR

PaddleOCR is an open source OCR tool developed by PaddlePaddle, which is Baidu's deep learning platform. The repository provides end-to-end OCR capabilities, including text detection, text recognition, and multilingual support (`https://github.com/PaddlePaddle/PaddleOCR`).

The PaddleOCR process has a lot of steps in place that can be seen in the following *Figure 13.13*:

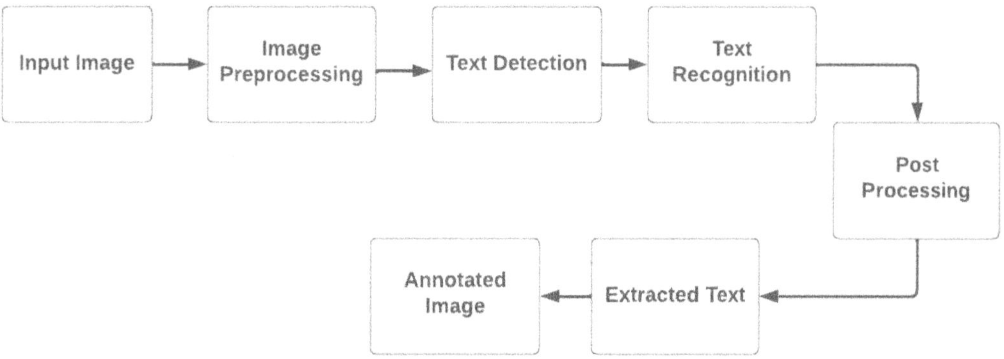

Figure 13.13 – OCR process step by step

Let's break down the process step by step:

1. The process begins with an input image that may contain text.
2. **Image preprocessing**: The image may undergo various preprocessing steps, such as resizing, converting to grayscale, and noise reduction, to enhance text visibility.
3. **Text detection**: The model detects regions of the image that contain text. This may involve algorithms such as **Efficient and Accurate Scene Text** (**EAST**) or **Differentiable Binarization** (**DB**) to find bounding boxes around the text.
4. **Text recognition**: The detected text regions are fed into a recognition model (often a **convolutional neural network** (**CNN**) followed by a **long short-term model** (**LSTM**) or a transformer) to convert the visual text into digital text.
5. **Post-processing**: The recognized text may be further refined through spell-checking, grammar correction, or contextual analysis to improve accuracy.

6. **Extracted text**: The final output consists of extracted digital text ready for further use.
7. **Annotated image**: Optionally, an annotated version of the original image can be generated, showing the detected text regions along with the recognized text.

It may seem complicated initially but luckily, most of these steps are abstracted away from the user and are handled by the PaddleOCR package automatically. Let's introduce a use case for OCR to extract text from YouTube video thumbnails.

> **YouTube thumbnails**
>
> **YouTube thumbnails** are small, clickable images that represent a video on the platform. They serve as the visual preview that users see before clicking to watch a video. Thumbnails are crucial for attracting viewers, as they often play a significant role in influencing whether someone decides to watch the content.

By analyzing the text present in thumbnails, such as video titles and promotional phrases, stakeholders can gain insights into viewer engagement and content trends. For instance, a marketing team can collect thumbnails from a range of videos and employ OCR to extract keywords and phrases that frequently appear in high-performing content. This analysis can reveal which terms resonate most with audiences, enabling creators to optimize their future thumbnails and align their messaging with popular themes. Additionally, the extracted text can inform **search engine optimization** (**SEO**) strategies by identifying trending keywords to incorporate into video titles, descriptions, and tags, ultimately enhancing video discoverability. In our case, we have provided a folder on the GitHub repository with YouTube thumbnails from the channel I am cohosting called **Vector Lab**, discussing Gen AI and ML concepts. Here's a link to the images folder on GitHub: `https://github.com/PacktPublishing/Python-Data-Cleaning-and-Preparation-Best-Practices/tree/main/chapter13/images`.

The images in the folder look like the following figure and the idea is to pass all these images and extract the text depicted on the image.

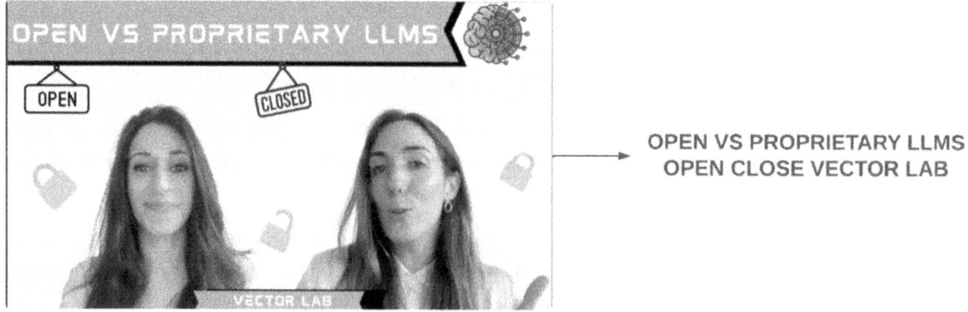

Figure 13.14 – Example YouTube thumbnail

Let's see how we can achieve that:

1. We'll start by initializing PaddleOCR:

```
ocr = PaddleOCR(use_angle_cls=True, lang='en')
```

The `use_angle_cls=True` flag enables the use of an angle classifier in the OCR process. The angle classifier helps improve the accuracy of text recognition by determining the orientation of the text in the image. This is particularly useful for images where text might not be horizontally aligned (e.g., rotated or skewed text).

The `lang='en'` parameter specifies the language for OCR. In this case, `'en'` indicates that the text to be recognized is in English. PaddleOCR supports multiple languages and sets the appropriate language in case you want to perform OCR in a language other than English.

2. Next, we define the path to the folder containing images to extract the text from:

```
folder_path = 'chapter13/images'
```

3. Then, we specify the supported image extensions. In our case, we only have `.png`, but you can add any image type in the folder:

```
supported_extensions = ('.png', '.jpg', '.jpeg')
```

4. Next, we get all paths to the images in the folder that we will use to load the images:

```
image_paths = [os.path.join(folder_path, file) for file in
os.listdir(folder_path) if file.lower().endswith(supported_
extensions)]
```

5. Then, we create an empty DataFrame to store the results:

```
df = pd.DataFrame(columns=['Image Path', 'Extracted Text'])
```

6. We use the following code to check if there are no image paths returned, which would mean that either there are no images in the folder or the images that exist in the folder don't have any of the supported extensions:

```
if not image_paths:
    print("No images found in the specified folder.")
else:
    for image_path in image_paths:
        process_image(image_path)
```

7. Now, let's expand on the `process_image` function. This function processes images and extracts text. For each thumbnail image, the function will extract any visible text, such as titles, keywords, or promotional phrases:

```
def process_image(image_path):
    result = ocr.ocr(image_path, cls=True)
```

```
    extracted_text = ""
    for line in result[0]:
        extracted_text += line[1][0] + " "
        print(f"Extracted Text from {os.path.basename(image_
path)}:\n{extracted_text}\n")
    df.loc[len(df)] = [image_path, extracted_text]
```

The `process_image` function performs OCR on an image specified by `image_path`. It starts by invoking the `ocr` method from the `PaddleOCR` library, which processes the image and returns the recognized text along with other details. The function initializes an empty string, `extracted_text`, to accumulate the recognized text. It then iterates through each line of text detected by the OCR process, appending each line to `extracted_text` along with a space for separation. After processing the entire image, it prints the accumulated text along with the filename of the image. Finally, the function adds a new entry to a DataFrame called `df`, storing `image_path` and the corresponding `extracted_text` in a new row, thus updating the DataFrame with the latest OCR results.

8. Optionally, we can save the DataFrame to a CSV file using the following code:

```
df.to_csv('extracted_texts.csv', index=False)
```

The results can be seen in *Figure 13.15*, where we have the path to the image on the left and the extracted text on the right.

```
                 Image Path                                       Extracted Text
0    chapter13/images/8.png  HOW TO MITIGATE SaCURITY RISKS IN AI AND ML SY...
1    chapter13/images/9.png  BUILDING DBRX-CLASS CUSTOM LLMS WITH MOSAIC A1...
2   chapter13/images/14.png  MPROVING TeXT2SO L PeRFORMANCe WITH EASE ON DA...
3   chapter13/images/15.png  HOW TO INTeRVIEW FOR A SOLUTIONS ARCHITaCT ROL...
4   chapter13/images/17.png  AGENTS AND CHAINS: HOW TO LOG, DEPLOY, AND DEB...
5   chapter13/images/16.png  MODEL TRAINING WITH DATABRICKS API - STaP BY-S...
6   chapter13/images/12.png               DaaP LEARNING WITH MLFLOW VECTOR LAB
7   chapter13/images/13.png  OPTIMIZING DATABRICKS LLM PIPaLINEs WiTH DsPy ...
8   chapter13/images/11.png               DaaP LEARNING WITH MLFLOW VECTOR LAB
9   chapter13/images/10.png  WhAT'S NaW - DATABRICKs VaCTOR SEARCH IS GA VE...
10  chapter13/images/21.png  HOW LONG SHOULD YOU TRAIN YOUR LANGUAGe MODeL?...
11  chapter13/images/20.png  CReATe FORaCAsTS WITH SQL AND AL FUNCTIONS ON ...
12  chapter13/images/22.png    TRAINING MOES AT SCALe WITH PYTORCH VECTOR LAB
13  chapter13/images/23.png           NEW MODEL: GPT-40 MINI NEW! VECTOR LAB
```

Figure 13.15 – OCR output

The results are great; however, we can see that there are some misspellings in certain cases, probably due to the font of the text in the images. The key thing to understand here is that we don't have to deal with the images anymore, but only with the text, thereby significantly simplifying our challenge. Based on what we learned in *Chapter 12, Text Preprocessing in the Era of LLMs* we can now manipulate and clean the text in various ways, such as chunking it, embedding it, or passing it through **large language models** (**LLMs**), as we will see in the next part.

Using LLMs with OCR

OCR technology, despite its advancements, often produces errors, especially with complex layouts, low-quality images, or unusual fonts. These errors include misrecognized characters and incorrect word breaks. So, the idea is to pass the OCR-extracted text through an LLM to correct these errors as LLMs understand context and can improve grammar and readability. Moreover, raw OCR output may be inconsistently formatted and hard to read; LLMs can reformat and restructure text, ensuring coherent and well-structured content. This automated proofreading reduces the need for manual intervention, saving time and minimizing human error. LLMs also standardize the text, making it consistent and easier to integrate into other systems, such as databases and analytical tools.

In this section, we will expand the thumbnail example to pass the extracted text through an LLM to clean it. To run this example, you need to do the following setup first.

Hugging Face setup

In order to run this example, you will need to have an account with Hugging Face and a token to authenticate. To do that, follow these steps:

1. Go to `https://huggingface.co`.
2. Create an account if you don't have one.
3. Go to **Settings**.
4. Then, click on **Access Tokens**. You should see the following page:

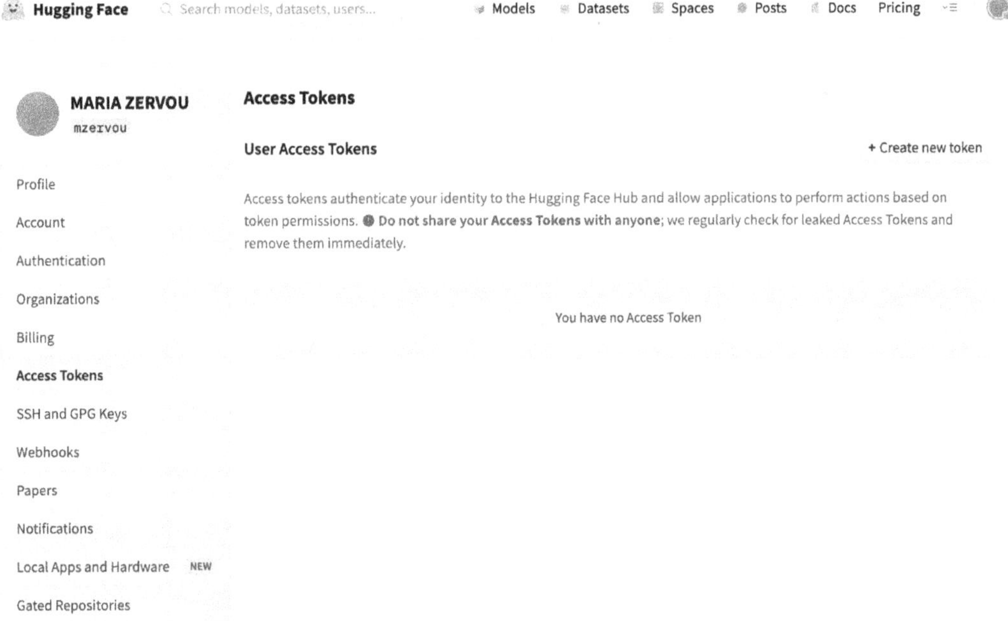

Figure 13.16 – Creating a new access token in Hugging Face

5. Click on the **Create new token** button to generate a new personal token.
6. Remember to copy and keep this token as we will need to paste it in the code file to authenticate!

Now, we are ready to dive into the code.

Cleaning text with LLMs

Let's have a look at the code that you can find in the GitHub repository: `https://github.com/PacktPublishing/Python-Data-Cleaning-and-Preparation-Best-Practices/blob/main/chapter13/3.ocr_with_llms.py`:

1. We start by reading in the OCR-extracted text that we wrote in the previous step:

```
df = pd.read_csv('extracted_texts.csv')
```

2. We then initialize the Hugging Face model. In this case, we are using `Mistral-Nemo-Instruct-2407`, but you can replace it with any LLM you have access to:

```
model_name = "mistralai/Mistral-Nemo-Instruct-2407"
```

   Hugging Face is a platform that provides a diverse repository of pretrained models, which can be accessed and integrated easily using user-friendly APIs. Hugging Face models come with detailed documentation and benefit from continuous innovation driven by a collaborative community. I see it as being similar to how GitHub serves as a repository for code; Hugging Face functions as a repository for ML models. Importantly, many models on Hugging Face are available for free, making it a cost-effective option for individuals and researchers.

   In contrast, there are many other paid models available, such as Azure OpenAI, which provides access to models such as GPT-3 and GPT-4. These models can be accessed on a paid model, and you have to manage the authentication process, which is different from authenticating with Hugging Face.

3. Add your Hugging Face API token, which was created in the previous setup section:

```
api_token = "add_your_token"
```

4. Here is the LangChain setup with some few-shot examples. Here, we use the `PromptTemplate` class, which helps create a prompt for the model:

```
prompt_template = PromptTemplate(
    input_variables=["text"],
    template='''
    Correct the following text for spelling errors and return
only the corrected text in lowercase. Respond using JSON format,
strictly according to the following schema:
    {{"corrected_text": "corrected text in lowercase"}}
     Examples: Three examples are provided to guide the model:
     Input: Shows the input text needing correction.
```

```
     Output: Provides the expected JSON format for the corrected
text. This helps the model learn what is required and encourages
it to follow the same format when generating its responses.

    Examples:
    Input: "Open vs Proprietary LLMs"
    Output: {{"corrected_text": "open vs proprietary llms"}}

    Input: "HOW TO MITIGATE SaCURITY RISKS IN AI AND ML SYSTEM
VECTOR LAB"
     Output: {{"corrected_text": "how to mitigate security risks
in ai and ml system vector lab"}}

    Input: "BUILDING DBRX-CLASS CUSTOM LLMS WITH MOSAIC A1
TRAINING VECTOR LAB"
     Output: {{"corrected_text": "building dbrx-class custom
llms with mosaic a1 training vector lab"}}

    Text to Correct: Placeholder {text} that will be replaced
with the actual input text when calling the model.

    Text to correct:

    {text}

    Final Instruction: Specifies that the output should be in
JSON format only, which reinforces the expectation that the
model should avoid unnecessary explanations or additional text.

    Output (JSON format only):
    '''
)
```

The code snippet defines a `PromptTemplate` class for use with a language model to correct spelling errors in text. The `PromptTemplate` class is initialized with two key parameters: `input_variables` and `template`. The `input_variables` parameter specifies the input variable as `["text"]`, which represents the text that will be corrected. The `template` parameter contains the prompt structure sent to the model. This structure includes clear instructions for the model to correct spelling errors and return the output in lowercase, formatted as JSON. The JSON schema specifies the expected output format, ensuring consistency in responses. The template also provides three examples of input text and their corresponding corrected output in JSON format, guiding the model on how to process similar requests. The `{text}` placeholder in the template will be replaced with the actual input text when the model is invoked. The final instruction emphasizes that the output should be strictly in JSON format, avoiding any additional text or explanations.

5. We initialize the model from Hugging Face using the model name and API token that we specified earlier:

```
huggingface_llm = HuggingFaceHub(repo_id=model_name,
huggingfacehub_api_token=api_token, model_kwargs={"task": "text-
generation"})
```

6. We then combine the prompt template and the model, creating a chain that will take input text, apply the prompt, and create output:

```
llm_chain = LLMChain(prompt=prompt_template, llm=huggingface_
llm)
```

7. We use `llm_chain` to generate a response:

```
response = llm_chain.invoke(text)
```

8. Finally, we apply text correction to the `Extracted Text` column:

```
df['Corrected Text'] = df['Extracted Text'].apply(correct_text)
```

Let's present some of the results:

```
Original: HOW TO MITIGATE SaCURITY RISKS IN AI AND ML SYSTEM VECTOR
LAB
Corrected: how to mitigate security risks in ai and ml system vector
lab

Original: BUILDING DBRX-CLASS CUSTOM LLMS WITH MOSAIC A1 TRAINING
VECTOR LAB
Corrected: building dbrx-class custom llms with mosaic a1 training
vector lab

Original: MPROVING TeXT2SO L PeRFORMANCe WITH EASE ON DATABRICKS 7
VECTOR LAB
Corrected: improving text2so l performance with ease on databricks 7
vector lab
```

As you can see, the results are quite good; we have managed to convert all the text to lowercase and fix most of the misspellings. The misspelled words that were not corrected and flagged in the preceding samples seem to be very technical content. For example, `text2so l` means `text2sql`, which is challenging for the model to fix unless it is fine-tuned on this type of correction data. Another approach you could try is to include these very technical cases that the model seems to miss in the few-shot examples in the prompt to "teach" the model how to interpret these words.

In the code file on the GitHub repository, you'll see that we have added error handling and parsing for the JSON output. This is necessary because we are asking the model to return the output in a specific format, but LLMs do not always follow these instructions precisely. There is currently ongoing work on enforcing the output of LLMs in a specific format, but at this point, it is experimental. You can find more information here: `https://python.langchain.com/v0.1/docs/integrations/llms/lmformatenforcer_experimental/`.

As of now, we have seen how we can use OCR to extract text from images and then fix the extracted text by passing it to an LLM. In the case of a thumbnail image, this extracted text can also be used as an image caption, as the video's title is usually depicted on the image. However, there are cases where the image doesn't contain any text, and we need to ask the model to infer the caption based on what it has seen and understood from the image. This will be the point of discussion for the next part.

Creating image captions

Creating accurate and meaningful captions for images involves not only recognizing and interpreting visual content but also understanding context and generating descriptive text that accurately reflects the image's content. The complexity arises from the need for models to process various elements in an image, such as objects, scenes, and activities, and then translate these elements into coherent and relevant language. This challenge is further compounded by the diverse and often subtle nature of visual information, which can include different lighting conditions, angles, and contexts.

To showcase the difference between the technique we demonstrated earlier, and the captioning based on image understanding, we will use the same images from the thumbnails and attempt to create captions for them based on the image understanding process instead of the text extraction process. You can find the code for this part here: `https://github.com/PacktPublishing/Python-Data-Cleaning-and-Preparation-Best-Practices/blob/main/chapter13/4.image_captioning.py`.

Now, let's dive into the code:

1. Let's start by importing the libraries we'll need for this example:

   ```
   import os
   import pandas as pd
   from PIL import Image
   import matplotlib.pyplot as plt
   from transformers import BlipProcessor,
   BlipForConditionalGeneration, AutoTokenizer,
   AutoModelForSeq2SeqLM
   from langchain import PromptTemplate, LLMChain
   from langchain.llms import HuggingFaceHub
   ```

2. We then define the folder containing the images:

   ```
   folder_path = 'chapter13/images'
   ```

3. Next, we make a list of supported image extensions:

   ```
   supported_extensions = ('.png', '.jpg', '.jpeg')
   ```

4. We then get all image paths for each image in the folder:

```
image_paths = [os.path.join(folder_path, file) for file in
os.listdir(folder_path) if file.lower().endswith(supported_
extensions)]
```

5. We create an empty DataFrame to store the results:

```
df = pd.DataFrame(columns=['Image Path', 'Generated Caption',
'Refined Caption'])
```

6. Then, we initialize the BLIP model and processor for image captioning:

```
blip_model = BlipForConditionalGeneration.from_
pretrained("Salesforce/blip-image-captioning-base")
blip_processor = BlipProcessor.from_pretrained("Salesforce/blip-
image-captioning-base")
```

The `BlipForConditionalGeneration` model is a pretrained model designed for image captioning. It generates descriptive text (captions) for given images by understanding the visual content and producing coherent and relevant descriptions. The model is based on the BLIP architecture, which is optimized for linking visual and textual information. `BlipProcessor` is responsible for preparing images and text inputs in a format suitable for the BLIP model. It handles the preprocessing of images (such as resizing and normalization) and any required text formatting to ensure that the data fed into the model is in the correct format.

7. Now, we initialize the LLM for text refinement. Once we create the caption with the BLIP model, we will then pass it to an LLM again to clean and optimize the caption:

```
llm_model_name = "google/flan-t5-small" # You can play with any
other model from hugging phase as well
tokenizer = AutoTokenizer.from_pretrained(llm_model_name)
model = AutoModelForSeq2SeqLM.from_pretrained(llm_model_name)
```

This piece of code specifies the name of the pretrained model to be used. Here, `"google/flan-t5-small"` refers to a specific version of the T5 model, called `FLAN-T5 Small`, developed by Google. The `AutoTokenizer` class from Hugging Face's Transformers library is used to load the tokenizer associated with the specified model. As we learned in *Chapter 12, Text Preprocessing in the Era of LLMs*, tokenizers are responsible for converting raw text into token IDs that the model can understand. They handle tasks such as tokenizing (splitting text into manageable units), adding special tokens, and encoding the text in a format suitable for model input. Finally, it loads the `google/flan-t5-small sequence-to-sequence` language model, which is suitable for tasks such as translation, summarization, or any task where the model needs to generate text based on some input text. The model has been pretrained on a large dataset, enabling it to understand and generate human-like text, and it is perfect for our use case of caption generation.

8. Next, we need to chain all our steps together and we will use functionality from LangChain to do so:

```
api_token = "add_your_hugging_face_token"
prompt_template = PromptTemplate(input_variables=["text"],
template="Refine and correct the following caption: {text}")
huggingface_llm = HuggingFaceHub(repo_id=llm_model_name,
huggingfacehub_api_token=api_token)
llm_chain = LLMChain(prompt=prompt_template, llm=huggingface_
llm)
```

The `PromptTemplate` object, which is used to define how prompts (input requests) are structured for the language model is created here. Here, we need a much simpler prompt than the one in the previous example as the task is simpler to explain to the model. This instructs the model to refine and correct the provided caption. The `{text}` placeholder will be replaced with the actual text that needs refinement. Then, an instance of `HuggingFaceHub` is created and finally, we create the LLMChain to connect the prompt with the language model.

9. We create a `refine_caption` function that accepts a generated caption as input and creates a prompt by formatting `prompt_template` with the input caption. It then uses `llm_chain` to run the prompt through the LLM, generating a refined caption, and it returns the refined caption:

```
def refine_caption(caption):
    prompt = prompt_template.format(text=caption)
    refined_caption = llm_chain.run(prompt)
    return refined_caption
```

10. We then create the `generate_caption` function, which accepts an image path as input:

```
def generate_caption(image_path):
    image = Image.open(image_path).convert("RGB")
    inputs = blip_processor(images=image, return_tensors="pt")
    outputs = blip_model.generate(inputs)
    caption = blip_processor.decode(outputs[0], skip_special_
tokens=True)
    return caption
```

This function performs the following:

- The function opens the image file and converts it to RGB format.
- It then processes the image using `blip_processor`, returning a tensor suitable for the BLIP model.
- The function generates a caption by passing the processed image to the BLIP model. It finally decodes the model's output into a human-readable caption, skipping special tokens, and returns the caption.

11. Finally, we process each image in the folder, generate an image caption, refine it, and append the final result to a DataFrame:

```
if not image_paths:
    print("No images found in the specified folder.")
else:
    for image_path in image_paths:
        caption = generate_caption(image_path)
        print(f"Generated Caption for {os.path.basename(image_
path)}:\n{caption}\n")
        refined_caption = refine_caption(caption)
        print(f"Refined Caption:\n{refined_caption}\n")
        df.loc[len(df)] = [image_path, caption, refined_caption]
```

Let's have a look at the captions generated by this process:

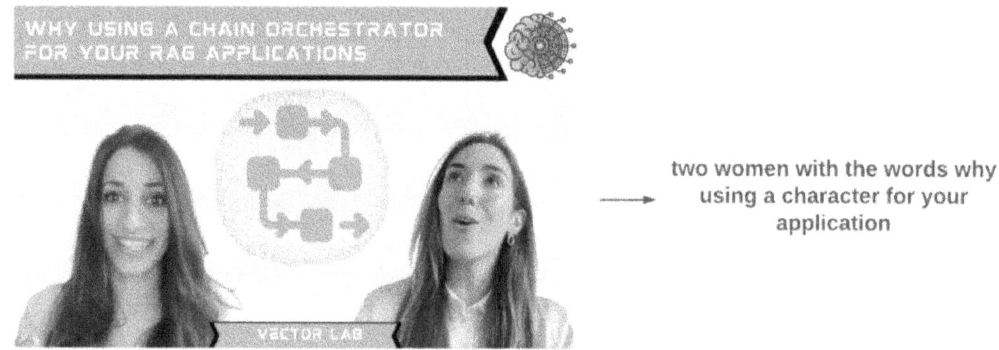

Figure 13.17 – Image caption creation

As we can see, this caption is poor compared to the previous method we demonstrated. The model attempts to understand what is happening in the image and grasp the context, but since the context is derived from a thumbnail, it ends up being quite inadequate. We need to understand that thumbnails often provide limited context about the video content; while they are designed to attract clicks, they may not convey enough information for the model to generate informative captions. The lack of context in combination with the fact that thumbnails are frequently visually cluttered with various images, graphics, and text elements makes it challenging for the model to discern the main subject or context. This complexity can lead to captions that are less coherent or relevant than we have experienced. So, in the case of dealing with thumbnails, the OCR process is best.

However, in cases where images do not contain text, unlike thumbnails that are often filled with written elements, the image understanding process becomes the primary method for generating captions. Since these images lack textual information, relying on the model's visual understanding is essential for creating accurate and meaningful descriptions. Here is some homework for you: Pass through the BLIP process an image that has no text and see what you get!

> **But what about videos?**
>
> To handle videos, the process involves reading the video file and capturing frames at specified intervals, allowing us to analyze each frame, so, *each image*, independently. Once we have the frames, we can apply techniques like those used for images, such as OCR for text extraction, or image understanding models, such as BLIP, for caption generation.

Next, we will move from image to audio data and discuss how we can simplify the audio processing.

Handling audio data

A lot of work is happening in the audio processing space with the most significant advancements happening in **automatic speech recognition** (**ASR**) models. These models transform spoken language into written text, allowing the seamless integration of voice inputs into text-based workflows, thereby making it easier to analyze, search, and interact with. For instance, voice assistants, such as Siri and Google Assistant, rely on ASR to understand and respond to user commands, while transcription services convert meeting recordings into searchable text documents.

This conversion allows the passing of text input to LLMs to unlock powerful capabilities, such as sentiment analysis, topic modeling, automated summarization, and even supporting chat applications. For example, customer service call centers can use ASR to transcribe conversations, which can then be analyzed for customer sentiment or common issues, improving service quality and efficiency.

Handling audio data as text not only enhances accessibility and usability but also facilitates more efficient data storage and retrieval. Text data takes up less space than audio files and is easier to index and search. Moreover, it bridges the gap between spoken and written communication, enabling more natural and intuitive user interactions across various platforms and devices. For instance, integrating ASR in educational apps can help students with disabilities access spoken content in a text format, making learning more inclusive.

As ASR technologies continue to improve, the ability to accurately and efficiently convert audio to text will become increasingly important, driving innovation and expanding the potential of AI-driven solutions. Enhanced ASR models will further benefit areas such as real-time translation services, automated note-taking in professional settings, and accessibility tools for individuals with hearing impairments, showcasing the broad and transformative impact of this technology.

In the next section, we will discuss the Whisper model, which is effective for transforming audio into text and performing a range of audio processing tasks.

Using Whisper for audio-to-text conversion

The **Whisper model** from OpenAI is a powerful tool for transforming audio to text and serves as a base for many modern AI and ML applications. The applications range from real-time transcription and customer service to healthcare and education, showcasing its versatility and importance in the evolving landscape of audio processing technology:

- Whisper can be integrated into voice assistant systems, such as Siri, Google Assistant, and Alexa, to accurately transcribe user commands and queries.
- Call centers can use Whisper to transcribe customer interactions, allowing for sentiment analysis, quality assurance, and topic detection, thereby enhancing service quality.
- Platforms such as YouTube and podcast services can use Whisper to generate subtitles and transcriptions, improving accessibility and content indexing.
- Whisper can be used in real-time transcription services for meetings, lectures, and live events. This helps create accurate text records that are easy to search and analyze later.
- In telemedicine, Whisper can transcribe doctor-patient conversations accurately, facilitating better record-keeping and analysis. Moreover, it can assist in creating automated medical notes from audio recordings.
- Educational platforms can use Whisper to transcribe lectures and tutorials, providing students with written records of spoken content, enhancing learning and accessibility.
- Security systems use direct audio processing to verify identity based on unique vocal characteristics, offering a more secure and non-intrusive method of authentication.

As a pretrained model, Whisper can be used out of the box for many tasks, reducing the need for extensive fine-tuning and allowing for quick integration into various applications. The model supports multiple languages, making it versatile for global applications and diverse user bases. While Whisper primarily focuses on transforming audio to text, it also benefits from advancements in handling audio signals, potentially capturing nuances, such as tone and emotion. Although direct audio processing (such as emotion detection or music analysis) might require additional specialized models, Whisper's robust transcription capability is foundational for many applications.

Using some audio from the **Vector Lab** (`@VectorLab`) videos, we will parse the audio through Whisper to get the extracted text.

Extracting text from audio

The following code demonstrates how to use the Whisper model from Hugging Face to transcribe audio files into text. It covers loading necessary libraries, processing an audio file, generating a transcription using the model, and finally decoding and printing the transcribed text. Let's have a look at the code, which you can also find here: `https://github.com/PacktPublishing/Python-Data-Cleaning-and-Preparation-Best-Practices/blob/main/chapter13/5.whisper.py`.

Let's begin:

1. We'll start by importing the required libraries:

```
import torch
from transformers import WhisperProcessor,
WhisperForConditionalGeneration
import librosa
```

2. We start by loading the Whisper processor and model from Hugging Face:

```
processor = WhisperProcessor.from_pretrained("openai/whisper-
large-v2")
model = WhisperForConditionalGeneration.from_pretrained("openai/
whisper-large-v2")
```

3. Next, we define the path to your audio file:

```
audio_path = "chapter13/audio/3.chain orchestrator.mp3"
```

You can replace this file with any other audio you want.

4. Then, we load the audio file:

```
audio, rate = librosa.load(audio_path, sr=16000)
```

Let's expand on this:

- `audio` will be a NumPy array containing the audio samples.
- `rate` is the sampling rate of the audio file. The `sr=16000` argument resamples the audio to a sampling rate of 16 kHz, which is the required input sampling rate for the Whisper mode.

5. Now, we preprocess the audio file for the Whisper model:

```
input_features = processor(audio, sampling_rate=rate, return_
tensors="pt").input_features
```

6. We then generate the transcription:

```
with torch.no_grad():
    predicted_ids = model.generate(input_features)
```

This line passes the preprocessed audio features to the model to generate transcription IDs. The model produces token IDs that correspond to the transcribed text.

7. Now, we decode the generated transcription:

```
transcription = processor.batch_decode(predicted_ids, skip_
special_tokens=True)[0]
```

This line decodes the predicted token IDs back into readable text. The `[0]` value at the end extracts the first (and only) transcription from the resulting list.

8. Finally, we print the transcribed text:

```
"As you can see, you need what we call a chain orchestrator
to coordinate all the steps. So all the steps from raising
the question all the way to the response. And the most popular
open source packages are Lama Index and LangChain that we can
recommend. Very nice. So these chains, these steps into the RAG
application or any other LLM application, you can have many
steps happening, right? So you need this chain to help them
orchestrate"
```

Is the transcription slow?

Depending on the model size and your hardware capabilities, the transcription process might take some time.

As we can see, the transcription is excellent. Now, in the use case we are dealing with, after transcribing the YouTube video, there are several valuable actions you can take on this project. First, you can create captions or subtitles to improve accessibility for viewers who are deaf or hard of hearing. Additionally, writing a summary or extracting key points can help viewers grasp the main ideas without watching the entire video. The transcription can also be transformed into a blog post or article, providing more context on the topic discussed. Extracting quotes or highlights from the transcription allows you to create engaging social media posts that promote the video. Utilizing the transcription for SEO purposes can improve the video's search engine ranking by including relevant keywords in the description. You can also develop FAQs or discussion questions based on the video to encourage viewer engagement. Additionally, the transcription can serve as a reference for research, and you might consider adapting it into a script for an audiobook or podcast. Incorporating the transcription into educational materials, such as lesson plans, is another effective way to utilize the content. Lastly, you can create visual summaries or infographics based on the key points to present the main ideas visually. How cool is that?

In the following section, we will expand the use case and do some emotion detection from the transcribed text.

Detecting emotions

Emotion detection from text, often referred to as sentiment analysis or emotion recognition, is a subfield of **natural language processing** (**NLP**) that focuses on identifying and classifying emotions conveyed in written content. This area of study has gained significant traction due to the growing amount of textual data generated across social media, customer feedback, and other platforms.

In our case, we will use the `j-hartmann/emotion-english-distilroberta-base` model, built upon the DistilRoBERTa architecture. The DistilRoBERTa model is a smaller and faster variant of the RoBERTa model, which itself is based on the Transformer architecture. This model is specifically fine-tuned for emotion detection tasks. It has been trained on a dataset designed to recognize various emotions expressed in text, making it adept at identifying and classifying emotions from written content. It is designed to detect the following emotions from text:

- **Joy**: This represents happiness and positivity
- **Sadness**: This reflects feelings of sorrow and unhappiness
- **Anger**: This indicates feelings of frustration, annoyance, or rage
- **Fear**: This conveys feelings of anxiety or apprehension
- **Surprise**: This represents astonishment or unexpectedness
- **Disgust**: This reflects feelings of aversion or distaste
- **Neutral**: This indicates a lack of strong emotion or feeling

These emotions are typically derived from various datasets that categorize text based on emotional expressions, allowing the model to classify input text into these predefined categories.

Let's have a look at the code, which is also available here: `https://github.com/PacktPublishing/Python-Data-Cleaning-and-Preparation-Best-Practices/blob/main/chapter13/6.emotion_detection.py`.

> **Memory check**
>
> The following code is memory intensive, so you may need to allocate more memory if working on virtual machines or Google Collab. The code was tested on Mac M1, 16 GB memory.

Let's start coding:

1. We first import the libraries required for this example:

```
import torch
import pandas as pd
from transformers import WhisperProcessor,
WhisperForConditionalGeneration,
AutoModelForSequenceClassification, AutoTokenizer
import librosa
import numpy as np
```

2. We then load the Whisper processor and model from Hugging Face:

```
whisper_processor = WhisperProcessor.from_pretrained("openai/
whisper-large-v2")
```

```
whisper_model = WhisperForConditionalGeneration.from_
pretrained("openai/whisper-large-v2")
```

3. Then, we load the emotion detection processor and model from Hugging Face:

```
emotion_model_name = "j-hartmann/emotion-english-distilroberta-
base"
emotion_tokenizer = AutoTokenizer.from_pretrained(emotion_model_
name)
emotion_model = AutoModelForSequenceClassification.from_
pretrained(emotion_model_name)
```

4. We define the path to your audio file:

```
audio_path = "chapter13/audio/3.chain orchestrator.mp3" #
Replace with your actual audio file path
```

5. Once the path is defined, we load the audio file:

```
audio, rate = librosa.load(audio_path, sr=16000)
```

6. We create a function called `split_audio` to split audio into chunks:

```
def split_audio(audio, rate, chunk_duration=30):
    chunk_length = int(rate * chunk_duration)
    num_chunks = int(np.ceil(len(audio)/chunk_length))
    return [audio[i*chunk_length:(i+1)*chunk_length] for i in
range(num_chunks)]
```

7. We also create a function to transcribe audio using Whisper:

```
def transcribe_audio(audio_chunk, rate):
    input_features = whisper_processor(audio_chunk, sampling_
rate=rate, return_tensors="pt").input_features
    with torch.no_grad():
        predicted_ids = whisper_model.generate(input_features)
    transcription = whisper_processor.batch_decode(predicted_
ids, skip_special_tokens=True)[0]
    return transcription
```

The function preprocesses the audio file for the Whisper model and generates the transcription. Once it's generated, the function decodes the generated transcription.

8. We then create a function to detect emotions from text using the emotion detection model:

```
def detect_emotion(text):
    inputs = emotion_tokenizer(text, return_tensors="pt",
truncation=True, padding=True, max_length=512)
    outputs = emotion_model(inputs)
```

```
        predicted_class_id = torch.argmax(outputs.logits, dim=-1).
    item()
        emotions = emotion_model.config.id2label
        return emotions[predicted_class_id]
```

This function begins by tokenizing the input text with `emotion_tokenizer`, converting it into PyTorch tensors while handling padding, truncation, and maximum length constraints. The tokenized input is then fed into `emotion_model`, which generates raw prediction scores (logits) for various emotion classes. The function identifies the emotion with the highest score using `torch.argmax` to determine the class ID. This ID is then mapped to the corresponding emotion label through the `id2label` dictionary provided by the model's configuration. Finally, the function returns the detected emotion as a readable label!

9. Then, we split the audio into chunks:

```
audio_chunks = split_audio(audio, rate, chunk_duration=30) #
30-second chunks
```

10. We also create a DataFrame to store the results:

```
df = pd.DataFrame(columns=['Chunk Index', 'Transcription',
'Emotion'])
```

11. Finally, we process each audio chunk:

```
for i, audio_chunk in enumerate(audio_chunks):
    transcription = transcribe_audio(audio_chunk,rate)
    emotion = detect_emotion(transcription)
    # Append results to DataFrame
    df.loc[i] = [i, transcription, emotion]
```

The output emotions are shown for each chunk of transcribed text, and in our case, all are neutral, as the video is just a teaching concept video:

```
   Chunk Index  Emotion
0            0  neutral
1            1  neutral
2            2  neutral
```

Now, we will expand our use case a bit further to demonstrate how you can take the transcribed text and pass it through an LLM to create highlights for the YouTube video.

Automatically creating video highlights

In the era of digital content consumption, viewers often seek concise and engaging summaries of longer videos. Automatically creating video highlights involves analyzing video content and extracting key moments that capture the essence of the material. This process saves time and improves content accessibility, making it a valuable tool for educators, marketers, and entertainment providers alike.

Let's have a look at the code. You can find it at the following link: `https://github.com/PacktPublishing/Python-Data-Cleaning-and-Preparation-Best-Practices/blob/main/chapter13/7.write_highlights.py`.

In this code, we will expand the Whisper example. We will transcribe the text, then join all the transcribed chunks together, and finally, we will pass all the transcriptions to the LLM to create the highlights for the entire video. Let's continue the previous example:

1. We start by initializing the Hugging Face model:

```
model_name = "mistralai/Mistral-Nemo-Instruct-2407" # Using
Mistral for instruction-following
```

2. Then, we add your Hugging Face API token:

```
api_token = "" # Replace with your actual API token
```

3. Here's the LangChain setup that we'll be using in this use case. Notice the new prompt that we added:

```
prompt_template = PromptTemplate(
    input_variables=["text"],
    template='''This is the transcribed text from a YouTube
video. Write the key highlights from this video in bullet
format.
    {text}
    Output:
    '''
    )

huggingface_llm = HuggingFaceHub(repo_id=model_name,
huggingfacehub_api_token=api_token, model_kwargs={"task": "text-
generation"})
llm_chain = LLMChain(prompt=prompt_template, llm=huggingface_
llm)
```

4. Next, we generate the transcription:

```
def transcribe_audio(audio_chunk, rate):
    input_features = whisper_processor(audio_chunk,
        sampling_rate=rate,
        return_tensors="pt").input_features
    with torch.no_grad():
        predicted_ids = \
            whisper_model.generate(input_features)
```

```
        transcription = whisper_processor.batch_decode(
            predicted_ids, skip_special_tokens=True)[0]
        return transcription
```

5. Then, we create a function to generate the key highlights from text using the LLM:

```
def generate_highlights(text):
    try:
        response = llm_chain.run(text)
        return response.strip() # Clean up any whitespace around
the response
    except Exception as e:
        print(f"Error generating highlights: {e}")
        return "error" # Handle errors gracefully
```

6. Next, we split the audio into chunks:

```
audio_chunks = split_audio(audio, rate, chunk_duration=30) #
30-second chunks
```

7. We then transcribe each audio chunk:

```
transcriptions = [transcribe_audio(chunk, rate) for chunk in
audio_chunks]
```

8. Then, we join all transcriptions into a single text:

```
full_transcription = " ".join(transcriptions)
```

9. Finally, we generate highlights from the full transcription:

```
highlights = generate_highlights(full_transcription)
```

Let's see the automatically created highlights:

```
Chain Orchestrator: Required to coordinate all steps in a LLM (Large
Language Model) application, such as RAG (Retrieval-Augmented
Generation).
Popular Open Source Packages: Lama Index and LangChain are recommended
for this purpose.
Modularization: Chains allow for modularization of the process, making
it easier to update or change components like LMs or vector stores
without rebuilding the entire application.
Rapid Advancements in JNNIA
```

As we can see, there are some minor mistakes, mainly coming from the Whisper process, but other than that, it is actually pretty good.

In the next part, we will quickly review the research happening in the audio space, as it is a rapidly evolving field.

Future research in audio preprocessing

There is a growing trend toward the development of multimodal LLMs capable of processing various types of data, including audio. Currently, many language models are primarily text-based, but we anticipate the emergence of models that can handle text, images, and audio simultaneously. These multimodal LLMs have diverse applications, such as generating image captions and providing medical diagnoses based on patient reports. Research is underway to extend LLMs to support direct speech inputs. As noted, "Several studies have attempted to extend LLMs to support direct speech inputs with a connection module" (`https://arxiv.org/html/2406.07914v2`), indicating ongoing efforts to incorporate audio processing capabilities into LLMs. Although not only relevant to audio, LLMs face several challenges with other data types, including the following:

- High computational resources required for processing
- Data privacy and security concerns

Researchers are actively exploring various strategies to overcome these challenges. To address the high computational demands, there is a focus on developing more efficient algorithms and architectures, such as transformer models with reduced parameter sizes and optimized training techniques. Techniques such as model compression, quantization, and distillation are being employed to make these models more resource-efficient without sacrificing performance (`https://arxiv.org/abs/2401.13601`, `https://arxiv.org/html/2408.04275v1`, `https://arxiv.org/html/2408.01319v1`). In terms of data privacy and security, researchers are investigating privacy-preserving ML techniques, including federated learning and differential privacy. These approaches aim to protect sensitive data by allowing models to learn from decentralized data sources without exposing individual data points. Additionally, advancements in encryption and secure multi-party computation are being integrated to ensure that data remains confidential throughout the processing pipeline. These efforts are crucial for enabling the widespread adoption of multimodal LLMs across various domains while ensuring they remain efficient and secure (`https://towardsdatascience.com/differential-privacy-and-federated-learning-for-medical-data-0f2437d6ece9`, `https://arxiv.org/pdf/2403.05156`, `https://pair.withgoogle.com/explorables/federated-learning/`).

Let's now summarize the learnings from this chapter.

Summary

In this chapter, we covered various image processing techniques, such as loading, resizing, normalizing, and standardizing images to prepare them for ML applications. We implemented augmentation to generate diverse variations for improved model generalization and applied noise removal to enhance image quality. We also examined the use of OCR for text extraction from images, particularly addressing the challenges presented by thumbnails. Additionally, we explored the BLIP model's capability to generate captions based on visual content. Furthermore, we discussed video processing techniques involving frame extraction and key moment analysis.

Finally, we introduced the Whisper model, highlighting its effectiveness in converting audio to text and its automatic speech recognition capabilities across multiple languages.

This concludes the book! You did it!

I want to express my sincere gratitude for your dedication to finishing this book. I've aimed to share the insights from my experience, with a focus on ML and AI, as these fields have been central to my career. I find them incredibly fascinating and transformative, though I might be a bit biased.

As you've seen in the later chapters, I believe LLMs are poised to revolutionize the field and the way we process data. That's why I dedicated the last chapters to building a foundation and showcasing how effortlessly different types of data can be transformed and manipulated using LLMs.

Please take my advice and spend some time diving into the code examples provided. Implement these techniques in your daily tasks and projects. If there's anything you find yourself doing manually or redoing frequently, *code it up* to streamline your process. This hands-on practice will help reinforce your learning. Experiment with the techniques and concepts from this book on your own projects, as real growth occurs when you adapt and innovate with these tools in practical scenarios.

I'm traveling the world speaking at conferences and running workshops. If you see me at one of these events, don't hesitate to say hello and talk to me! Who knows, our paths might cross at one of these events! In any case, I'd also love to hear about your progress and see what you've learned and built. Feel free to share your experiences with me—your insights and developments are always exciting to see. So, let's stay connected! Connect with me on LinkedIn (`https://www.linkedin.com/in/maria-zervou-533222107/`) and you can subscribe to my YouTube channel (`https://www.youtube.com/channel/UCY2Z8Sc2L0wQnTOQPlLzUQw`) for ongoing tutorials and content to keep you informed about new developments and techniques in ML and AI.

Thank you once again for your time and effort. Remember, learning is just the beginning; real growth comes from practicing and applying your knowledge. I'm excited to see where your journey leads you and hope our paths cross again in the future.

Index

C

D

E

N

Q

R

S

T

V

W

Y

Z

Other Books You May Enjoy

If you enjoyed this book, you may be interested in these other books by Packt:

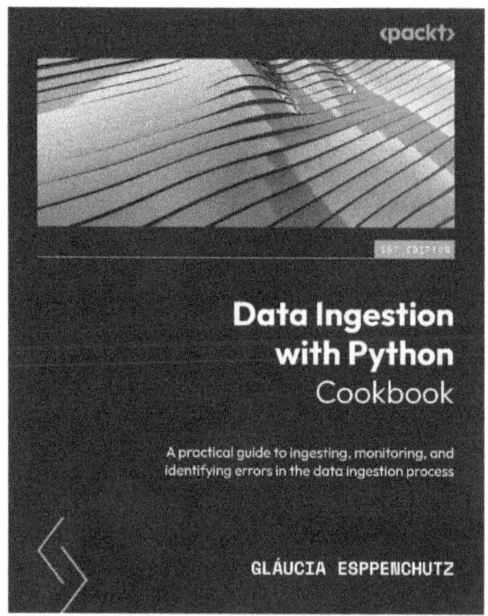

Data Ingestion with Python Cookbook

Gláucia Esppenchutz

ISBN: 978-1-83763-260-2

- Implement data observability using monitoring tools
- Automate your data ingestion pipeline
- Read analytical and partitioned data, whether schema or non-schema based
- Debug and prevent data loss through efficient data monitoring and logging
- Establish data access policies using a data governance framework
- Construct a data orchestration framework to improve data quality

Python Data Cleaning Cookbook

Michael Walker

ISBN: 978-1-80323-987-3

- Using OpenAI tools for various data cleaning tasks
- Producing summaries of the attributes of datasets, columns, and rows
- Anticipating data-cleaning issues when importing tabular data into pandas
- Applying validation techniques for imported tabular data
- Improving your productivity in pandas by using method chaining
- Recognizing and resolving common issues like dates and IDs
- Setting up indexes to streamline data issue identification
- Using data cleaning to prepare your data for ML and AI models

Packt is searching for authors like you

If you're interested in becoming an author for Packt, please visit `authors.packtpub.com` and apply today. We have worked with thousands of developers and tech professionals, just like you, to help them share their insight with the global tech community. You can make a general application, apply for a specific hot topic that we are recruiting an author for, or submit your own idea.

Share your thoughts

Now you've finished *Python Data Cleaning and Preparation Best Practices*, we'd love to hear your thoughts! Scan the QR code below to go straight to the Amazon review page for this book and share your feedback or leave a review on the site that you purchased it from.

`https://packt.link/r/1-837-63474-2`

Your review is important to us and the tech community and will help us make sure we're delivering excellent quality content.

Download a free PDF copy of this book

Thanks for purchasing this book!

Do you like to read on the go but are unable to carry your print books everywhere?

Is your eBook purchase not compatible with the device of your choice?

Don't worry, now with every Packt book you get a DRM-free PDF version of that book at no cost.

Read anywhere, any place, on any device. Search, copy, and paste code from your favorite technical books directly into your application.

The perks don't stop there, you can get exclusive access to discounts, newsletters, and great free content in your inbox daily

Follow these simple steps to get the benefits:

1. Scan the QR code or visit the link below

`https://packt.link/free-ebook/9781837634743`

2. Submit your proof of purchase
3. That's it! We'll send your free PDF and other benefits to your email directly

www.ingramcontent.com/pod-product-compliance
Lightning Source LLC
Chambersburg PA
CBHW081225220326
41598CB00037B/6877

* 9 7 8 1 8 3 7 6 3 4 7 4 3 *